Vincent L. Zelick
May – 1975

CHEMISTRY OF WATER SUPPLY, TREATMENT, AND DISTRIBUTION

CHEMISTRY OF WATER SUPPLY, TREATMENT, AND DISTRIBUTION

ALAN J. RUBIN
Associate Professor of Civil Engineering
Water Resources Center
The Ohio State University

ann arbor science PUBLISHERS INC.
POST OFFICE BOX 1425 • ANN ARBOR, MICHIGAN 48106

P. O. Box 1425, Ann Arbor, Michigan 48106

Library of Congress Catalog Card Number 73-82270
ISBN 0-250-40036-7

Printed in the United States of America

PREFACE

In the past few years the tendency has been strong to think of "ecology" when speaking of environmental protection. There has been little enough concern with human health; and correspondingly, water supply problems have been deemphasized if not outrightly ignored in recent federal legislation.

Does the nation need a typhoid epidemic or the outbreak of another waterborne disease to reverse the trend? The possibility is not so remote. Cholera has taken lives in Italy this year, and typhoid has been reported in Florida. Yet aside from these potentially acute situations, of greater significance in the long run are the chronic effects of trace contaminants. The chemistry of even the common minerals in aquatic systems is very complex and still much is unknown. This is further complicated by the continued pollution of our waterways leading to the introduction of numerous synthetic compounds of unknown persistence and undetermined reactivity. Needless to say, many health implications are almost completely unknown.

To what extent, if any, are these contaminants being removed during conventional treatment? For that matter, what trace constituents are being added as a result of treatment? Do we have the analytical capability to adequately monitor our water supplies? Without official interest in such questions, and this includes research support at the federal level, our young scientists and engineers will turn to areas of greater personal and professional reward.

Fortunately, there is a body of young as well as experienced professionals still concerned with the problems of delivering potable water. Evidence for this was the very gratifying attendance and participation in the *Symposium on the Chemistry of Water Supply, Treatment, and Distribution* held by the Environmental Chemistry

Division of the American Chemical Society at its national meeting in April of 1973. This book takes its name from that symposium and is a direct result of it, being comprised of about half the papers presented.

Organization of *Chemistry of Water Supply, Treatment, and Distribution* is quite simple. Topics are laid out in the sequence paralleling the flow of water from supply, through a treatment plant, and on to distribution. Thus, the leading paper on setting the new drinking water standards is followed by reports dealing with raw water quality, analytical methods, removal processes and finally, the quality of the finished water. Some of the unit operations dealt with are coagulation, filtration, adsorption and disinfection. The latter includes studies on viruses as well as disinfectants other than chlorine. The chemistry of cadmium, aluminum, arsenic, sulfur species and the products of active carbon reacting with chlorine are dealt with at length. The papers for the most part contain details of techniques and results not usually described in journal articles.

The efforts of the individual authors in contributing to the success of this enterprise and the support of the Civil Engineering Department and the College of Engineering of The Ohio State University are gratefully acknowledged. Sponsorship of the original symposium in Dallas by the Environmental Chemistry Division is also appreciated and acknowledged. Special thanks are extended to friends and colleagues, too many to name, who assisted and encouraged me.

Alan J. Rubin
Columbus, Ohio
November 1973

TABLE OF CONTENTS

Chapter | Page

CHAPTER 1

CRITERIA DEVELOPMENT FOR NATIONAL DRINKING WATER STANDARDS

Gordon G. Robeck*

For approximately 60 years the Federal Government, through the U. S. Public Health Service (PHS), has had the responsibility of certifying the safety of water supplied to and used on interstate carriers. The U. S. Environmental Protection Agency has recently been given this responsibility and is attempting to update the standards. The actual process of developing criteria and setting standards has traditionally been through combined expertise of Federal staff members and nonfederal advisors. Both groups rely principally on scientific literature, legal precedence, and practical experience for selecting constituent limits. Advisory committee representation, drawn from the most involved and affected public and professional groups, has resulted in the development of acceptable criteria for most constituents. However, some limits have a rather inadequate scientific base. The selection of criteria and examples of adequate and inadequate bases for standards are the subjects of this paper.

In 1962, PHS introduced standards for fluoride, nitrate, and certain organics as represented by a carbon chloroform extract. Each had a different degree of scientific and field data to support its limit; however, the committee thought there was enough justification in each case to warrant inclusion. Pesticides were not included at that time for specific reasons.

*G. G. Robeck, Director of the Water Supply Research Laboratory, National Environmental Research Center, Cincinnati, Ohio.

While there are many harmful substances in natural runoff, as well as in man's waste discharges to streams, only a few are known to occur in water consistently and in sufficient concentrations to have either health or esthetic significance. Thus, a certain abundance and general widespread occurrence are necessary before a specific constituent is considered for inclusion in the Drinking Water Standards. If this policy were not followed, analytical and enforcement chores would be unduly burdensome.

Knowing of the occurrence of a chemical in water implies that there is an analytical method for quantifying the substance. However, as there is usually no quick, reasonable way to measure certain compounds, especially organics, a sensible standard cannot be set for certain specific compounds. Thus, analytical ability is an indirect factor in establishing enforceable standards.

Protecting human health is the primary reason for setting standards. Several considerations go into selecting health limits for pathogens or chemicals. Among these considerations are the public health statistics related to morbidity and mortality, the population exposed, the physical and chemical state of the substance, the toxicity of the substance to man or to experimental animals that can be related to man, and the amount of the substance likely to be found in other sources, such as food and air. For example, because the organic mercury content is known to be relatively high in certain marketable fish taken from the sea, the limit for mercury in water was adjusted downward to keep the total daily intake for man well below a hazardous level.

A brief discussion of the mercury case may give insight into how criteria are developed and standards are selected. No limit was set for mercury in the 1962 Drinking Water Standards because it was not considered a very likely contaminant of drinking water. Since the inorganic mercurials are neither very soluble in water nor easily absorbed, they are not considered to be as toxic in water as other forms of mercury.

Nevertheless, during this decade there has been a decided change in attitude about the hazard of mercury in the environment. For example, the experience at Minamata Bay, Japan, and more recently, the occurrence of a relatively large concentration of organic forms of mercury in foodstuff marketed in this country have caused great concern. In 1970, PHS set a tentative limit of 0.005 mg/l of mercury in drinking water though there was limited specific data upon which to base the selection of this particular exposure level. At that time, there was considerable difficulty with analytical methods for quantifying mercury. Much resampling and splitting of samples between laboratories was necessary to establish the amount of mercury present in various media and in which chemical form it existed. However, because of the apparent general distribution of mercury, the Federal Technical Committee for Reviewing Standards proceeded to develop a standard for drinking water.

Review of current information indicates that, among populations exposed to fish containing methyl mercury, the minimum effect level occurs when the blood methyl mercury level reaches 0.2 μg of mercury per gram of blood. This value corresponds to prolonged, continuous exposure to approximately 0.3 mg of methyl mercury per day for a human. Applying a safety factor of 10 to this minimum effect exposure level, the maximum allowable exposure by all routes would be 30 μg/day. Thus, the current FDA guideline of 0.5 mg of mercury per kg allows for a maximum daily consumption of 60 g of fish with no additional exposure.

In addition to a health criteria for total intake, there is also a need to allocate a certain sensible portion to all routes of exposure -- food, water, and air. In 1970, the technical committee decided that 0.005 mg/l could be allotted to water. A 2-liter consumption would amount to a third of the maximum allowable intake of 30 μg/day.

In the meantime, more has been discovered about the occurrence of mercury in drinking water and its removal rate by conventional water treatment. Of 273 water supplies sampled, only one exceeded 0.005 mg/l; thus mercury in drinking water is apparently not a major problem. The single exceeder was extensively resampled and the elevated concentration did not occur again. In addition, several of the conventional water treatment processes reduce the mercury content of water. Inasmuch as mercury concentrations are relatively high in certain foods and low in drinking water (now measurable below 1 μg/l), the Public Advisory Committee set the mercury standard for water at 0.002 mg/l. This limit was based on total mercury intake and is considered to be a screening level. More elaborate analyses would be performed on exceeders to determine the proportions of alkyl and inorganic forms.

Toxicological data will help determine if a factor of safety should be 2, 10, or even as high as 500. A safety factor of 2 or 10 may be satisfactory for certain inorganics with a long history of human health observations. However, in the case of chlorinated hydrocarbon pesticides where the use has been recent and only experimental animal data are available, a factor of 500 may be necessary. In certain cases, toxicological data and epidemiological observations do not necessarily support the same conclusion; therefore, wherever there is reasonable doubt, the conservative limit is used until more conclusive evidence is accumulated. Because the arsenics presently fall into this category, a discussion of this particular circumstance may serve to illustrate how the committee copes with such a situation.

In 1962, though the mandatory limit of 0.05 mg/l for arsenic was not changed, a recommended or preferred limit was set at 0.01 mg/l because it was thought that arsenic may be carcinogenic. During a review in 1970, the technical committee used several consultants, one of whom felt that the public suffers from arsenophobia. He contended that toxicity studies have established the usefulness of arsenic as a poultry feed additive without adverse effects and felt that the overly restrictive limits could be relaxed. A considerable body of animal toxicological data exists on the effects of inorganic arsenite and arsenate, and minimum-effect and no-effect levels in dogs, rats, and mice have been determined. Three generations of breeding mice were exposed to 5 mg/l of arsenite in the diet with no observable effects on reproduction. The fact that little of the organic form of arsenic found in shrimp meat was retained was demonstrated in rats.

Nevertheless, in considering a health effect limit for arsenic, there are conflicting data to consider. Admittedly, attempts to demonstrate through animal studies that arsenic is tumorigenic have met with failure; chronic toxicity studies indicate that mg/l concentrations can be consumed without effects; and, occupational health studies have demonstrated that man can be exposed to rather high concentrations without adverse effects.

On the other hand, clinical and epidemiological studies conducted in several locations in the world have demonstrated the adverse health effects of arsenic in drinking water. For example, in an area on the southwest coast of Taiwan where the population uses water from wells with high arsenic content, some individuals develop hyper-pigmentation, keratosis, and finally skin cancer. These conditions increase with age, which indicates longer exposure, and the effect is more pronounced in males. The prevalence rates for skin cancer show an ascending gradient according to the arsenic content of the well water. The conditions of melanosis, keratosis, and skin cancer were not noted in control villages where the arsenic content was 0.02 mg/l or less. In the villages studied, these conditions were noted, but were much less prevalent when the arsenic content was less than 0.3 mg/l.

Another example is from Antofagasta, Chile, where the drinking water supply from 1958 to 1970 contained 0.8 mg/l arsenic. Dermatological manifestations were noted, especially in children. The water supply of this city of 130,000 inhabitants was changed, and Dr. Jose Manuel Borgono of Chile stated that these adverse conditions were not observed when the water supply contained less than 0.15 mg/l.

A third field example is from a ranch in Nevada where well water with an arsenic content that ranged from 0.5 to 2.75 mg/l was reported to have resulted in a skin condition in at least one child and possibly in another.

In such studies total intake and retention are usually difficult to estimate. Because arsenic is a geochemical pollutant, it can be expected to occur in the air, food, and water of those sections of the country where it occurs naturally in soil and water in higher concentrations. Therefore, much additional research is needed on distribution and availability of arsenic in various media before any major change is made in the standards. Nevertheless, because of the interpretation given to field observations by epidemiologists, the present recommendation is to limit arsenicals in water to 0.1 mg/l, which is approximately 20 per cent of the average total intake of arsenic.

Another factor that influences limits is attainability. This is clearly demonstrated in the area of coliform bacteria, which can be easily controlled with a disinfectant without knowing precisely the factor of safety for a certain occurrence or level of coliform. General organic content or odorous limits may also be determined by the available technology and by the amount of money the community wants to spend for a high quality water. For example, before the 1962 Standards were published, small activated carbon filters were used by researchers in the USPHS in Cincinnati[1] to capture and characterize some organics in water that were retained on the surface of this type of carbon. Thus, they were able to demonstrate that in certain surface sources there were many potentially dangerous or odorous chemicals. Based on a large national sampling, a limit of 0.2 mg/l was selected as the limit for organics. Such limits were detected by extracting the carbon with chloroform (CCE) and as a means of curbing excessive odors or organic concentrations of an ill-defined nature. A vast majority of supplies sampled were below that limit.

Inasmuch as the limit in the 1962 Standards was not a mandatory limit, but merely a recommended one, few water works supplies have been monitored routinely for this parameter. Obviously, because a more practical method of gaging for harmful organics is necessary, improvements in yield and operation have recently been made in the CCE determination. Consequently, the actual limit will be changed to 0.7 mg/l, which represents a slightly more stringent limit than the present one of 0.2 mg/l.

A Carbon Alcohol Extract (CAE) limit of 3.0 mg/l is also proposed for the next edition of Drinking Water Standards. When McCabe[2] analyzed old data from Heuper and Payne,[3] he indicated that life-shortening had occurred in rats challenged with this type of extract. Further, Rosen *et al.*[4] had also shown the presence of a synthetic organic, ABS, in some CAE's. Here again, the limit was selected to minimize the intake of excessive quantities of this somewhat unknown material in water. Most supplies comply with this limit according to preliminary sampling in various major basin sources. A few highly polluted sources are known to exceed 3 mg/l in certain seasons of the year. A variance can be granted if a more specific analysis accounts for the content of the extract and the component parts are judged harmless. For example, if color bodies dominate the CAE, an individual investigation can be made to determine whether there is an excessive health-related component in the extract.

Sometimes a criterion for a constituent limit is such that the analytical work can be reduced by including the effect of the substance with some other generalized limit. This was the case with phenol where a limit of 1 μg/l was set in 1962 because more than that amount tended to react adversely with chlorine to form odorous components. The 1970 EPA Technical Review Committee recognized that the analytical methods for that concentration have not been very practical for most water works concerns; therefore, the committee decided to drop phenols from the Standards. The odor limit serves as the means to control phenol inasmuch as it is that indirect effect that is significant.

The historical development of limits for pesticides may also serve to illustrate how care and caution have to be used in setting limits for materials added to the environment for presumably beneficial purposes. In 1962, considerable effort was made to examine the logic of and need for including selected limits for specific or general groups of pesticides in the Drinking Water Standards. A decision was made to postpone the establishment of limits until more information on simplified analyses and mammalian experiments had been accumulated. In 1965, another committee, formed to advise PHS on effective implemention of Drinking Water Standards, suggested limits for pesticides; however, these limits were never formally adopted for enforcement. Although the recommended limits were based mainly on fish toxicity studies, they did serve as guidelines for evaluating a water supply and controlling excessive use of herbicides in reservoirs. Since that time many investigators have developed experimental animal data and made observations of human experiences, and, as a result, a different set of limits has been proposed in the current revision. Nonetheless, in some cases the factors of safety must be as high as 500 due to incomplete information regarding long-range effects of chlorinated hydrocarbons. Table 1.I gives the old guidelines as compared with the proposed limits.

The limit for heptachlor was lowered because most of the permissible body intake can be found in food; thus, the usual 20 per cent allotted to water could not be used. Fortunately, most of the concentrations in public water supplies are well below these limits. Furthermore, fish and wildlife considerations may cause the general pesticide limits in raw water to be even lower than that required for drinking water.

Generally speaking, any standard is made from criteria plus an action plan to implement the desired goal; therefore, developing and carrying out a monitoring or surveillance program is an integral part of the standard. In the case of pesticides, regional characterization of sources, types, and levels by state or Federal programs may help minimize the analytical work required of the local water utility. Variances in this part of the standards will probably be permitted on a case-by-case basis. Certainly historical constituent level patterns, manufacturing operations, or land-use practices will help state and Federal engineers decide what the continuing monitoring schedule should be for compounds that man has added to the environment.

Many other examples of criteria, especially those established for esthetic purposes, could be described. However, the reasons for selecting a constituent and its limit for the Drinking Water Standards have been illustrated by a discussion of the basis for including mercury, arsenic, CCA, CAE, phenol, and several pesticides. Additionally, these are the constituents that may be either changed or

Table 1.I

Proposed Limits for Pesticides in EPA Drinking Water Standards

Pesticide	*Old Guideline* *mg/l*	*Proposed 1973* *mg/l*
Chlorinated Hydrocarbons		
Aldrin	0.017	0.001
Chlordane	0.003	0.003*
DDT	0.042	0.05
Dieldrin	0.017	0.001
Endrin	0.001	0.0005
Heptachlor	0.018	0.0001
Heptachlor epoxide	0.018	0.0001
Lindane	0.056	0.005
Methoxychlor	0.035	0.1
Toxaphene	0.005	0.005*
Chlorophenoxyl Herbicides		
2,4-D	0.1	0.002
2,4,5-TP	0.1	0.03
2,4,5-T	0.1	(Dropped because of disuse)
Organophosphate Insecticides	0.1**	0.1**

*Limit selected on basis of odor although the toxic limits are similar.

**Expressed in terms of parathion equivalent cholinesterase inhibition.

added to the next revision of the Drinking Water Standards. Obviously, many of the criteria are based on limited data and analytical methods; therefore, research on the physiological effects, measurement techniques, and treatment of contaminants should be increased to develop creditable standards. EPA, through its Water Supply Research Laboratory at the National Environmental Research Center in Cincinnati, is planning to gradually expand its present activity in the area. Passage of legislation pending before Congress would undoubtedly speed up this research.

REFERENCES

1. Ettinger, M. B. *J. Amer. Water Works Assoc.*, *52*, 689 (1960).
2. McCabe, L. J. "Life Table Analysis of Heuper's Data," Interoffice Memo, Robert A. Taft Sanitary Engineering Center, Cincinnati, Ohio (January 1964). (now National Environmental Research Center, Cincinnati).
3. Heuper, W. C. and W. W. Payne. *Am. J. Clin. Pathol.*, *39*, 475 (1963).
4. Rosen, A. A., F. M. Middleton, and N. W. Taylor. *J. Amer. Water Works Assoc.*, *48*, 1321 (1956).

CHAPTER 2

NUTRIENT INPUTS TO A LAKE AND THEIR EFFECTS UPON WATER QUALITY

Donald B. Aulenbach and Nicholas L. Clesceri*

LAKE SYSTEM COMPONENTS

Water

There are two criteria which differentiate a lake from any other body of water. These criteria are: (1) The flow of water is essentially negligible. A specific rate of flow through a lake may be stated in order to differentiate between certain lakes and streams; however, no specific flow rate is offered here. Other influences, which may create a flow of water within a lake but do not come under the same category as downhill flow in a stream, are wind induced currents and seiches within a lake that may be caused by numerous factors. (2) There is a constant water level throughout its entire area. This factor differentiates a lake from a flowing stream in which the water is always moving downhill. However, a lake may be a specific portion of a stream in which the flow is very slow and the water level is essentially constant. Obviously, any small flow within a lake is induced by some downhill gradient, though this is normally very slight. Again, no specific value for a gradient is offered here due to various conditions, primarily the size of a lake, which would affect the definition. A very few lakes have absolutely no flow, such as Crater Lake in Oregon, which has no inlet and no outlet and whose level is maintained by precipitation and evaporation. The distinction between a lake and a pond is primarily a function of size. In this case, size is considered from the standpoint of surface area rather than the volume of the water.

---

*D. B. Aulenbach and N. L. Clesceri, Rensselaer Fresh Water Institute at Lake George, New York and Environmental Engineering, Rensselaer Polytechnic Institute, Troy, New York.

Obviously, the value used to differentiate between a lake and a pond may be quite arbitrary. New York State uses 10 acres (4 hectares) as the surface area above which size a pond would be considered a lake.

Lakes may be subdivided into size categories by various parameters. One parameter would be the total water volume, *i.e.*, the amount of water contained in the lake. This is usually considered to some mean level since fluctuations in water level may make a significant change in the total volume of water within a lake. Depth is an important parameter and should be included in any description of a lake, giving both the mean and maximum depth. The area of a lake is also an important parameter in describing a particular lake. A combination of the volume and area are frequently given by the ratio of surface area to volume. A similar ratio would be that of the shallow areas to the deep areas. The shallow areas are frequently defined as those down to the point of summer thermocline or as areas in which photosynthesis takes place (euphotic zone). Whenever the terms "shallow" or "deep" are used, these must be further defined.

Some consideration must be given to the source or sources of the water in the lake. Ultimately, all water is due to precipitation. A lake with a small drainage basin area may have as its prime source of water direct precipitation. Or, the major source of water for a lake with a very large drainage basin area may be surface runoff. In other instances, depending upon the ground conditions and the perviousness and/or imperviousness of the soil at various levels within the earth, groundwater may be the prime source of water for a particular lake.

Basin

The description of the lake bottom, or morphology of the lake, is a function of the original form of the valley or basin in which the lake initially formed, as modified by both erosion and silting. In general, a lake tends to fill in with both organic and inorganic debris until it ultimately ceases to exist in the form of a lake. The time required to reach extinction is a function of many factors, though initial depth of the lake is a prime factor. Only a few large lakes (*e.g.*, Lakes Baikal and Tanganyika) extend back beyond Quaternary times into the Tertiary; most lakes in existence today originated in the Pleistocene Era.[1] At the other extreme of the time scale is one of the most recently formed natural lakes, Crater Lake in Oregon, which is only approximately 1,000 years old. Of course there have been many new artificial lakes created recently due to man's dam building activities. The artifically formed lakes deserve the same consideration as naturally occurring lakes because the factors affecting a lake are similar whether it be of natural or artificial origin.

Of prime concern is the perviousness, or conversely, the imperviousness of the lake bottom. Generally, the least pervious bottom is most desirable because it maintains maximum retention of the water within the lake. This is generally achieved by solid rock or fine, compacted clay. Rock, however, has a tendency to crack and form channels that would allow the water to flow through the fissures that have been created. Where soluble rocks such as limestone occur, these may be dissolved by the water, creating channels that may carry the water either into or out of a lake. Such conditions exist in Florida where the first layer of solid material under the sand is limestone. Earthquakes and tectonic movements may change the bottom of the lake and may either open or close fissures, changing the rate of inflow or outflow of groundwater in the lake.

Further, the quality of the water within the lake is a function of the composition of the material that comprises the bottom of the lake. In many instances, the minerals dissolved from the bedrock may constitute sufficient trace nutrients to support prolific biological life. Limestone, for example, may contribute calcium and magnesium carbonates to the water, whereas salt formations may liberate large quantities of sodium and chloride to the water of the lake. The contribution of minerals to the lake water is as varied as the mineral content of the rock on the bottom of the lake itself.

A consideration is that the bottom ultimately becomes the receptacle of all the solid and precipitated material in a lake. Inorganic materials may accumulate and merely fill in the lake physically. On the other hand, decomposition products of biological systems (including both waste materials and dead cells) normally settle to the bottom where they undergo decomposition. Though this decomposition is primarily bacterial, there may be other higher forms of life that contribute. In general, organic benthal deposits become anaerobic at all depths except the extreme surface/ water interface when there is oxygen in the adjacent water phase. In deep lakes the bottom temperature is generally near 4° C, at which temperature bacterial decomposition is relatively slow. Thus, a gradual accumulation of undecomposed organic matter may occur. If, on the other hand, the organic debris settles in shallow areas where the temperature of the bottom is quite warm, bacterial decomposition may be much more rapid, and recycling of the dissolved materials of the decomposition back into the water phase may occur at a much more rapid rate. In the latter case, more nutrients occur for additional biological growth within the overlying lake water, increasing the biomass production and hence dead cells, which in turn settle to the bottom and increase the organic sediment. Thus, although there is less decomposition of the settled organic matter in the deep portions of a lake, the total amount of accumulation in shallow water may be greater due to the recycling of the nutrients that have been settled to the bottom of the lake.

Watershed

Of significant importance in the consideration of the water quality of a lake is the description of the watershed. The size and shape of the watershed greatly influence the types and quantities of materials carried into the lake. A large portion of the nutrients contained within a lake are contributed in some form or other from the watershed.

The first consideration is the size of the watershed. In general, the larger the watershed, the more opportunity there is for nutrients to be leached out of the soil or be contributed by other factors, including both natural forms of vegetation and human activities. Another parameter related to size is the ratio of the size of the watershed to the size of the lake itself. In general, the larger the ratio, the greater the problem of excessive nutrients becomes.

The slope of the watershed is also a significant factor. There are two extremes in slope. In a very gradual slope, the runoff is in contact with the soil or rock particles for a longer period of time, affording greater opportunity for the materials to dissolve out of the soil and get into the water tributary to the lake. With a steep slope, there is less contact time and the water from the rainfall reaches the lake in a shorter period of time, thereby dissolving fewer nutrients from the soil. However, as the rate of flow increases, so does the sediment transport. Thus, at high rates of flow, silt carriage can be significant. The ideal flow would be somewhere in between these two extremes.

The composition of the soil or rock through which the runoff passes on its way to the lake is also an important factor. This may be divided into both inorganic and organic components. In general, the inorganic components are the minerals of the rock or soil itself. If these minerals are relatively soluble they will readily dissolve in the runoff and be carried into the lake. Sodium and potassium salts of nitrates and sulfates are particularly soluble and easily dissolved in heavy runoff. Other materials such as calcium and magnesium carbonates (limestone) are slightly less soluble but will dissolve slowly with sufficient contact with the flowing water. Other materials such as heavy metals and silicates are relatively insoluble and will not be easily dissolved in the flowing water. The organic portion is generally the result of the vegetation on the watershed. Trees, brush, and grasses generally will grow where there is adequate water. Depending upon the latitude and the climate, these may have a cyclic nature. In temperate climates, vegetation will grow during the summer months and die during the winter months. In general, the dead leaves fall to the ground, where the leaching of internal material can be rapid, and are carried to the lake via runoff. In more tropical climates, this process is not as periodic as in temperate climates, but rather tends to be continuous. At very high altitudes there is little vegetation and,thus, the organic material is considerably less. Wildlife activities may also be a significant factor where such wildlife exists. Where herds of animals persist, such as is the case in parts of Africa, the contribution of organic animal waste may be significant.

The extent of human activities is also significant in the contribution of nutrients to a lake. In general, the human factor reduces the contribution from wildlife, but changes the composition, in that in addition to normal bodily wastes, wastes from garbage, trash, and laundry are also contributed to the watershed. The type of waste disposal is very significant in the contribution of nutrients to the lake. In urban areas, street runoff can be a significant source of both nutrients and silt. During construction periods, silt can be a serious problem. The total number of persons living in a watershed and the concentration of persons in a certain area (*e.g.*, around a small bay) are significant. Caution must always be taken to avoid taking average values over too large an area and failing to observe local concentrations which may result in the creation of local problems.

NUTRIENTS

In general, there are ten elements that are considered to be the major nutrients required to support biological life. Not necessarily in order of decreasing need, these include carbon, hydrogen, oxygen, nitrogen, phosphorus, sulfur, calcium, magnesium, potassium, and iron. Although silicon is not included, it is an essential element for the growth of diatoms. Sodium, also not included, is omnipresent. In addition to the list of major nutrients, there are many minor nutrients that must be present in trace amounts in order to support certain specific biological systems. These include cobalt, copper, boron, molybdenum, and others.[2] The amount of each trace element required is unknown and may vary for each specific organism. In addition, there may be other trace elements that are essential in extremely small amounts but have not been determined by the analytical techniques presently available. Mackenthun[3] has compiled a bibliography of trace materials that are essential to certain biological systems. In addition to the above elements, it includes such things as vitamins, thiamin, biotin, hormones, and organic chelators.[4]

There have been many estimates as to the ratio of the essential elements needed to support biological life in general. Vollenweider[4] has cited numerous references to studies that have attempted to evaluate this ratio. The ratio of the macronutrients required comes in the range of N:P = 15:1. Few researchers have gone any farther into the exact amount of the other nutrients necessary for biological growth. Generalities are difficult to make inasmuch as each individual species, and possibly each individual organism within a species, may have differing nutrient requirements. Even less is known about the specific requirements for some of the trace nutrients. Thus, only general statements of the amounts of each of the nutrients required to support biological growth can be made.

BIOSYSTEMS

The need for nutrients is related to the biosystems in existence. In general, the biosystems related to a lake have been divided into three main categories: (1) primary producers, (2) secondary producers, and (3) decomposers. Though there may be numerous other categories, depending upon the reason for the categorization, these three types will be sufficient to describe most situations.

Primary Producers

The primary producers may be defined as those relying upon photosynthesis as a source of energy for growth and reproduction. This group is responsible for converting sunlight into organic matter as cell material, which is a form of storing of solar energy. Approximately 5.3 per cent of the solar energy falling on the earth's surface is converted into organic matter through the process of photosynthesis.[5] Photosynthesis normally involves chlorophyll; however, other pigments may also be responsible for photosynthetic action in certain specific organisms.

The process is called primary production because it represents the conversion of solar energy into organic matter, which is the first process in the food web. In the terrestrial system, this primary production occurs from leaves and grass; in the aquatic system, algae are primarily responsible, though in some instances aquatic plants may also be important; in the marine system, the diatoms, a specific type of algae, predominate.

Secondary Producers

The organisms that depend upon the primary producers as a source of energy are commonly called secondary producers or consumers. An example of an aquatic food web may be described as phytoplankton, which are consumed by zooplankton, which in turn may be consumed by small fish, which may in turn be consumed by larger, and in some cases, still larger fish, which ultimately may be consumed by man. As the organic matter is transferred through each link of this food web, approximately 10 per cent of the energy of the consumed cell is converted into organic matter in the produced cell. The remaining 90 per cent is used for energy for existence, locomotion, reproduction, or other energy-consuming processes. It may be seen that by the time the system reaches man, large amounts of organic matter, solar energy, and nutrients have been consumed to provide for the needs of man.

Decomposers

In discussing the complete biological cycle one must consider the decomposers. These are the organisms which normally stabilize dead organic matter, converting it back into basic elements and some organic matter for the cell material of the decomposers themselves. A stable biological system requires a balance of the producers and the decomposers. If this were not so there would be an accumulation of primary producers which would soon deplete the nutrients, unless external sources were sufficient. The decomposers maintain this balance and allow for recycling of the nutrients available. Most frequently, the decomposers consist of bacteria, but may also include fungi.

SOURCES OF NUTRIENTS

Precipitation

In determining the balance of the sources of nutrients to a lake, consideration must be given to the nutrients that are contributed directly by precipitation. While many consider rainwater to be similar to distilled water, it can contain numerous impurities, some of which may be considered as nutrients. These accumulate as precipitation, primarily in the form of rain, falls through the atmosphere dissolving some of the gases as well as scavenging particles that occur in the atmosphere. These consist primarily of carbon dioxide but may also include oxides of nitrogen and sulfur, and traces of phosphorus. Obviously, nitrogen gas and oxygen would also be dissolved in rainwater. Of particular concern are nitrogen and sulfur. Measurements of the total nitrogen in the precipitation (including both wet and dry fallout) at Lake George, New York, indicate an average value of 110 μg/l of nitrogen.[6] Studies made at Hubbard Brook, New Hampshire,[7] and other locations,[4] have confirmed that total nitrogen concentrations in the range of 0.1 to 2 mg/l have occurred in rainfall. There has been no direct identification of the source of this nitrogen. However, oxides of nitrogen are contributed by most combustion processes including home heating and automobile exhausts. These oxides of nitrogen may be carried great distances through the atmosphere before they are scrubbed out by the rainfall. It must be pointed out that in the Lake George Basin the largest source of nitrogen comes from precipitation. Thus, in order to establish the levels of nutrients contributed to a lake, measurements of the nitrogen in the precipitation must be made.

Phosphorus concentrations as high as 10 μg/l have been found in the combined wet and dry fallout in the Lake George area. Though this is not the major source of phosphorus to Lake George, it is a significant contributor and again, must be evaluated in establishing a balance of nutrient inputs to a lake. Vollenweider[4] indicates concentrations of total phosphorus in precipitation may reach as high as 150 μg/l in some locations.

Sulfur is also an essential nutrient. At the present time very little information is available as to the total sulfur content of precipitation. However, inasmuch as sulfur oxides do gain access to the atmosphere, it is very probable that there is some sulfur contributed by precipitation.

It is likely that other nutrients may gain access to a lake system from the total wet and dry fallout. Further measurement and research would be required in this area in order to establish the exact levels of these nutrients that are available and/or are significant in a lake system.

Runoff

Measurements in the Lake George Basin have shown that the nutrient content of runoff is generally lower than that of the rainfall that produced the runoff. This is particularly true throughout the growing season during which time the growth of terrestrial plants may be affecting rain as well as nutrient throughfall. Some nutrients may be sorbed onto the forest litter either when the water percolates through the litter or when it drains as overland flow. However, in temperate climates consideration must be given to the dying plant material and the falling leaves that accumlate on the forest floor and from which nutrients may be leached out during either the fall runoff or spring runoff with the snow melt. As of the present time, the data resulting from studies that have been made to trace the variation in nutrient content with time of year and with extent of runoff are not sufficient to enable general conclusions to be drawn.

There is another factor that should be considered as affecting the quality of the runoff. Whereas some of the nutrients may be taken up by the plants growing on the watershed, depending upon the type of rock and soil material, there may also be added nutrients dissolved from the soil and rock and carried to the lake. The nutrients contributed by street drainage have been shown to be appreciable, particularly in urban areas.

Thus it may be seen that the effects of carriage of nutrients into a lake via the watershed are very much related to the type of soil and the development of human activities on that watershed.

Groundwater

The contributions of both quantity of groundwater and the nutrients contained in that water are two parameters that are more difficult to evaluate in a lake system. The contribution of groundwater may be either positive or negative; that is, there may be an inflow of groundwater to the lake or an outflow of groundwater from the lake. Depending upon the specific situation, it is even possible that the flow may be positive or negative depending upon the amount of precipitation on the watershed. During and for a period of time after heavy rainfall, water absorbed into the ground may percolate into the lake due to the higher gradient of the water in the ground relative to that in the lake. However, after periods of extended lack of precipitation the groundwater level may be considerably reduced and the water may proceed from the lake, supplying the water to the ground.

Contributions of groundwater to a lake may be of two general types. The first type is intra-basin, or the result of precipitation that falls within the basin, infiltrates into the ground, is carried to the first impervious layer, and then percolates in a downward gradient toward the lake. In areas where there is very little ground cover, or the impervious layer is near the surface, this groundwater represents only a very shallow entry into the lake. This situation provides a minimum amount of time of contact with the soil and rock particles and, therefore, results in relatively less dissolution of the minerals and nutrients into the aqueous phase. In instances where this groundwater is very near the surface, the nutrients may be taken up by the roots of plants that penetrate into the groundwater table. Thus, there may be some actual removal of nutrients from the groundwater into the plant material. In general, intra-basin groundwater is relatively low in dissolved minerals and nutrients.

On the other hand, inter-basin groundwater may come from an adjacent or even a remote area, flowing great distances through the soil and emerging generally at relatively great depths within the lake. This water is frequently quite high in nutrients. A prime example of the nutrient content of waters that travel great distances through the ground is the famous Saratoga Springs of New York. The recharge area for this water is approximately 100 miles west of Saratoga Springs. The water, as it emerges from various wells in the Saratoga Springs area, is high in nutrient content, and in some cases is even naturally carbonated. These waters have varying qualities, *i.e.*, some are considered very highly palatable while others are considered unpalatable, and certain are renowned for their medicinal qualities, or for use as therapy in hot baths. However, in many instances, the nutrients contained therein may stimulate algal growths in a lake system. Thus, in general, inter-basin groundwater will be higher in both mineral content and nutrients than water that travels shorter distances within the soil. It is frequently very difficult to determine the recharge area for a particular water that enters a lake.

The determination of inflow or outflow of groundwater in a lake is often very difficult. It is usually measured by hydrologic balances in a watershed. The amount of precipitation is relatively easy to measure, as is the outflow from the surface outlet of a lake. Major stream inputs to the lake may also be measured with good accuracy, and reasonable estimates made of the evaporation of water from a lake surface. This would be combined with evapotranspiration from the watershed itself. In general, it is considered that any significant discrepancies between the input and the outgo of a lake system are attributed to groundwater inflow or outflow.

Even more difficult to evaluate than the total inflow or outflow of groundwater is the quality of the inflow. In many cases, the inflow may be dispersed over a wide area and at several depths within the lake. It is difficult to obtain a sample of water from such an aquifer system without having it mixed (the equivalent of contamination) with the lake water itself. Underwater wells penetrating the aquifer just before its discharge into the lake provide the best information on the quality of the groundwater prior to its entrance into the lake. Thus, even if the location of the groundwater influent to the lake can be determined, it is still difficult to obtain a representative sample that can be used to establish a nutrient balance in a lake.

Human Development

There is an old statement that says "a lake is born to die." However, human development has been shown in many instances to speed the rate of dying of a lake. Thus, the extent and nature of human development on a lake must be evaluated as a very important factor of the quality and character of the waters within a lake. Changes may be of two forms. One is the physical changes that man applies directly to the lake, and the other is the result of the nutrients that are contributed by the various activities of man, generally in and around a lake.

Physical

There are many things that man has done to change the total amount of water within a lake. In some instances, he has diverted streams into a lake so that the total supply of water may be greater. In other instances, man has used the lake as a water supply reservoir and has diverted much of the water out of the lake to provide for human needs, including drinking water, industrial use, and irrigation of agricultural crops. Extreme variations in surface water level may make significant changes in the amount and type of shoreline exposed. It will also make a significant change in the direction of groundwater flow into or out of the lake. With a smaller amount of water within a lake the nutrients that reach the lake may be more concentrated due to the lesser amount of dilution available. In instances where a lake is designed primarily as a water supply reservoir the secondary effects of the fluctuation in water level are of less concern, inasmuch as the prime concern is to provide the amount of water needed for man's uses. However, if the change in the water level significantly affects the quality of the water provided, it may impair the use of the water for certain demands. Thus, a consideration must be made of any alterations in quality of the water due to the changing water level.

Man also frequently changes the shoreline. There are many varied uses of a lake, each of which demands a different shoreline. For swimming, for example, a gently sloping sandy bottom is preferable, whereas boating and shipping require steep-sloped, deep shorelines. In other cases, shallow areas of lakes are filled in to provide more area for housing. Many proposals have been made to build structures out over the water due to the shortage of land available; in fact, such a proposal has already been undertaken in some instances. Each shoreline change has some effect upon the quality of the water within the lake and the biosystem associated with that shoreline. For example, rooted aquatic plants generally occur only in shallow areas. Thus, any change in the amount of shallow areas available will significantly change the amount of rooted aquatic plant growth. Otherplants such as *Nitella sp.*, which may occur at only certain depths within a lake, are primarily governed by the depths of light penetration and/or nutrients.

Nitella has an optimum light absorbance at which it can best survive in a lake. The depth of this optimum absorbance is a function of the turbidity of a specific lake. Dredging to create harbors may remove contact areas for bottom growing organisms. The construction of docks or piers may change the shoreline current or may cover areas which are optimum for the growth of certain specific biological systems. Where boats enter shallow areas, the propellers frequently stir up the bottom muds, displacing them to other areas within the lake. The turbidity itself may affect the breathing systems of fish, or the displacement of the silt may cover either breeding areas for fish or the areas where fish food would normally grow at an optimum rate.

Dredging and filling to change conditions within a lake often create a problem similar to those resulting from changes in the shoreline. Where dredging is conducted in order to maintain shipping, the removal of the dredging materials normally creates great turbidity which can interfere with either fish or fish food organisms. Proper disposal sites must be prepared for the dredging materials so that these materials do not ultimately get back into the lake. Very frequently, the material dredged is used to fill a swampy area to provide for housing developments. In many instances, these wetlands are the prime areas for the spawning of fish; thus, filling swamp areas may destroy the natural fish productivity of a lake.

A major concern is the problem of construction within the lake drainage basin area. Terrestrial vegetation tends to slow the flow of water to the lake and withholds the soil at its original location. Normally, one of the first phases of construction is the clearing of vegetation from the surface of the land. Often the vegetation in adjacent areas is destroyed by construction machinery. Barren of vegetation, the soil is washed away by rainfall and is carried into the lake at a much more rapid rate. This results in the filling in of certain areas of the lake, and of course, the depositing of additional nutrients and turbidity in the lake. Provisions should be made to slow the passage of the runoff to the lake during construction periods by the use of horizontal curbing or ridges that would hold the water back and allow the silt particles to settle out before they reach the lake.

Street runoff is a major concern. In areas where snow occurs during the winter, salt is frequently applied to both streets and highways. Studies in the Irondequoit Bay near Rochester, New York,[8] have indicated a significant increase in the salt content of the bay since the start of salting of the roads in the area. Inasmuch as this bay is already meromictic in the spring and the fall overturn is delayed by approximately a month, the addition of this salt prevents to an even greater extent the exchange of bottom waters with the outflow of the bay into Lake Ontario. In urban areas, street runoff can also include the same products of transportation that occur in rural areas. These include rubber, petroleum products, and asbestos from brake linings. In urban areas of temperate climate, the organic material from fallen leaves may be leached into the storm drains and to the streams having access to the lakes. Street runoff has been shown to contain significant amounts of both nitrogen and phosphorus, thus possibly contributing significant nutrients to a lake.[9] These nutrients are very difficult to control inasmuch as the significant volume after a heavy rainfall would preclude its being treated directly in a waste treatment system. However, one scheme is to hold the storm water in any one of many different systems until it can be treated in a normal treatment system. Holding facilities include large surface water tanks and, in at least one instance, large plastic or rubber bags that are suspended beneath the water in a bay area. Handling of storm water presents a specific problem which must be considered for each lake system.

Waste Discharges

There are numerous ways of differentiating the types of wastes and their discharges. From the standpoint of a lake, the two types of categories of concern are point sources and diffuse sources. In general, all types of wastes may be from a point or a diffuse source, but in many instances, one or the other is more common to a specific type.

One of the most significant diffuse sources is fertilizer used for agriculture, and for domestic purposes for lawn care in urban areas. Unless the runoff from such fertilized areas is collected in a stream, it usually occurs as a diffuse source and frequently enters the lake system through the groundwater. There is great variation in the type of fertilizer used and, therefore, the same variation in the nutrients that gain access to the lake. Domestic wastes may fall under both categories. Septic tank systems, which are frequently used in resort areas surrounding a lake, represent a diffuse source. If the septic tanks are operating properly, any nutrients that reach the lake do so by passage through the soil and occur as groundwater inflow to the lake. In some instances there may be some improperly operating septic tanks that have a surface runoff directly to the lake or a nearby stream. On the other hand, where municipal sewage treatment facilities are provided, the nutrients are concentrated at a point. The ultimate discharge of the treated wastes will determine whether or not the point discharge is a greater problem than the distributed discharge. The greatest potential advantage of a centralized waste treatment system is that the wastes can be controlled more readily. It may be determined, for example, that the best system would be to divert the collected and treated waste into another watershed, particularly to a flowing stream, possibly downstream from the lake in question. In general, nutrients in a stream are less of a problem than nutrients in a lake. On the other hand, if the treated effluent with all its nutrients is discharged into a small bay or a small area of a lake, it may create an even greater problem than if the waste had been discharged into individual septic tanks. Soil conditions in the area also govern the desirability of septic tank systems versus collected and treated sewage systems. One of the main difficulties here is that traditionally the prime concern was for the removal of the water involved. Thus, a coarse, open, sandy soil would be the most desirable for discharge of a septic tank effluent. However, in a lake the prime concern is generally the nutrients rather than the total volume of water distributed by sewage systems. In this case, a coarse, open soil will allow both the water and its contained nutrients to reach the lake in a shorter period of time, thereby providing less contact time with the soil for either ion exchange or uptake by biological systems. In general, the optimum type of soil for the removal of nutrients is a fine clay-like soil. This type of soil frequently has exchange capacities for phosphorus, though nitrogen is generally scarcely removed by any type of soil system. A properly operating septic tank system in which the seepage area is aerobic will result in conversion of the reduced nitrogen compounds to nitrate. Thus, in order to utilize the ion exchange capacities of the soil, fine clay-like soil would be more desirable for the removal of nutrients. It may be seen that this is converse to the type of soil desired for the removal of the water. Another advantage of the collection and treatment of the waste is that with the waste all in one point it is more amenable to removal of the nutrients by chemical or biological means.

The other significant type of waste is industrial waste. In general, these are point sources, although in many instances, there may be numerous point source discharges from any one industry. Industrial wastes are so varied and diversified that space cannot be allotted here to discuss each one individually. Suffice it to be said that any of the raw materials, products, or by-products of an industry may gain access to a lake if the industry is on that watershed. Measurements must be made of the specific materials in the wastes reaching the lake, and from this, an evaluation can be made of the effects of the discharges from that industry upon the lake.

WATER QUALITY STANDARDS

Initially, water quality standards were established from the standpoint of protecting the health of the water consumer. The first concern was the bacterial quality, the second, the chemical quality. There was also concern for the physical parameters such as turbidity, color, and odor, which were primarily esthetic problems. There are two lists of chemical water quality parameters.[10] Substances present in concentrations greater than recommended in the first list would be justification for seeking another water supply. The second list includes chemicals which, if present in concentrations above the stated limits, would be cause for rejection of the water supply. In general, the first list affects the esthetic quality of the water, whereas the second list represents potential health hazards. These lists are updated periodically and additional elements or compounds are added as necessary. For example, the nitrate concentration recommended for drinking water purposes has been reduced from the original recommended limit of 45 mg/l to the present level of 10 mg/l.[11] Nitrates are included in the list of recommended concentrations, as indicated above; however, phosphates are not on any list.

The U.S. Environmental Protection Agency (EPA) has established guidelines for the phosphorus content of lakes and streams. In its latest guidelines,[12] EPA has stated that phosphorus shall not exceed 25 μgP/l in lake waters. Further, phosphorus should not exceed 50 μgP/l in a stream at a point at which the stream enters a lake or reservoir, and should not exceed 100 μgP/l in flowing water at any time. Though it is impossible to directly compare drinking water standards and stream and lake standards since different elements are cited, the stream and lake standards are more stringent than the drinking water standards.

As has been discussed by Vollenweider,[4] phosphorus concentrations in excess of 10 μgP/l and nitrogen concentrations greater than 300 μgN/l in a lake may allow excessive algal growth. It may be seen that the latest EPA guideline of 25 μgP/l for phosphorus in a lake is 2.5 times the accepted limit above which excessive algal growth may occur. If this same factor were applied to the total nitrogen content, nitrogen concentrations over 750 μg/l (2.5 times 300 μgN/l) would not be acceptable in a lake system. This value of 750 μgN/l is considerably less than the 10 mgN/l which is acceptable for drinking water. Thus, it may be concluded that water quality standards for the control of eutrophication are considerably more stringent than those for human consumption of water.

Another factor must be considered. As reported previously,[4] phosphorus concentrations of up to 130 μgP/l have been recorded in rainfall. Thus, the importance of precipitation as a source of nutrients and other chemicals must not be overlooked in many instances.

The important fact is that drinking water quality standards cannot be used to evaluate the quality of water in a lake. The control of nuisance algae in a lake requires much more stringent water quality standards than drinking water quality standards.

CASE STUDY: LAKE GEORGE

As an application of a balance of nutrients in a lake, the various concentrations of nitrogen and phosphorus in tributary waters of Lake George, New York, were measured and compared with the nitrogen and phosphorus content of the waters of the lake. Lake George can be described by the parameters shown in Table 2.I. Basically, the lake is long and narrow, having been formed from two river valleys by the receding glacier approximately 10,000 years ago. It may be classified as a deep lake and it is subject to thermal stratification during the period from about May to October. Because it is an oligitrophic lake, the minimum dissolved oxygen in the hypolimnion at the end of the summer is seldom less than 65% saturation. The area of the tributary watershed is relatively small, being approximately four times the area of the surface of the lake itself.

Table 2.I

Description of Lake George[13]

Length	51 km	32 mi
Mean Breadth	2.3 km	1.4 mi
Max. Breadth	4.0 km	2.4 mi
Area	114 km^2	44 mi^2
Max. Depth	58 m	191 ft
Mean Depth	18 m	59 ft
Length of Shoreline	209.6 km	131 mi
Volume	2.1 km^3	0.5 mi^3
Watershed Area (land)	492 km^2	190 mi^2
Watershed Area (including lake)	606 km^2	234 mi^2

An estimate of the annual nitrogen and phosphorus sources and sinks in Lake George is presented in Table 2.II. It must be emphasized that the concentrations of these elements vary in all their sources and that concentrations in the lake also vary to a degree. Thus, the values used are the best estimates that could be obtained from the approximately four years of data available at the present time. It may be seen that about 79% of all of the nitrogen contributed to the lake on an annual basis comes from precipitation directly on the lake. The amounts of nitrogen falling onto the surrounding land of the basin are not added to this value inasmuch as the values for the stream runoff represent the actual quality of the precipitation runoff water as it reaches the lake. There is a significant reduction in the total nitrogen content between the precipitation and the runoff. Since most of the analyses were made during the growing season, this very likely represents the uptake of nitrogen by the terrestrial vegetation. Data are presently being collected in order to evaluate the nitrogen content over the entire season, with particular reference to spring runoff during periods of snow melt. The values of flow from the Lake George Village Sewage Treatment Plant are quite accurate because they are taken from the flowmeters at the Treatment Plant. The 10 mg/l of nitrogen in wastewater is also a reasonable value inasmuch as it represents both the nitrogen content of the effluent from the treatment plant before it is applied to the sandbeds, and the concentration as the effluent emerges from the ground and into a nearby stream. At present, data are not available from the Bolton Landing Sewage Treatment Plant, the only other sewage treatment plant in the basin. Values from the remainder of the lake are based on population and reasonable values of concentration from septic tank systems. The discharge at Ticonderoga is measured by means of a recording gauge, and the concentration of nitrogen is obtained from frequent measurements of the nitrogen content of the water near the discharge. It may be seen that only 11% of the total annual nitrogen added to Lake George passes out of the lake through the discharge stream. It must be assumed that the other 89% of the nitrogen added to the lake annually remains in the lake and is most likely precipitated to the bottom sediments since the nitrogen is not observed in the water of the lake. For comparison, the standing amounts of nitrogen in the entire lake are also indicated. It may be seen that the total amount of nitrogen at any one time is somewhat greater than the annual contribution to the lake. Therefore, the detention time of the nitrogen in this lake is approximately 10 years, which is the duration of the hydraulic residence.

In the case of phosphorus, the 10 μgP/l in the total annual precipitation represents approximately 10% of the total addition to the lake. The runoff in the streams is somewhat higher than the precipitation since runoff from urban areas is included in this figure. Streams in non-urban areas indicated a concentration of 7 μgP/l, which represents a slight reduction in the total phosphorus content. This reduction could be attributed to plant uptake. The increase in the phosphorus content in the stream runoff is due largely to the high concentration of phosphorus in urban runoff. Thus, stream runoff contributes approximately 25% of the phosphorus to Lake George. The Lake George Village Sewage Treatment Plant discharges its secondary treated effluent onto natural sandbeds with no collection of the water that has passed through the sand. During the spring of 1973 an outcropping of water approximately 0.5 miles away from the treatment plant was found. This outflow ultimately reaches West Brook, which is tributary to Lake George. Although positive tracer studies have not been conducted, it is assumed that this outcropping of water represents the treated sewage discharge onto the sandbeds. At the point of the outcropping a few measurements of the phosphorus concentration are on the order of 150 μgP/l. Thus, it may be considered that the soil does accomplish a significant reduction in the total phosphorus concentration of the sewage effluent. Further studies are being conducted to confirm that this is actually the water from the treatment

Table 2.II

Estimated Annual Nitrogen and Phosphorus Sources and Sinks in Lake George

		Nitrogen			Phosphorus		
	Quantity km^3/yr	Conc. $\mu g/l$	Tot.wt. kg/yr	% of Total Add'n. to Lake	Conc. $\mu g/l$	Tot.wt. kg/yr	% of Total Add'n. to Lake
Sources							
Precipitation[a]							
Directly on lake	0.105	1,350	142,000	79	10	1,050	10
Surrounding land	0.442	1,350	596,000	-	10	4,420	-
Entire basin	0.547	1,350	738,000	-	10	5,470	-
Runoff streams	0.181	80	14,450	8	15	2,715	25
Wastewater							
Lake George Vil. STP[b]	8.198×10^{-4}	10,000	8,198	4	150	123	1
Bolton Landing STP	Values not yet available				Values not yet available		
Remainder of Lake	-	10,000	13,130	7	c	6,770	64
Total	-	-	21,328	11	-	6,893	65
Total	0.286	-	177,780	100	-	10,658	100
Sinks							
Discharge at Ticonderoga	0.225	92.5	20,812	11	6	1,350	13
Bottom Sediments[d]	-	-	156,968	89	-	9,308	87

	Quantity km^3	Nitrogen		Phosphorus	
		Conc. μg/l	Tot.wt. kg	Conc. μg/l	Tot.wt. kg
Standing Amount					
Water of Lake George	2.1	100	210,000	7	14,700

[a] Includes wet and dry fallout

[b] Sewage Treatment Plant

[c] Based on 2g P/cap-day

[d] By difference

plant and to affirm that the value of 150 μg/l of phosphorus is valid.

The Bolton Landing Sewage Treatment Plant also utilizes a natural sand bed; however, this bed is not so deep as that at Lake George Village and several outcroppings of water have been reported in the area around the treatment plant. Thus, it is possible that there is less phosphorus removal in the treated Bolton Landing Sewage effluent because there is less passage through soil. Up to the present time, measurements have not been made of the flow or the concentration at the Bolton Landing Sewage Treatment Plant and, thus, no estimates can be made of the contributions to the lake. However, this source is included to indicate that it may be a significant contributor to the nutrient loading of the lake. On the other hand, the diffuse potential sources of phosphorus around the lake from septic tank discharges represent 64% of the potential input of phosphorus. Since flow data are not available and little is known as to the extent of phosphorus removal by the soil in individual septic tank systems, the contribution of phosphorus was based on 2 grams of phosphorus per capita per day. While it is realized that the soil does have a certain capacity for removing phosphates, it must also be pointed out that in Lake George some of the septic tanks are located very close to the lake and there is little opportunity for removal of the phosphorus by the soil before the liquid effluent reaches the lake. It is for this reason that no reduction in phosphorus is assumed for the septic tank discharges. The lake discharge at Ticonderoga represents less than 13% of the total annual input of phosphorus to Lake George. This indicates that approximately 87% of the annual phosphorus added to Lake George remains in the lake and is probably deposited in the bottom sediments. The 14,700 kilograms of total phosphorus in the standing waters of the lake at any one time is slightly greater than the annual contribution from all sources. The residence time for the phosphorus is on the order of 11 years.

In determining the nitrogen and phosphorus balances in Lake George, some other factors were taken into account although they are not shown in the table. The contributions from fertilizer were considered to be negligible since less than 2% of the entire basin is actively farmed. On the other hand, numerous persons living around the lake do fertilize their lawns. Although this may have some effect upon local bays, it is not considered to be a significant factor for the entire lake. Other inputs could include swimmers and the effects of boating; again, however, these sources are considered to be insignificant when compared with the total overall input. Likewise, removals from the lake of fish and organisms, such as flying insects, which spend their developing stages in the water and then emerge, cannot be considered significant when compared to the total amount of nutrients within the lake.

It becomes immediately evident that there is a large discrepancy between the amount of nutrients flowing into the lake and the discharge through the outflow at Ticonderoga. Apparently, nearly 90% of the added nitrogen and phosphorus is retained in the lake. Although direct measurements have not been made of the nitrogen and phosphorus content of the sediments, it is assumed at the present time that this is the ultimate sink for these nutrients. In order to confirm that these nutrients are being deposited in bottom sediments, extensive studies would be needed to measure the accumulation or changes in these sediments over a period of a year. Aerial distributions of the sediments could make the interpretation of results difficult.

Some additional unusual results were found during a sampling period at Lake George in 1972. In order to observe any changes during a shorter period of time, samples were taken every four hours at two different stations in Lake George during July, 1972. The results as shown in Figures 2.1 and 2.2 indicated that the phosphate concentration went through a low value and returned to a high point over a 48-hour period. At both station 5A, which is somewhat subject to human stress, and station 6, which is little subject to human stress, the concentration of orthophosphate was similar at all depths. The values varied from less than 0.2 μg/l, the minimum detectable limit by the analytical methods used, to 2 μgP/l, which is considered a significant variation. In order to confirm this potential orthophosphate cycle, a second series of sampling was conducted in August, 1972, this time selecting station 1, which is the location most subjected to human stress, and station 6, to compare the results with the previous study. Samples were secured every six hours at only the five-meter depth since the results of the previous studies indicated little significant variation with depth. Also, the study was continued for 72 hours to see if the trend was similar to that observed previously. Results of this study are shown in Figure 2.3. It is obvious that the same cycle that was apparent in July was repeated in August. Stations 1 and 6 are approximately 25 miles apart and the cycle was in the same phase at both

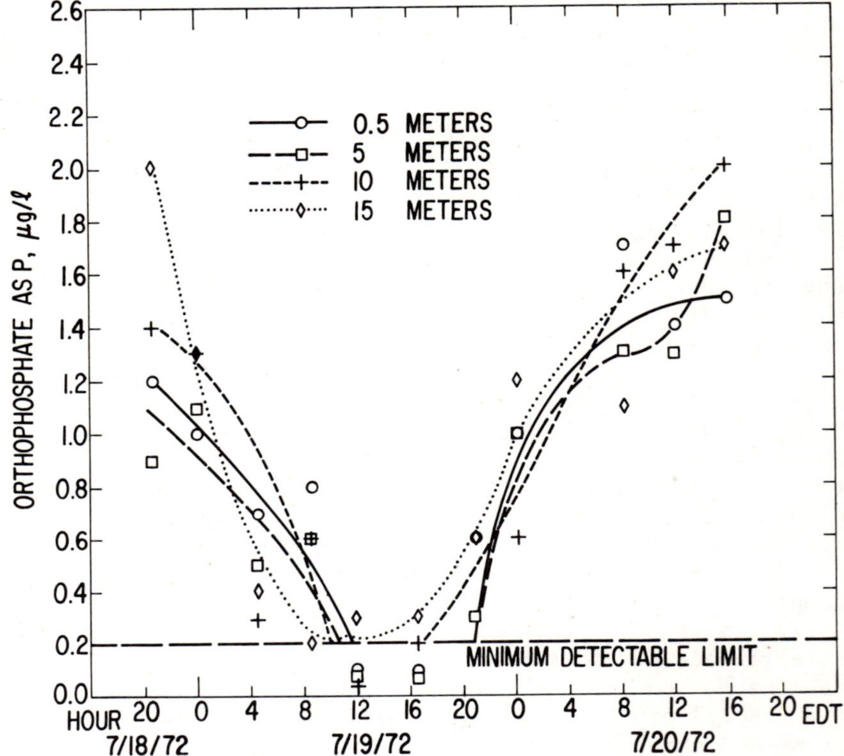

Figure 2.1. Orthophosphate variation with time, Station 5A, Lake George.

locations; the August samples were in phase with the 48-hour cycle from the July samples. These results seem to confirm that there was a cycle that extended over approximately 48 hours in the orthophosphate concentration within Lake George.

The observation of a cycle of orthophosphate concentration may be significant in reviewing the correlation in the literature between nitrogen and phosphorus content of a lake and excessive algal production. In many instances the results correspond, but in other instances they appear to be contradictory. Obviously, there are cases where some other nutrients may have been limiting, and thus, the nitrogen and phosphorus would not be the only parameters which could control excessive algal growth. However, there

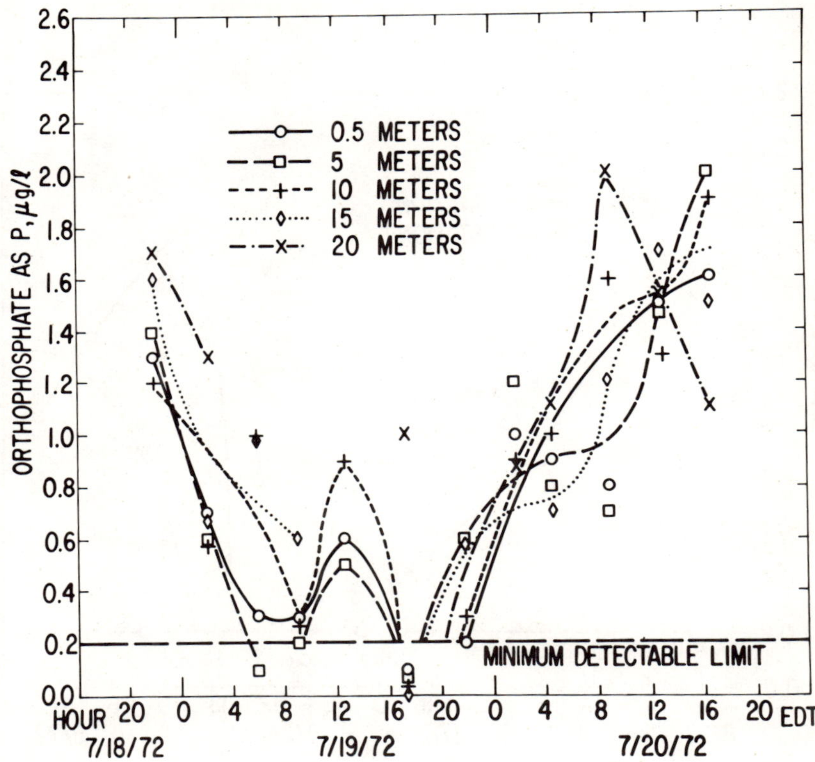

Figure 2.2. Orthophosphate variation with time. Station 6, Lake George.

are numerous instances in the literature where the same nitrogen and phosphorus content resulted in no excess algal productivity in one lake but a large excess productivity in another. Applying this orthophosphate cycle to these results may give an indication of at least one of the reasons for this discrepancy. The orthophosphate concentration varied by a factor of 10. If, as is usually the case, the orthophosphate samples were secured on a random time basis, and an insufficient number of samples had been obtained for a statistical analysis, the values recorded for the orthophosphate content of any body of water could potentially vary by a factor of 10 from that value. Thus, what may be designated as a high or low value could be just the opposite depending upon the exact time of the sampling within this 48-hour cycle.

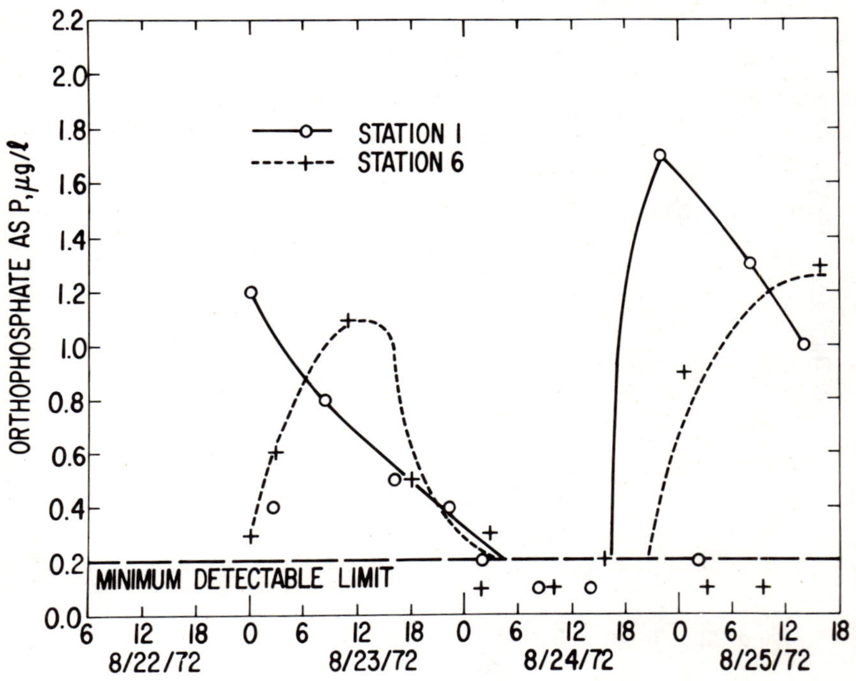

Figure 2.3. Orthophosphate variation with time. At 5-meter depth, Lake George.

From the data available, the limiting concentration of phosphorus for excessive algal growth cannot be determined, nor is it possible to explain the cause for the cycle. However, it is evident that additional analytical work will have to be performed on a lake in order to determine whether a phosphorus concentration as measured represents the total picture within a lake. Further information must also be gathered to determine whether the high point, the low point, or a mid-point on the cycle could represent the limiting amount of phosphorus in a lake system.

SUMMARY

For a complete evaluation, a lake must be described not only according to its physical characteristics, but also according to its chemical and biological parameters. In general, certain biological activity is controlled by the input of chemicals to a lake. Therefore, determining the inputs of nutrients to a lake is the first step toward a better understanding of biological growth. The fate of the nutrients must also be known in order to maintain a complete balance within any lake system. In order to prevent excessive algal growth, the quality of the water in a lake must be superior to that required for human drinking water consumption. Orthophosphate appears to fluctuate for a 48-hour cycle by a factor of approximately 10, which makes the evaluation of the phosphorus content of a lake very difficult to establish. It may also help to explain some of the discrepancies that have existed in establishing a limiting phosphorus level for the production of excess biological growth within a lake.

The evaluation of the nutrients in a lake and their effects upon water quality is not a simple undertaking. It is not something that can be established by taking one sample within each lake. It requires sampling at various locations and at different depths within a lake. Such a sampling should continue for a period of at least one year in order to get best interpretation of the results. At least one extended period of sampling should be conducted for a minimum of 60 hours, with sampling at four or six hour intervals to determine any cycles of nutrients during this period. With sufficient information of the proper type, a correlation can be established between nutrient content and biological productivity. With knowledge of the sources of inputs of these nutrients, means of control can be established to maintain the quality of lakes and waters in a manner satisfactory for the best uses of all.

ACKNOWLEDGMENT

Research supported in part by the New York State Science and Technology Foundation and in part by the Eastern Deciduous Forest Biome Project, U.S.-International Biological Program, funded by the National Science Foundation under Interagency Agreement AG-199,40-193-69, with the Atomic Energy Commission--Oak Ridge National Laboratory. EDFB Memo Report 73-14. FWI Report 73-8.

REFERENCES

1. Ruttner, F. *Fundamentals of Limnology*. University of Toronto Press, Toronto, Canada, 1969.
2. Mackenthun, K. M. "Nitrogen and Phosphorus in Water. An Annotated Selected Bibliography of their Biological Effects," U.S. Department of Health, Education and Welfare, Public Health Service, Washington, D.C., 1965.
3. Mackenthun, K. M. "Algal Growth. Aqueous Factors Other than Nitrogen and Phosphorous. Selected Bibliographical References," U.S. Department of the Interior, Federal Water Pollution Control Administration, Washington, D.C., 1966.
4. Vollenweider, R. A. "Scientific Fundamentals of the Eutrophication of Lakes and Flowing Waters, with Particular Reference to Nitrogen and Phosphorous as Factors in Eutrophication," Organization for Economic Cooperation and Development, Directorate for Scientific Affairs, Paris, France, 1968.
5. Gates, D. M. *Scientific American, 224(3)*, 89 (September, 1971).
6. Williams, S. L., E. M. Colon, R. Kohberger, and N. L. Clesceri. *Bioassay Techniques and Environmental Chemistry*, G. E. Glass, Ed., Ann Arbor Science Publishers, Ann Arbor, Michigan, 1973.
7. Fisher, D. W., A. W. Gambell, G. E. Likens and F. H. Borman. *Water Resources Research, 4*, 1115 (1968).
8. Diment, W. H., R. C. Bubeck, and B. L. Deck. "Some Effects of De-Icing Salts on Irondequoit Bay and its Drainage Basin," Highway Research Records No. 425, Highway Research Board, National Research Council, 1973, pp. 23-25.
9. McCarty, P. L. *J. Amer. Water Works Assoc., 59*, 344 (1967).
10. Public Health Service. "Drinking Water Standards," Department of Health, Education and Welfare, Publication No. 956, Washington, D.C., 1962.
11. Federal Water Pollution Control Administration. "Report of the Committee on Water Quality Criteria," Department of the Interior, Washington, D.C., 1968.
12. Environmental Protection Agency. *Clean Water Report, 11(10)*, 93 (1973).
13. Aulenbach, Donald B. and Nicholas L. Clesceri. "Sources of Nitrogen and Phosphorus in the Lake George Drainage Basin: A Double Lake," Proceedings of the 19th Annual Technical Meeting of the Institute of Environmental Sciences, April 2-5, 1973. p. 424.

CHAPTER 3

APPLICATION OF GAS CHROMATOGRAPHY OF VOLATILE METAL CHELATES TO WATER ANALYSIS

Roger A. Minear and Cheryl M. Palesh*

INTRODUCTION

The toxic nature of many metals, even at the trace level, has been known for many years with respect to public health. Lead, cadmium, arsenic, mercury, antimony and beryllium all fall into this category and have been reported as causative agents in accidental industrial deaths.[1] In addition, many metals have been demonstrated to be toxic to aquatic life above certain "threshold" levels. Maintenance of metals below these levels was assumed adequate. Recently, a "no threshold concept" has been proposed, and subsequently challenged,[2] which suggests that detrimental effects might be experienced at all concentrations.

Avoiding the argument over the validity of the "no threshold concept," the nondegradable nature of heavy metals in the environment and their failure to be completely locked into sediments are documented facts. Biological transformation and concentration magnification in water systems have been well documented for cadmium[1] and mercury[3] and concern has been expressed regarding other metals such as arsenic, selenium and tellurium.[4,5]

As concern continues to develop with respect to the long term effects of low level exposure to many metals, the need

---

*R. A. Minear, Department of Environmental Engineering, Illinois Institute of Technology, Chicago, Illinois.
C. M. Palesh, Illinois Water Treatment Company, Rockford, Illinois.

for inexpensive multi-element analysis techniques grows. A wide range of methods is available for the analysis of trace metals in water, and their respective sensitivity, selectivity, cost, sample handling and multi-element capabilities have been discussed.[6] The use of volatile metal chelates shows promise of becoming the method of ultimate sensitivity for some metals when total available mass of the element becomes a consideration. Such is the case of multi-element analysis on natural water samples. While atomic absorption spectrophotometry provides high selectivity, simultaneous analysis of more than one element is not yet a practical reality. For samples containing analyte metals below direct detection limits, concentration techniques must provide sufficient solution volume for repeated analyses, frequently resulting in the processing of large volumes of water.

Within the last ten years, the successful determination of trace metals by gas chromatography has received much attention for analysis in a wide variety of sample matrices[7-22] most of which ultimately were converted to an aqueous system prior to the analysis procedure. The simplicity of such procedures, the potential for sample concentration by extraction into an organic phase, and the sensitivity of gas chromatography make trace analysis for some metals a fast and simple procedure. Extension of this technique to direct analysis of natural water systems seems logical.

CHELATING AGENTS

Chelates of β-diketones, $R\text{-}\overset{O}{\overset{\|}{C}}\text{-}CH_2\text{-}\overset{O}{\overset{\|}{C}}\text{-}R'$, have been most extensively studied owing to the combination of their volatility, and their thermal and solvolytic stability. Of the eleven or more chelating agents forming volatile metal chelates,[6] trifluoroacetylacetone (tfa) ($R = CF_3$, $R' = CH_3$) has received the widest use as an analytical technique for determination of beryllium, chromium, copper and aluminum (References 13 through 19, 21, 22).

The chelating agent, 2,2,6,6-tetramethylheptane-3,5-dione (dipivaloylmethane or piv) (R = R' = t-butyl) has been used successfully in the extraction and gas chromatographic analysis of the rare earths.[23] When applied to the analysis of alkali[11] and alkaline earth[24] metals, vapor phase exchange reactions resulting in mixed complexes prevented successful and unambiguous evaluation of the components. Piv has also been studied as an extraction agent for

Fe^{3+}, Al^{3+}, Cu^{2+}, Ni^{2+}, Zn^{2+}, Cd^{2+}, Mn^{2+}, Co^{2+} and Pd^{3+}.[25] Iron, copper, and aluminum were quantitatively extracted while zinc, nickel, manganese, cadmium and cobalt ranged from 53 to 85% for single extractions and were strongly pH dependent. No prior work appears to have been done on the gas chromatography of $Pb(piv)_2$ or its reaction from aqueous solution. Metal ions are complexed by the enolate form of the chelating agent with subsequent loss of hydrogen ion by the β-diketone

$$n\ \mathrm{R{-}\underset{\underset{O}{\|}}{C}{-}CH = \underset{\underset{OH}{|}}{C}{-}R'} + M^{n+} \rightarrow$$

$$\mathrm{M(R{-}\underset{\underset{O}{\|}}{C}{-}CH = \underset{\underset{O}{|}}{C}{-}R')_n} + n\ H^{+}$$

The effect of structure of the ligand on the volatility of the chelate was evaluated by Berg and Truemper,[26] with the systematic substitution of alkyl, fluorinated alkyl, aromatic and heterocyclic groups for R and R'. The degree of volatility is related to symmetry and polarity of the molecule, with spherical molecules being more volatile than planar molecules. This is the factor responsible for the success of GC analysis of a metal chelate. A spherical molecule is approximated by a tetrahedral or octahedral arrangement of the metal coordination bonds. Of the metal piv chelates, nickel and copper are planar, while the rest are tetrahedral.[27] Similarly, the tfa chelates take a tetrahedral or octahedral configuration, when fluorine rather than methyl groups are attached to the alpha carbon.

The greater volatility of the fluorocarbon chelates appears to be a general phenomenon.[26,28] The reason for this is not exactly known, but several possibilities have been advanced.[29] This phenomenon may be due to a reduction in van der Waals forces, with a decrease in intermolecular hydrogen bonding. Molecular models of octahedral hexafluoroacetylacetone (hfa) chelates demonstrate that a larger portion of the periphery is occupied by the fluorine; this means that the metal ion is then enclosed in a highly electronegative fluorocarbon shell. The resulting intermolecular attracting forces in the above are weaker in comparison to nonfluorinated β-diketones. Another explanation is that the somewhat bulkier fluorine atoms could prevent close packing in the crystal lattice, a factor that would greatly enhance volatility.

The property of volatility is of importance in considering GC analysis. The elution of tfa and hfa chelates is accomplished at temperatures between 50 and 150° C; thus, thermal decomposition is much less a problem.[30]

The octahedral complexes of the unsymmetrical bidentate ligands form trans and cis isomers, *e.g.* $Cr(tfa)_3$. In the cis isomer, the three trifluoromethyl groups are adjacent in the upper front face of the octahedron. The trans isomer has one of these groups reversed. While one of the ligands is still cis to one of the CF_3 groups, it is trans to the other; this is shown in Figure 3.1. Due to the polarity of the cis isomer, it is less volatile than the trans form and is eluted from the GC column prior to the cis isomer.

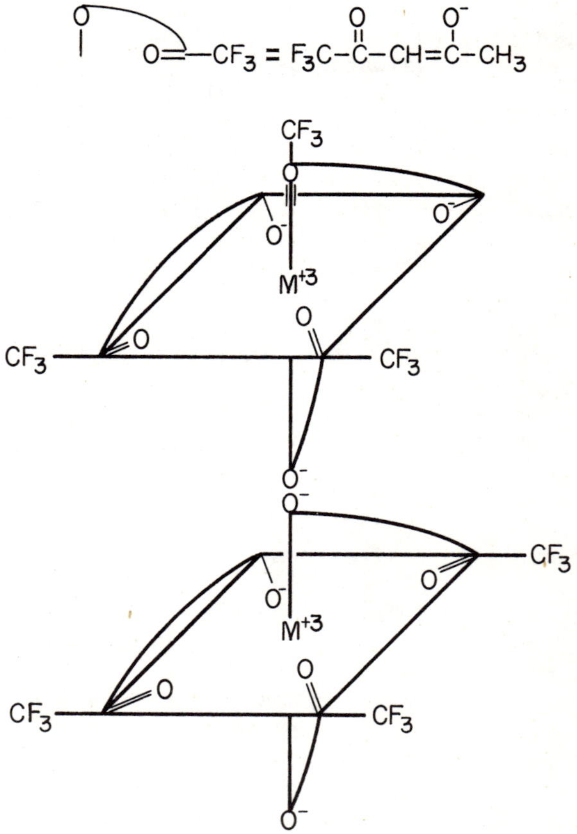

$$F_3C-\overset{O}{\overset{\|}{C}}-CH=\overset{O^-}{\overset{|}{C}}-CH_3$$

Figure 3.1. Cis and trans isomers of trivalent acetylacetonates.

EXPERIMENTAL

Instrumentation

Gas chromatographic analyses were conducted with a Varian Aerograph model 1740, gas chromatograph equipped with either a flame ionization detector (FID) or a tritium electron capture detector (ECD). All-pyrex systems were used unless otherwise specified.

Atomic absorption analyses were conducted with a Varian AA120 atomic absorption spectrophotometer.

Reagents

Chelating agents, trifluoroacetylacetone and 2,2,6,6-tetramethylheptane-3,5-dione were purchased from Pierce Chemical Co., Rockford, Illinois, and redistilled prior to use. The tfa fraction at 105-106° C was collected at atmospheric pressure. The piv fraction was collected at 48-49° C under reduced pressure of 0.3 to 0.5 mm Hg. Chelating agent solutions were made up in pesticide or nanograde benzene. Aqueous solutions were made from double distilled water, the second distillation from pyrex.

Preparation of Chelate Standards

Standard $Pb(piv)_2$ solutions were prepared by an anhydrous reflux of powdered lead metal with the 0.1 *M* piv for 24 hours. Two identical samples were prepared. After filtration, washing with 0.1 *M* NaOH, and drying over $CaCl_2$, the samples were diluted to 100 ml, yielding concentrations of 1.50 and 1.64×10^{-3} *M* $Pb(piv)_2$. Chromium metal was similarly reacted. The final concentration was 1.04×10^{-2} *M* $Cr(piv)_3$.

The tfa chelates of aluminum, chromium, copper and iron were prepared by adding the aqueous metal plus acetate buffer to 0.1 *M* tfa in benzene. This mixture was shaken for approximately one hour, after which the organic layer was separated, evaporated to dryness and stored in a desiccator. The colored product that remained was purified by sublimation, weighed and dissolved in benzene.

Tfa Studies

Aqueous samples were treated in a manner similar to the preparation of the chelate standards exclusive of the evaporation step. Back washing for removal of excess tfa from the benzene layer was accomplished with 0.125 *N* NaOH.

Field samples were filtered through borosilicate glass fiber filters (Reeve Angel, grade 934AH) to remove particulate matter. The filtrate was then passed through chelating resins Dowex A-1, an iminodiacetic acid resin, or Duolite ES-63, a phosphonic acid resin, at 8.8 and 3.5 ml min^{-1} cm^{-2}, respectively. Five-liter water samples were used and the solutions passing the columns were collected for further analysis after concentration by freeze drying.

Metal ions were eluted from the resin columns with 2.0 *N* HNO_3 at 1.8 ml min^{-1} cm^{-2}. Solutions were adjusted to pH 4.5 with NH_4OH prior to metal extraction and analysis by gas chromatography. Particulate analyses were conducted by similar procedure after wet ashing. Freeze dried samples were analyzed by gas chromatography and atomic absorption spectrophometry. Duplicate samples were processed in each case.

The glass fiber filters were wet ashed, followed by pH adjustment with NH_4OH. The tfa extraction was then carried out.

Piv Studies

Preliminary studies with piv involving lead, nickel, copper and zinc showed promise only for lead in aqueous solution. Consequently, further studies were restricted to aqueous lead solutions to evaluate the potential of the method for analytical application.

These studies involved examination of direct extraction of aqueous solutions of lead with piv in benzene. Variables examined were pH, nature of buffer, time of extraction, and temperature. Also, sealed tube studies were undertaken at elevated temperatures with varied temperature and time of extraction.

RESULTS AND DISCUSSION

Tfa Studies

Standards

Standard chelates formed from aqueous solutions exhibited characteristics shown in Table 3.I and extraction efficiencies ranged from 96 to 99.9 per cent. Sensitivity limits

Table 3.I.

Physical Characteristics of the tfa Chelates

Chelate	*Color*	*Observed mp (°C)*	*Reported mp (°C)**
$Al(tfa)_3$	White	117 - 117.5	117 (26)
$Cu(tfa)_2$	Blue	189 - 190	200 (26)
$Cr(tfa)_3$	Purple	145 - 147	cis 112 - 114 trans 154 - 155 (30)
$Fe(tfa)_3$	Red	114 - 115	115 (26)

* Reference in parentheses.

for the chelates were found to be 10^{-10} to 10^{-9} g (for 1 μl samples) for Cu, Cr, Fe and Al species using an electron capture detector. These values are somewhat higher than reported by other authors and likely reflect less than optimal settings of the instrument (See Table 3.II. Flame ionization detection provides less sensitivity than ECD as illustrated in Figure 3.2, but partial decomposition of $Fe(tfa)_3$ and $Cu(tfa)_2$ inside the column diminished ECD sensitivity and decreased background current due to fouling of the detector foil.

Table 3.II

Sensitivity of GC and AA Analyses for Metals

Metal	*Method*	*Sensitivity (μg/l) (Original Water Conc.)*	*Reference*
Aluminum	AA (1% Absorption)	1000	33
	GC-ECD	4	34
		48	31
		3	32
	GC-FID	4150	31
		820	This Study
Chromium	AA (1% Absorption)	150	33
	GC-ECD	20	34
		0.09	32
	GC-FID	200	This Study
Copper	AA (1% Absorption)	200	33
	GC-ECD	471	31
	GC-FID	2870	31
		150	This Study

Column Selection

Of several columns evaluated, five per cent QF-1 on Varaport 30 (100/120 mesh) and 15 per cent Carbowax 20 M on Chromosorb W (80/100 mesh), each being a 6' x 2 mm id pyrex coil, provided separations that allowed determination of Cu, Cr and Al but not Fe in the presence of the other three.

The Carbowax 20 M column provides excellent resolution of Cu from Cr (Figure 3.3) but not Fe, Al and Cu from one another (Figure 3.4). Variations in carrier gas flow and temperature programing did not improve this condition. Separation of Al, Cu and Cr, the latter of which is resolved into cis and trans isomers, was achieved on QF-1

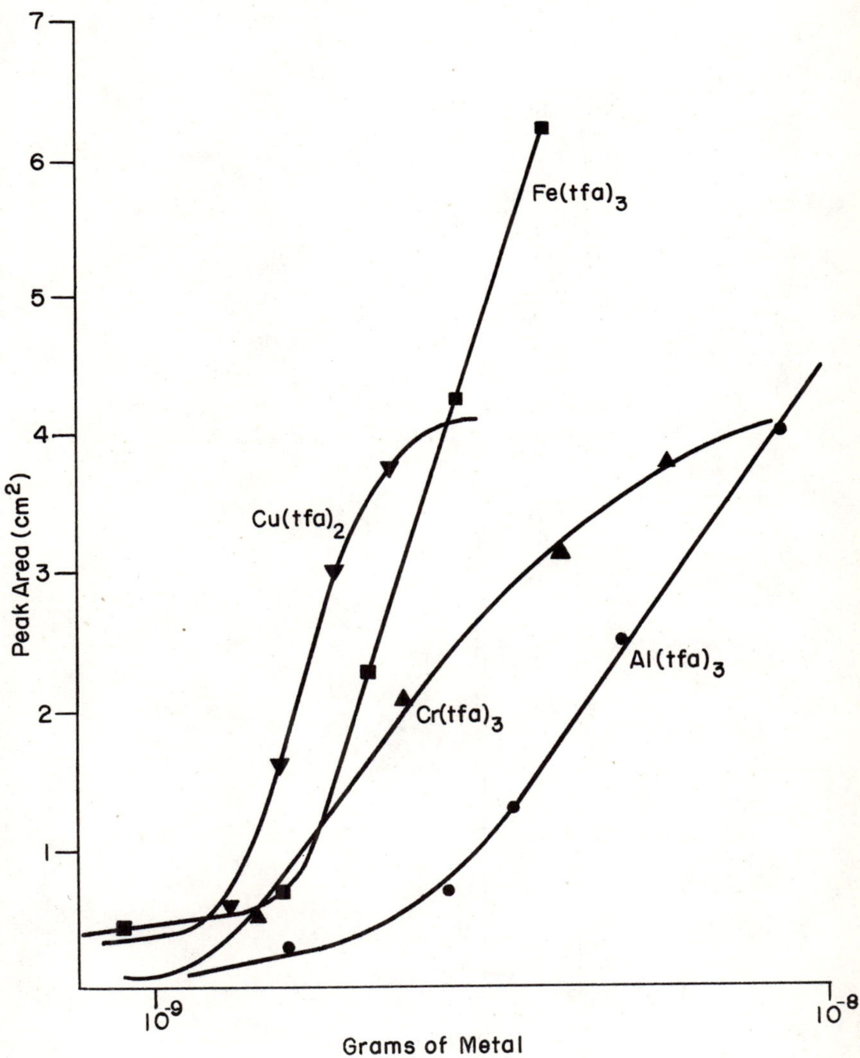

Figure 3.2. Standard curves for the tfa chelates of iron, copper, chromium and aluminum using flame ionization detection.

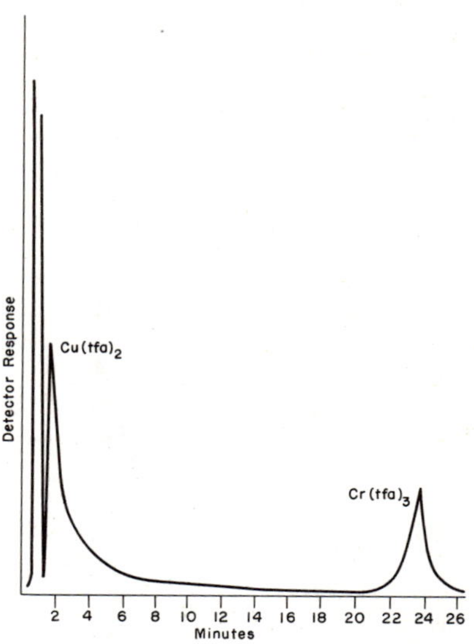

Figure 3.3. Separation of $Cu(tfa)_2$ and $Cr(tfa)_3$ using Carbowax 20M as the liquid phase.

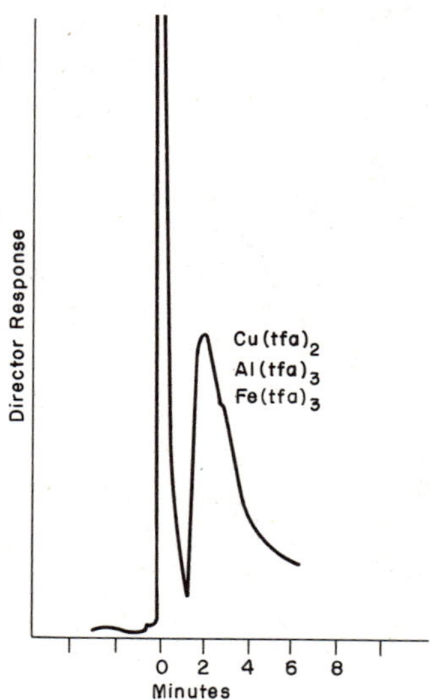

Figure 3.4. Chromatogram resulting from analysis of $Al(tfa)_3$, $Fe(tfa)_3$, and $Cu(tfa)_2$ using Carbowax 20M as the liquid phase.

(Figure 3.5). At sufficiently slow flow rates $Fe(tfa)_3$ decomposes on the column and fails to elute (as was the case in Figure 3.5). At higher carrier gas flow, iron elutes simultaneously with copper and again temperature programing offers no improvement. Adjustment of reaction conditions and use of a NaOH backwash can be used to eliminate $Fe(tfa)_3$ in the organic phase.

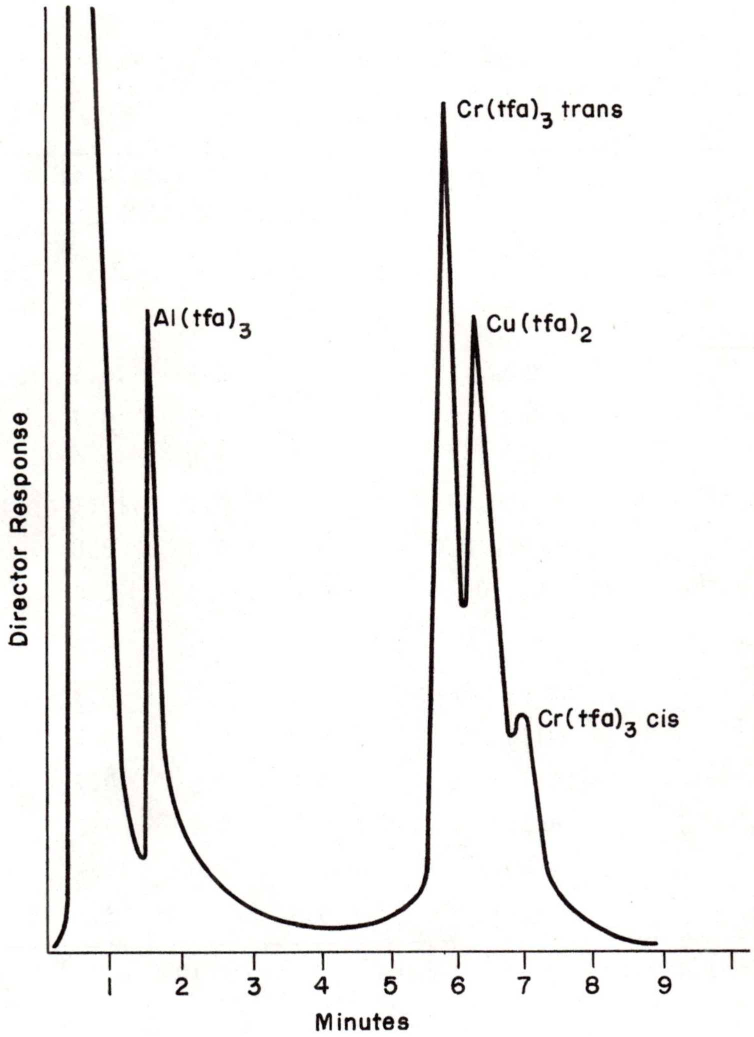

Figure 3.5. Separation of aluminum, chromium and copper tfa chelates by QF-I as the liquid phase.

Interferences

The effect of cation and anion interference upon copper and chromium recoveries was evaluated. As demonstrated in Table 3.III, the only strong interference resulted from

Table 3.III

*Inorganic Interference Studies for the Extraction and Gas Chromatographing of Copper and Chromium tfa Chelates**

Ions	*Source*	*Concentration of Metal Based on Peak Area (ppm)*			
		Copper		*(trans) Chromium*	
Anions					
NO_3^-	KNO_3	8.7	8.9	10.0	10.1
CL^-	KCL	10.4	9.5	9.7	9.9
F^-	NaF	10.0	11.4	9.0	9.7
SO_4^{-2}	Na_2SO_4	11.0	9.1	10.5	11.8
PO_4^{-3}	$NaH_2PO_4\ H_2O$	9.6	9.9	8.8	9.6
$P_2O_7^{-4}$	$Na_4P_2O_7\ 10H_2O$	0.0	0.0	0.0	0.0
Cations					
Al^{+3}	$Al_2(SO_4)_3\ 18H_2O$	10.7	12.4	9.3	9.0
Ni^{+2}	$NiSO_4$	9.7	9.1	9.4	8.7
Zn^{+2}	$Zn(SO_4)\ 7H_2O$	8.8	9.5	10.7	9.7
Mg^{+2}	$MgSO_4$	11.1	9.8	9.2	9.9
Pb^{+3}	$Pb(NO_3)_3$	9.4	9.0	11.4	9.8
NH_4^+	NH_4Cl	9.9	10.7	9.5	8.9

* Concentration of each Anion or Cation: 1000 ppm
Concentration of each Metal Ion: 10 ppm

pyrophosphate. When added to the water containing the metal ions, a light green color resulted. Upon addition of tfa and acetate buffer, the color was present in both phases indicating incomplete extraction of the metal. Separation of the organic layer and subsequent backwashing with aqueous NaOH resulted in transfer of color to the aqueous phase. Formation of a mixed phospho complex would be consistent with observed behavior.

The results given in Table 3.IV for the Dowex A-1 and Duolite ES-63 resins show that copper uptakes on the two resins range from 37 to 82% and 29 to 91%, respectively. The behavior of the two resins did not always correspond for the same sample. Copper as the free ion would be expected to be completely removed.[35,36] It is no surprise that this was not observed since complexation of soluble copper has been reported in fresh water,[37] sea water,[38,39] and waste waters.[37,40]

Aluminum is not expected to be picked up by the resin, and clearly was not, thus underscoring the need for developing other methods for determination of its concentration. Previous work[35] has indicated chromium is only partially removed from solution by Dowex A-1 and then is not completely recovered from the resin. The data (Table 3.IV) are consistent with earlier observations. The ES-63 resin demonstrated no affinity for chromium.

The points to be made here are the following:

1. Complexation of metal ions by natural organics obviates effective ion exchange concentration if it is even applicable to a given ion.
2. Other *in situ* concentration techniques such as anodic stripping may not respond to total metal in solution.
3. Concentration of the water sample to allow AA analysis requires sufficient sample for multiple determinations.

The tfa technique requires very little absolute mass of metal since final sample volumes of 1-2 ml can be utilized to simultaneously analyze for more than one metal ion.

Quite obviously, the technique is not completely satisfactory if only Cu, Al and Cr can be determined. What is desired is a chelating agent that is capable of reacting with several metals present in aqueous solution to form thermally stable, volatile chelates which can be

Table 3.IV

Results of GC Analysis on Natural Samples

Sample Location	pH	Metal	Collected by Dowex A-1	Passed by Dowex A-1	Concentration (ppm) Collected by Duolite ES-63	Passed by Duolite ES-63	Particulate
Calumet River							
Approx. 500 ft from Industrial Discharge	7.5	Al	0.003	20.1	ND*	18.7	70.6
		Cu	0.030	0.050	0.024	0.060	0.001
		Cr	0.500	0.002	ND	0.57	ND
Calumet River							
Non-industrial Area	8.0	Al	ND	5.130	ND	5.040	18.40
		Cu	0.050	0.460	0.380	0.060	ND
		Cr	0.220	ND	ND	0.250	ND
North Shore Sanitary							
Drainage Canal	6.8	Al	ND	1.1	ND	1.000	2.77
		Cu	0.018	0.004	0.020	0.002	ND
		Cr	0.011	0.003	ND	0.020	ND

* ND, below detection limits.

extracted quantitatively (or in constant proportion to their aqueous concentration) into an organic solvent and subsequently be resolved one from the other by gas chromatography.

Piv Studies

In the process of seeking such a reagent for heavy metals, it was found that the 2,2,6,6-tetramethylheptane-3,5-dionate of lead could be chromatographed (Figure 3.6) and provided

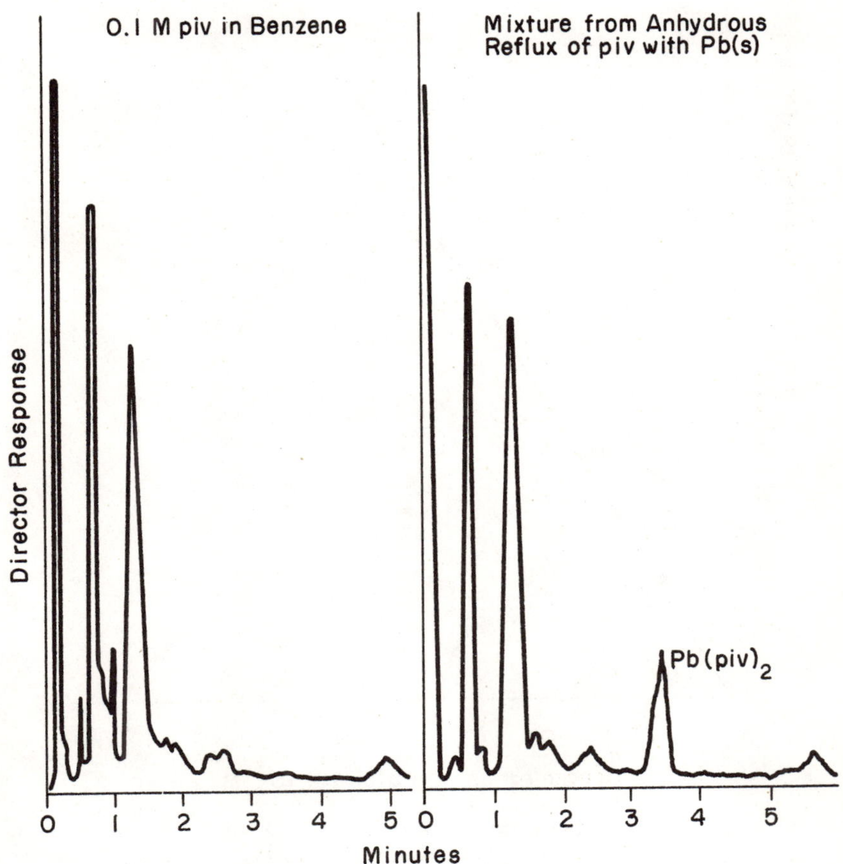

Figure 3.6. Gas chromatographic analysis of $Pb(piv)_2$.

a linear relationship between metal concentration and peak area as shown in Figure 3.7. At the same time nickel and chromium chelates were prepared. Chromium piv was also chromatographed but the nickel chelate decomposed on the column.

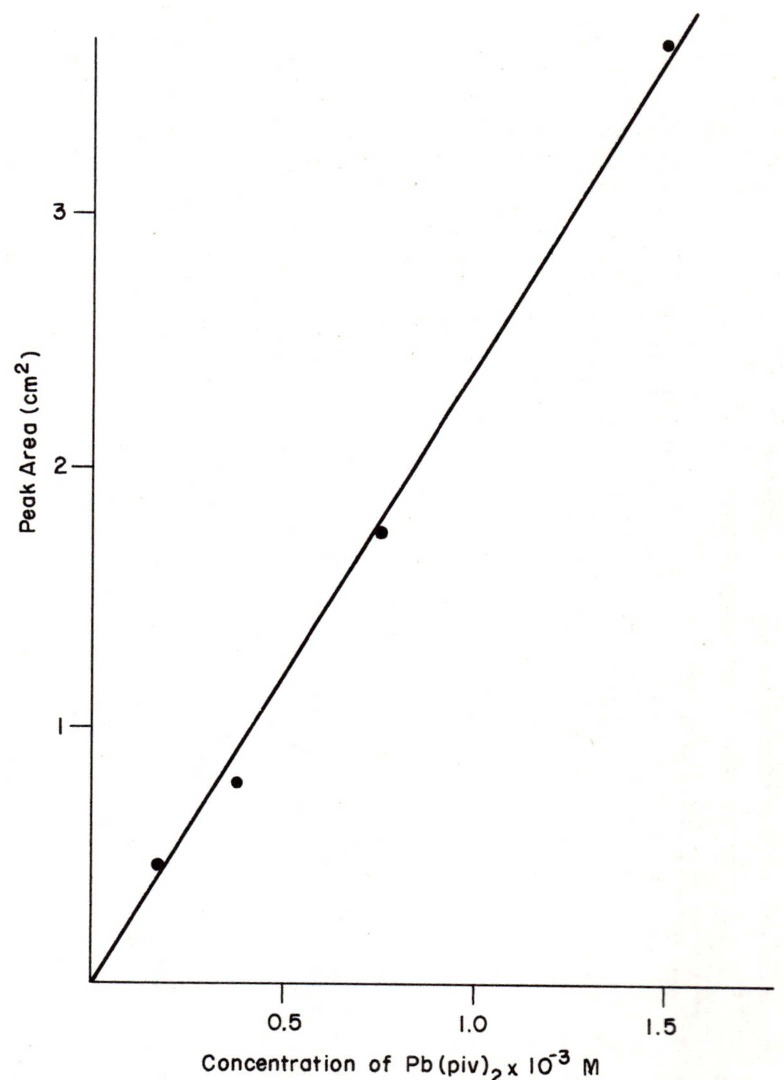

Figure 3.7. Standard curve for $Pb(piv)_2$ as concentration vs. peak area, using electron capture detection.

Reaction from aqueous lead solutions with extraction into benzene provided a product whose gas chromatographic behavior was identical to that of the anhydrous reflux product. However, extraction efficiency was found to be quite low and seemingly dependent upon pH and aqueous lead concentration as illustrated in Tables 3.V and 3.VI.

Table 3.V

*Effect of pH on Extraction of Lead from Aqueous Solution with piv**

pH	*Concentration of $Pb(piv)_2$ (M) $x\ 10^{-4}$ in Benzene Layer*	*Per Cent Extraction*
4	2.4	8.8
7	5.0	18.5
10	7.6	27.8
12.5	1.8	2.9
Unbuffered 6.5 before, 4.1 after	5.88	21.6

* 5 ml 200 ppm Lead; 5 ml 0.1 M piv Solution; 5 ml Buffer

Precipitation of the lead with the buffer anion was suspected. Attempts to improve the reaction efficiency by varying the buffer at pH 10 were not successful as demonstrated in Table 3.VII. The tris and pyridine buffers produced products which could not be gas chromatographed and were possibly mixed chelates.[41,42,43]

Unbuffered solutions shaken at room temperature did not provide a quantitative relationship (Figure 3.8a), nor did reaction in sealed tubes at elevated temperatures in the absence of shaking (Figure 3.8b). When sealed tubes of sufficient size to provide vigorous mixing were used, a reasonable linear relationship was found between aqueous lead concentration and gas chromatograph peak area (Figure 3.8c). Lack of uniformity of mixing from sample

Table 3.VI

*Per cent of Lead Recovered as a Function of Lead Concentration in the Aqueous Solution**

Concentration of Pb in ppm (25 ml)	*% Pb Extracted*	*Mean*	*Standard Deviation*
100	24.6	21.4	3.33
	22.4		
	17.4		
50	29.7	26.2	2.15
	27.5		
	22.2		
10	1.1	0.8	
	1.3		
	ND		
5	ND	ND	
	ND		
	ND		

* 10 ml of 0.01 M piv Solution; 1 ml of 0.01 M Borate Buffer

to sample may be responsible for the considerable variability exhibited, especially at the lower concentrations. Though extraction efficiency was still only 40 per cent, the linear relationship warrants further investigation to determine variation due to aqueous matrix composition.

The low extraction efficiency of the lead piv is not likely related to hydrate formation which occurs when the metal ion coordination number is greater than twice its charge[41] since the lead complex is tetrahedral. This phenomenon has been related to the thermal instability of some divalent metal ion complexes[42] which are octahedral. What is apparent is that additional work is needed to test the analytical potential of this lead chelate and the variability of the per cent extraction.

Table 3.VII

Effect of Buffer on Extraction of Lead with piv

Buffer	*Volume (ml)*	*Per cent Pb(piv)$_2$ Formed*
Borate	2	26.7
	10	24.4
Cyanic	2	29.4
	10	
Phosphate	2	28.8
	10	26.2
TRIS	2	Reaction Occurred which Made Determination Impossible
Pyridine	2	"
	10	

For other metals, adduct formation has been exploited to improve extraction efficiencies with β-diketone chelating agents,[41] and more recently this approach has been examined as a means of producing thermally stable diketonates using Di-n-butylsulfoxide[42] and organophosphorus triesters.[43]

SUMMARY

The use of trifluoroacetylacetone for the analysis of aqueous copper, aluminum, chromium and beryllium solutions is well established and has been demonstrated with respect to natural water analysis. What is clear in this application is that only a limited number of metals can be determined simultaneously. Other chelating agents are available that form chelates with metal ions but they either do not react from aqueous solution or do not possess sufficient thermal stability for gas chromatography.

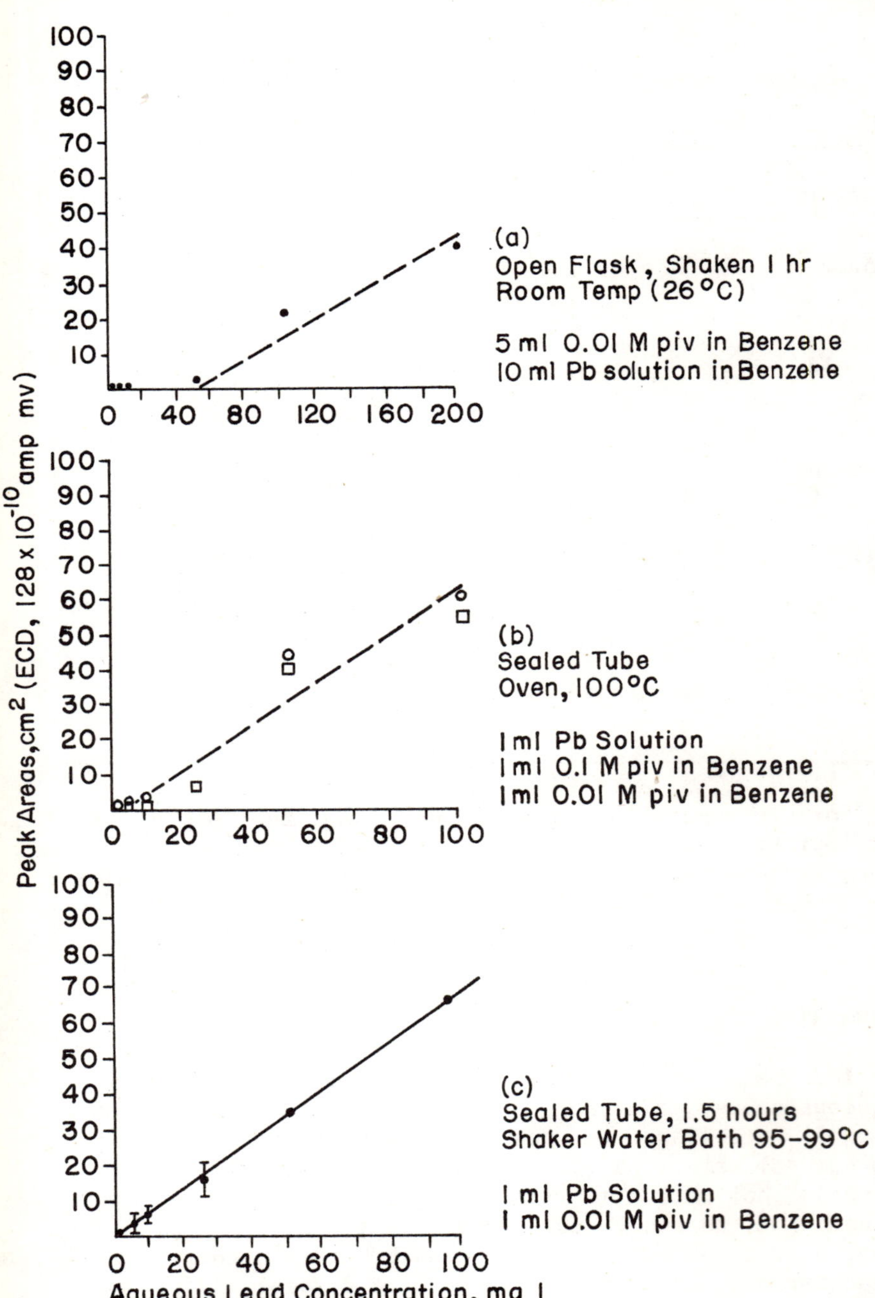

Figure 3.8. Aqueous lead extraction studies.

The need for multi-element analysis is clear since existing analysis techniques require sample preconcentration which is limited to evaporation or freeze drying of natural water samples. Use of ion exchange or chelating resins is limited to those ions for which the resin has sufficient affinity and even then not all the aqueous ion present in solution is removed from natural waters due to complexation with natural organic species. Very sensitive techniques such as anodic stripping voltammetry suffer from similar problems and, furthermore, are not applicable to more than a few electroactive metals. Existing multi-element analysis methods require expensive instrumentation and, frequently, considerable sample preparation for reliable quantitative results. Usually, this means reducing the sample to a solid matrix.

Gas chromatography of volatile metal chelates offers a potential workable alternative which may provide sensitive analysis on very small absolute quantities for several elements simultaneously and on less expensive equipment that is common to most laboratories performing environmental analyses for pesticides. High sensitivity can be accomplished and mass spectrometers interfaced with gas chromatographs can also provide absolute specificity for metals even in the absence of gas chromatographic peak resolution.[22] Admittedly this increases the cost of equipment but many such units are now finding routine application to pesticide analysis and could serve a dual function.

Currently, gas chromatography of volatile metal chelates shows promise provided true multi-element application can be developed. The development of new chelating agents and/or use of mixed chelates formed through use of stable adducts may well bring about this development. Work is currently underway in the authors' laboratory to evaluate a new ligand for water analyses. In conjunction with this activity, mixed ligand formation is planned for future piv and other β-diketone investigations.

REFERENCES

1. *Chem. Eng. News*, July 19, 1971, p. 29.
2. Dinman, B. P. *Sci., 175,* 495 (1972).
3. *Chem. Eng. News*, July 5, 1971, p. 22.
4. McBride, B. C., and R. S. Wolfe, *Biochem., 10,* 4312 (1971).
5. Ferguson, J. F., and J. Gavis. *Water Res., 6,* 1259 (1972).
6. Minear, R. A., and B. B. Murray. In *Trace Metals and Metal-Organic Interactions in Natural Waters,* P. C. Singer, Ed., Ann Arbor Science Publishers, Ann Arbor, Michigan, 1973.
7. Ross, W. D., R. E. Sievers, and G. Wheeler, Jr. *Anal. Chem., 37,* 598 (1965).
8. Morie, G. P., and T. R. Sweet. *Anal. Chem., 37,* 1552 (1965).
9. Moshier, R. W., and J. W. Schwarberg. *Talanta, 13,* 445 (1966).
10. Ross, W. D., and R. E. Sievers. *Talanta, 15,* 87 (1968).
11. Belcher, R., J. R. Majer, R. Perry, and W. I. Stephen. *Anal. Chim. Acta, 45,* 305 (1969).
12. Noweir, M. H., and J. Cholak. *Environ. Sci. Technol., 3,* 927 (1969).
13. Sabory, J., P. Mushak, F. W. Sunderman, Jr., R. H. Estes, and N. O. Roszel. *Anal. Chem., 42,* 294 (1970).
14. Foreman, J. K., T. A. Gough, and E. A. Walker. *Analyst, 95,* 797 (1970).
15. Hansen, L. C., W. G. Scribner, T. W. Gilbert, and R. E. Sievers. *Anal. Chem., 43,* 349 (1971).
16. Ross, R. D., L. C. Hansen, and W. G. Scribner. ARLP-71-0285, Wright-Patterson Air Force Base, 1971.
17. Booth, G. H., Jr., and W. J. Darby. *Anal. Chem., 43,* 831 (1971).
18. Taylor, M. L., and E. L. Arnold. *Anal. Chem., 43,* 1328 (1971).
19. Eisentraut, K. J., D. S. Griest, and R. E. Sievers. *Anal. Chem., 43,* 2003 (1971).
20. Bayer, E., H. P. Muller, and R. E. Sievers. *Anal. Chem., 43,* 2012 (1971).
21. Ross, W. E., and R. E. Sievers. *Environ. Sci. Technol., 6,* 155 (1972).
22. Wolf, W. R., M. L. Taylor, B. M. Hughes, T. O. Tiernan, and R. E. Sievers. *Anal. Chem., 44,* 616 (1972).
23. Eisentraut, K. J., and R. E. Sievers. *J. Am. Chem. Soc., 87,* 5254 (1965).

24. Schwarberg, J. E., R. E. Sievers, and R. W. Moshier. *Anal. Chem., 42,* 1828 (1970).
25. Koshimura, H., and T. Okubo. *Anal. Chim. Acta, 49,* 67 (1970).
26. Berg, E. W., and J. T. Truemper. *J. Phys. Chem., 64,* 487 (1960).
27. Gerlach, D. H., and R. H. Holm. *Inorg. Chem., 8,* 2292 (1969).
28. Sievers, R. E., B. W. Ponder, M. L. Morris, and R. W. Moshier. *Inorg. Chem., 2,* 693 (1963).
29. Moshier, R. W., and R. E. Sievers. *Gas Chromatography of Metal Chelates*, Pergamon Press, New York, 1965.
30. Scribner, W. E., W. J. Treat, J. D. Weis, and R. W. Moshier. *Anal. Chem., 39,* 1136 (1965).
31. Albert, K. D. *Anal. Chem., 36,* 2034 (1964).
32. Ross, R. D. *Anal. Chem., 35,* 1596 (1963).
33. *Standard Methods*. 13th ed., American Public Health Association, 1971.
34. Genty, C., C. Houin, P. Malherbe, and R. Schott. *Anal. Chem., 43,* 235 (1971).
35. Riley, J. P., and D. Taylor. *Anal. Chim. Acta, 40,* 479 (1968).
36. Powers, F. G. Unpublished Masters Thesis, Illinois Institute of Technology, Chicago, Illinois, 1973.
37. Allen, H. E., W. R. Matson, and K. H. Mancy. *J. Water Pollut. Control Fed., 42,* 573 (1970).
38. Williams, P. M. *Limnol. Oceanogr., 14,* 156 (1969).
39. Slowey, J. R., L. M. Jeffery, and D. W. Hood. *Nature, 214,* 377 (1967).
40. Bender, M. E., W. R. Matson, and R. A. Jordan. *Environ. Sci. Technol., 4,* 520 (1970).
41. Scribner, W. G. ARLP-68-0193, Wright-Patterson Air Force Base, 1968.
42. Burgett, C. A. Paper No. 87, 24th Pittsburgh Conference, Cleveland, Ohio, March 1973.
43. Mitchell, J. W., and C. V. Banks. *Anal. Chim. Acta, 57,* 415 (1971).

CHAPTER 4

NEW ANALYTICAL TECHNIQUES FOR THE STUDY OF WATER DISINFECTION

George P. Whittle and Alexander Lapteff, Jr.*

INTRODUCTION

For the proper assessment of the disinfection of water and wastewater, the concentrations of all active forms of the disinfectant must be rapidly and accurately determined. This is particularly true as related to the use of chlorine as the disinfectant predominantly employed in this country in water and wastewater treatment. Differences in efficacy of disinfection between the free chlorine forms of hypochlorous acid and hypochlorite ion have been well documented. Similarly, differences in disinfection ability of free chlorine and the combined form of monochloramine have been established in numerous investigations. The relative disinfection power of dichloramine, although reported to be greater than that of monochloramine, is not as well established as is that of the previously mentioned chlorine forms. While the remaining principal combined form, nitrogen trichloride, is known to be a strong oxidizing agent, its disinfection properties are largely unknown. Rather than cite a supportive, voluminous bibliography, the reader is referred to the recent publication by White[1] for an excellent review of the state of knowledge of the disinfection process.

In general, the extent of knowledge concerning the disinfection characteristics of a particular chlorine form may be directly related to the certainty with which that form

---

*G. P. Whittle, Professor of Civil Engineering, University of Alabama, University, Alabama. A. Lapteff, Jr., The Ralph M. Parsons Company, Washington, D.C.

may be analytically determined. Thus, conclusions drawn from disinfection studies involving free chlorine and monochloramine are strengthened by the fact that good analytical techniques, *e.g.*, amperometric titration, have been available for analysis of these forms. The conclusions drawn from dichloramine studies are not as clearly defined and disinfection data for nitrogen trichloride are essentially nonexistent. This lack of knowledge may be related to and caused, in part, by the increasing uncertainty of analysis in proceeding from that found for dichloramine to that available for nitrogen trichloride.

In recent years, several new methods for chlorine residual determination have been developed which have shown considerable improvement over the traditional orthotolidine procedures. The amperometric titration procedure, while not new, represented a significant advance in chlorine analysis and currently serves as the method of choice, particularly for free, monochloramine, and total chlorine, in calibration procedures for development of new methods. The DPD methods, first proposed some sixteen years ago and subsequently modified and improved, continue to provide reliable procedures for residual chlorine analyses. In recent years several new or modified methods have been proposed including the ferrous DPD, the stabilized neutral orthotolidine (SNORT), the methyl orange, and the leuco crystal violet methods. These latter five methods together with previously accepted standard methods were evaluated in two collaborative studies[2,3] conducted by the Analytical Reference Service of the U. S. Department of Health, Education, and Welfare. The analyses in these studies were restricted to determining free, total, and combined chlorine (by difference). While considerable variability of performance was noted among all methods evaluated, the five new methods mentioned previously were found to produce acceptable results and were published as tentative standard methods in the 13th edition of Standard Methods.[4] In a separate study,[5] the U. S. Army Medical Environmental Engineering Research Unit reported that the syringaldazine method proposed by Bauer and Rupe[6] was the most specific for free chlorine of five commercially available test kits evaluated.

In Table 4.I, a list is presented of the methods published in the 13th edition of Standard Methods together with the chlorine forms to which each method is currently applicable. The syringaldazine method, although not accepted as a standard method, is included in Table 4.I since it shows considerable promise for free chlorine determinations.

Table 4.I

Applicability of Analytical Methods to Specific Chlorine Forms

Method	*Free*	*Total*	*Combined*	*Monochloramine*	*Dichloramine*	*Nitrogen Trichloride*
Iodometric		X				
Amperometric	X	X	X	X	X	
Orthotolidine		X				
Orthotolidine-Arsenite	X	X	X			
Stabilized Neutral Orthotolidine (SNORT)	X	X	X	X	X	X
Ferrous DPD	X	X	X	X	X	X
Leuco Crystal Violet	X	X	X			
Methyl Orange	X	X	X			
Syringaldazine	X					

The leuco crystal violet (LCV) method first proposed by Black and Whittle[7] in 1967 has subsequently been employed by several investigators[8-11] in disinfection and other types of studies. However, the method as currently proposed is limited to the determination of free, total, and combined chlorine. The method is also limited to the determination of a maximum concentration of 2.0 mg/l of free chlorine in the absence of a sample dilution procedure.

OBJECTIVES OF STUDY

In an effort to expand the capabilities of the LCV method, this study was conducted with the objectives of determining free chlorine, monochloramine, dichloramine, and nitrogen trichloride concentrations as high as 10 mg/l. This higher concentration value was selected in anticipation of more stringent future requirements related to chlorination practice for the disinfection of both potable waters and wastewaters. In addition, the expanded method would be of value to the researcher in elucidating the disinfecting ability of all forms of active chlorine.

MODIFIED FREE CHLORINE METHOD

As presently proposed in Standard Methods, the leuco crystal violet (LCV) method is based on the reaction of chlorine (or iodine) with the indicator 4,4',4''-methylidynetris(N,N-dimethylaniline) at sample pH 4.0 to produce a bluish-violet color. The reaction product with free chlorine has been postulated previously to be a chlorinated crystal violet dye.[7] For total chlorine determination, the reaction is carried out through an iodide ion addition step in which the reaction product has been demonstrated to be crystal violet dye formed by reaction with hypoiodous acid. With added iodide ion and at sample pH 4.0, free chlorine and all forms of combined chlorine will respond for a total determination. For both free and total chlorine samples, the absorbance of the developed color is measured at 592 nm and the value obtained is compared with a previously prepared calibration curve.

Experimental

Preliminary studies revealed that by allowing the free chlorine to react first with the indicator at sample pH 7.0 prior to final adjustment to pH 3.7, free chlorine could be determined without sample dilution up to a concentration of 10 mg/l. The absorbance values in relation to chlorine concentration followed Beer's law up to a concentration of 5 mg/l. The details of the free chlorine method presented below as well as the methods for combined chlorine forms that follow evolved from numerous investigations of test parameters and experimental conditions. For a detailed account of these investigative procedures and other experimental conditions, reference is made to the study by Lapteff.[12]

The major apparatus employed in this study, other than standard laboratory glassware and equipment, included a research-type spectrophotometer, a recorder, and an amperometric titrator. For routine analytical work, a suitable spectrophotometer or colorimeter with a wavelength capability near 592 nm may be employed.

Reagents

All reagents and test solutions were prepared using water obtained from a Continental water conditioning system consisting of two mixed bed cation-anion exchange resin columns in series with a carbon column. This system produced deionized, chlorine-demand-free water and, unless otherwise noted, the term "water" employed in this study will denote this specially treated water. Chlorine-demand-free water may be prepared by other procedures outlined in Standard Methods. All glassware was initially treated with a chromic acid solution and rinsed with water prior to use.

LCV indicator solution: In a brown-glass bottle, acidify approximately 500 ml of water with 3 ml of concd H_2SO_4. Add 3.0 g of 4,4',4''-methylidynetris(N,N-dimethylaniline) (Eastman No. 3651), stir to dissolve, and dilute to 1 liter. Add 2.7 g of $HgCl_2$ and continue stirring until dissolution is complete. For maximum stability, store in a brown-glass bottle or an opaque plastic container.

Phosphate buffer, pH 7.0: Dissolve 76.2 g anhydrous KH_2PO_4 and 194.7 g of $Na_2HPO_4 \cdot 7\ H_2O$ in water and dilute to 1 liter.

n-Propanol, reagent grade.

Buffer-alcohol solution: To 1 liter of water add 150 ml of phosphate buffer, pH 7.0, and 750 ml of n-propanol. With stirring, add 10-15 mg of free chlorine by the addition of a small volume of a stock chlorine solution. Expose the solution to sunlight to dissipate the excess chlorine.

Hydroxylamine hydrochloride solution: Dissolve 40 g of $NH_2OH \cdot HCL$ in water and dilute to 1 liter.

Acetate buffer, pH 3.7: Dissolve 725 g concd (glacial) acetic acid and 143.5 g of sodium acetate, $NaC_2H_3O_2 \cdot 3H_2O$.

Procedure

Free chlorine solutions were prepared in the concentration range of 0.10 mg/l to 10.0 mg/l and standardized by amperometric titration.

Free chlorine, as well as the total and the various combined forms, may be determined by utilizing either 100-ml volumetric flasks in which the treated samples are diluted to the mark or 250-ml beakers in which no final dilution is performed. For the dilution method, a 50-ml sample was pipeted into a 100-ml volumetric flask and treated with the following reagents:

1. Add 10 ml of buffer-alcohol solution and swirl to mix.
2. Add 1.0 ml of mixed indicator solution, mix and allow at least a 1-min reaction time.
3. Add 1.0 ml of hydroxylamine hydrochloride solution and allow at least a 1-min reaction time.
4. Add 2.0 ml of acetate buffer, pH 3.7, dilute to the mark, and measure the absorbance at 592 nm using 1-cm cells.

For the beaker procedure, a 100-ml sample volume was measured by a graduated cylinder into a 250-ml beaker. The sample was gently agitated on a magnetic stirrer and treated with the reagents in the manner described previously with the exceptions that twice the reagent volumes were employed and no final dilution was performed.

Discussion of Results

Standard absorbance or calibration curves are shown in Figure 4 .1 for both the flask and beaker procedures for sample chlorine concentrations as high as 10 mg/l. For the flask procedure, Beer's law is apparently followed up to 5.0 mg/l of free chlorine. The molar absorptivity determined for the linear portion of the absorbance curve was 1.487 $\times 10^4$ l, cm^{-1}, m^{-1} (Cl_2) with a correlation coefficient of 0.9980. The per cent error of analysis was calculated to be $\pm$ 4.50 per cent.

At pH 7.0 the indicator, leuco crystal violet, is relatively insoluble in aqueous solution. The addition of n-propanol promoted solubility of the indicator for an improved reaction with aqueous free chlorine. The manner of addition of the indicator had no significant effect on the final absorbance values which offers a distinct advantage over the presently adopted standard LCV method. In addition, reduction of the final sample pH to 3.7, rather than to pH 4.0 reported previously,[4,7] produced better solubility of the final reaction mixture in the new test procedures.

In the presently accepted LCV method for free chlorine no significant interference from concentrations of combined forms as high as 4.0 mg/l Cl_2 had been reported.[7] The method proposed in this study was designed to determine free chlorine coexisting with combined chlorine concentrations as high as 10 mg/l Cl_2. It was found necessary to incorporate a suitable reductant in the procedure to prevent interference at these higher values of combined chlorine concentrations. The addition of hydroxylamine hydrochloride at sample pH 7.0 after completion of the free chlorine reaction with the indicator effectively reduced any combined forms and allowed the specific determination of free chlorine. Since the oxidation of leuco crystal violet is irreversible with respect to common aqueous reductants, the reaction time with $NH_2OH \cdot HCl$ is critical only in regard to allowing a sufficient time for the reduction of any coexisting combined chlorine. In the absence of the addition of hydroxylamine hydrochloride, only monochloramine at a high concentration produced a serious interference in the free chlorine determination.

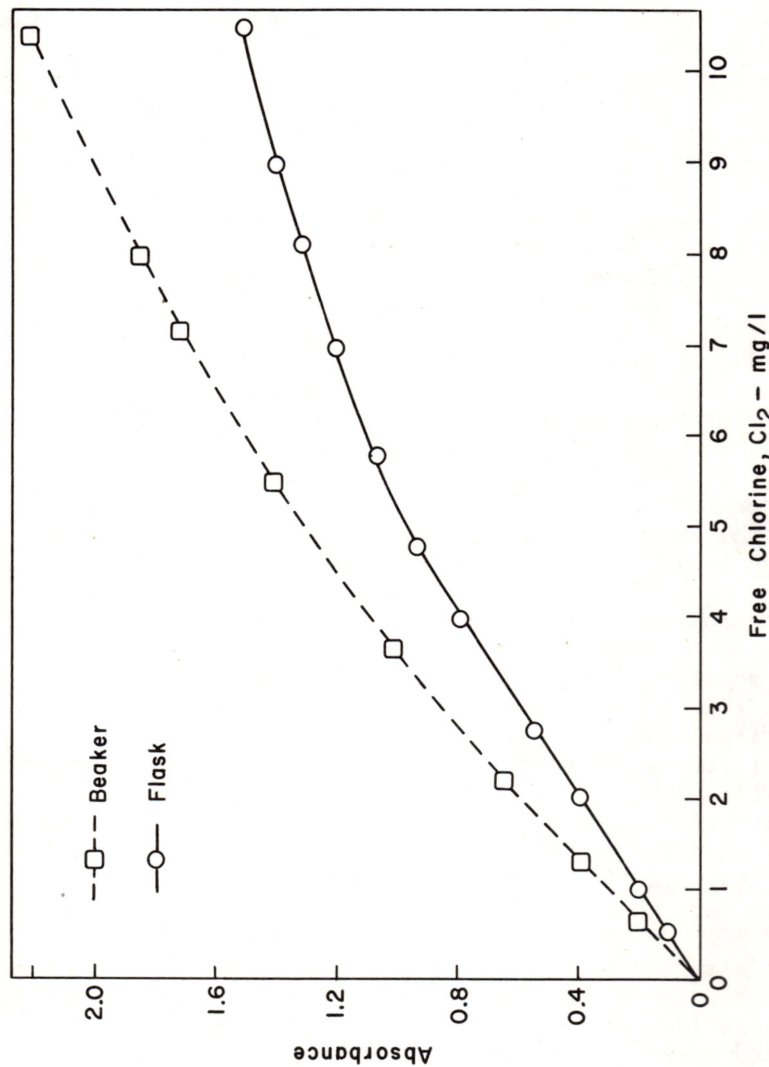

Figure 4.1. Standard absorbance curves for free chlorine.

MONOCHLORAMINE METHOD

The proposed test for monochloramine was based on the well-known reaction with iodide ion at pH 7.0 to produce iodine. The iodine produced was reacted with leuco crystal violet at pH 7.0 to form a completed reaction product. The sample was subsequently treated with $NH_2OH \cdot HCL$ to remove other coexisting combined forms before final sample pH adjustment to 3.7 for maximum absorbance values.

Experimental

A stock solution containing only free chlorine and monochloramine was prepared by reacting chlorine with an ammonium sulfate solution at pH 9.0 such that the ammonia to chlorine ratio was at least 15 to 1. Since solutions of other combined forms were required for interference studies as well as for establishing calibration curves for each form, the preparation procedures for each combined species are presented in this section.

Monochloramine

Add a free chlorine solution to an ammonium sulfate solution in an ammonia to chlorine weight ratio of at least 15 to 1. To approximately 3 liters of demineralized water add 36.6 ml of 5 per cent $(NH_4)_2SO_4$. Slowly add, with mixing, 90 ml of stock chlorine solution (1 ml = 0.35 mg as Cl_2) and adjust final solution pH to 9.0 with 1 *N* NaOH. Allow mixture to react until tests indicate only monochloramine or monochloramine plus free chlorine is present.

Dichloramine

Solutions of dichloramine were prepared in a similar manner to that for monochloramine, except the final solution was adjusted with 1 + 19 sulfuric acid to the pH range of 3.7 to 4.5. This solution is allowed to react 12 to 17 hours.

Nitrogen Trichloride

Solutions of nitrogen trichloride were also prepared in a manner similar to that for monochloramine except the solution pH was adjusted to 2.0 with 1 + 19 sulfuric acid. A reaction time of two to three hours is allowed after addition of all reagents.

Each chloramine stock test solution, after complete reaction, contains approximately 10 mg/l as Cl_2 of the particular combined chlorine species. Mono- and dichloramine solutions were analyzed amperometrically. For the nitrogen trichloride solution, a qualitative test to determine if other combined forms were absent was performed by first extracting one liter of the solution with three 20-ml portions of carbon tetrachloride. The extract was optically scanned from 230 to 350 nm using 10-cm absorption cells and measured against a carbon tetrachloride blank. Absorption spectra exhibiting a maximum absorption at 340 nm indicated the presence of trichloramine.[13] The absence of absorbance peaks at 245 and 297 nm indicated mono- and dichloramine did not coexist with nitrogen trichloride at pH 2.0. Free chlorine was analyzed using the leuco crystal violet method proposed by Black and Whittle.[4,7] Total chlorine was determined amperometrically. The nitrogen trichloride solution concentration was calculated by the difference between the total chlorine and free chlorine test results.

In preparing these combined chlorine test solutions, the analyst should satisfy himself that the test solutions contain the desired residual. In this study, it was frequently necessary to re-prepare solutions several times and subject them to various independent test procedures, *e.g.* amperometric titration and spectrophotometric examination, in order to be certain of the test solution composition.

Reagents

In addition to the reagents previously described, the following buffer solution containing potassium iodide was prepared.

Buffer-alcohol-KI solution: To 1 liter of the previously prepared buffer-alcohol solution, add 50 ml of a solution containing 3 g of potassium iodide per liter. (Note: If a new solution is to be prepared, prepare the buffer-alcohol solution first, treat with chlorine to satisfy any chlorine demand, and allow chlorine residual to dissipate <u>before</u> the addition of KI.)

Procedure

A calibration curve for monochloramine was prepared by first measuring a suitable aliquot of the stock solution into a 100-ml volumetric flask or a 250-ml beaker and adding water to obtain a 50- or 100-ml total volume, respectively. For the determination of combined forms of chlorine using the beaker method it is desirable to dilute the final treated sample to 200 ml. The following quantities of reagents are added based on a 50-ml total sample volume:

1. Add 1.0 ml of pH 7.0 buffer solution. (Note: This is required only for these samples having initial pH values less than 7.0. The analyst should determine if the 1.0-ml volume is sufficient to adjust the pH to 7.0; if not, add additional buffer solution.)
2. Add 10 ml of buffer-alcohol-KI solution, mix, and allow a reaction time of at least 1 minute.
3. Add 1 ml of the mixed indicator, swirl to mix, and allow a 1-minute reaction time.
4. Add 1.0 ml of $NH_2OH \cdot HCl$ solution and wait 1 minute.
5. Add 2.0 ml of pH 3.7 buffer, dilute to mark, and measure absorbance at 592 nm.

Discussion of Results

The absorbance of the final solution was related to the sample chlorine concentration and a calibration curve was prepared as shown in Figure 4.2. Solutions containing only free chlorine were treated as described above for this monochloramine procedure and a standard absorbance curve is also shown in Figure 4 .2. The molar absorptivity for the monochloramine curve was calculated to be 3.75×10^4 l, cm^{-1}, m^{-1} (Cl_2) as compared to 3.35×10^4 l, cm^{-1}, m^{-1} (Cl_2) for free chlorine. Monochloramine and free chlorine regression curves exhibited correlation coefficients of 0.9996 and 0.9995, and per cent errors in analyses of $\pm$ 1.04 and $\pm$ 1.45 per cent, respectively. The differing response of free chlorine and monochloramine in the monochloramine test necessitated a slightly different correction procedure for free chlorine in a solution of coexisting forms than would be the case if the absorbances of each form were simply additive. To accomplish this correction, free chlorine was determined by the method described previously employing

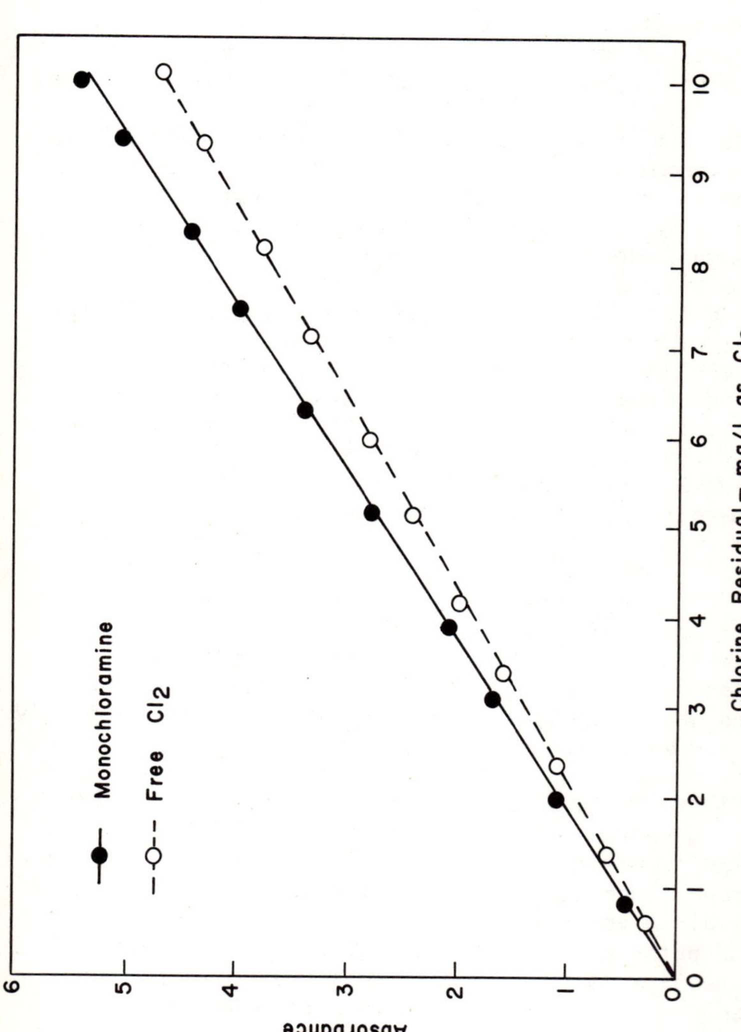

Figure 4.2. Standard absorbance curves for monochloramine and for free chlorine in monochloramine method.

the standard absorbance curve shown in Figure 4.1. The absorbance produced by the same free chlorine concentration in the monochloramine test was obtained by reference to the free chlorine curve in Figure 4.2. This free chlorine absorbance value was subtracted from the total absorbance measured in the monochloramine test to obtain the monochloramine standard absorbance curve.

For solutions with absorbances in 1.0-cm cells greater than 2, cells having a lesser light path may be employed, or the developed sample may be diluted with water buffered at pH 3.7. Since Beer's law was found to be followed with respect to dilution and/or changes in light-path length, these absorbance values may be corrected to the undiluted state by the appropriate dilution or cell-thickness factor. Similarly, aliquots of the stock or unknown solution may be taken such that the final absorbance value is less than 2. In this case, it is necessary to dilute the aliquot to an approximate 50-ml volume before performing the test.

Under the conditions of the test, dichloramine did not respond and nitrogen trichloride responded very slowly allowing the determination of monochloramine when hydroxylamine hydrochloride was added at the completion of the monochloramine reaction at pH 7.0. It is essential that the sample pH be 7.0 or above before addition of the mixed reagent containing potassium iodide in order to prevent interference from dichloramine and nitrogen trichloride.

NITROGEN TRICHLORIDE METHOD

The monochloramine procedure was modified to include the determination of nitrogen trichloride by allowing a sufficient reaction time with potassium iodide for both forms to respond.

Experimental

A stock nitrogen trichloride solution was prepared as described previously. These stock solutions were invariably found to contain some trace amounts of free chlorine. The previously described method for free chlorine was employed to determine these trace quantities. An absorbance curve for free chlorine solutions employing the nitrogen trichloride procedure corresponded to that found in the monochloramine procedure. As discussed previously, values from this absorbance curve were used to correct the values obtained for the free Cl_2 plus NCl_3 solutions in preparing a standard absorbance curve for NCl_3 only.

Procedure

Nitrogen trichloride may be determined by the flask or beaker method. The reagent quantities employed below are based on a 50-ml sample volume.

1. Add 3.0 ml of pH 7.0 buffer solution or a sufficient quantity to adjust sample pH to 7.0. (Note: This large volume of buffer is required in calibration procedures since the stock nitrogen trichloride solution has a pH in the range of 1.8 to 2.2. For potable or wastewaters having a pH near or above 7.0, this buffer addition step may be eliminated.)
2. Add 10 ml of buffer-alcohol-KI solution and allow a reaction time of 10 minutes.
3. Add 1.0 ml of mixed indicator solution and allow a reaction time of at least 1 minute.
4. Add 1.0 ml of $NH_2OH \cdot HCl$ solution and wait one minute.
5. Add 2.0 ml of pH 3.7 buffer, dilute to mark, and measure absorbance at 592 nm.

A standard absorbance curve for nitrogen trichloride was prepared as shown in Figure 4.3.

Discussion of Results

The nitrogen trichloride absorbance curve exhibited a molar absorptivity of 1.53×10^4 l, cm^{-1}, $mole^{-1}$ (Cl_2), a correlation coefficient of 0.9987, and a ± 3.64 per cent error in analysis. The 10-min reaction period with KI at pH 7.0 was selected after it was determined that the reaction was essentially complete under these conditions. Reaction time periods up to 15 minutes produced no significant increase in absorbance of the final color. The slowness of this reaction will permit the determination of monochloramine in the presence of nitrogen trichloride. This was indicated by the observation that a concentration of nitrogen trichloride as high as 10 mg/l as Cl_2 produced an apparent monochloramine concentration of only 0.02 mg/l as Cl_2 in the monochloramine test.

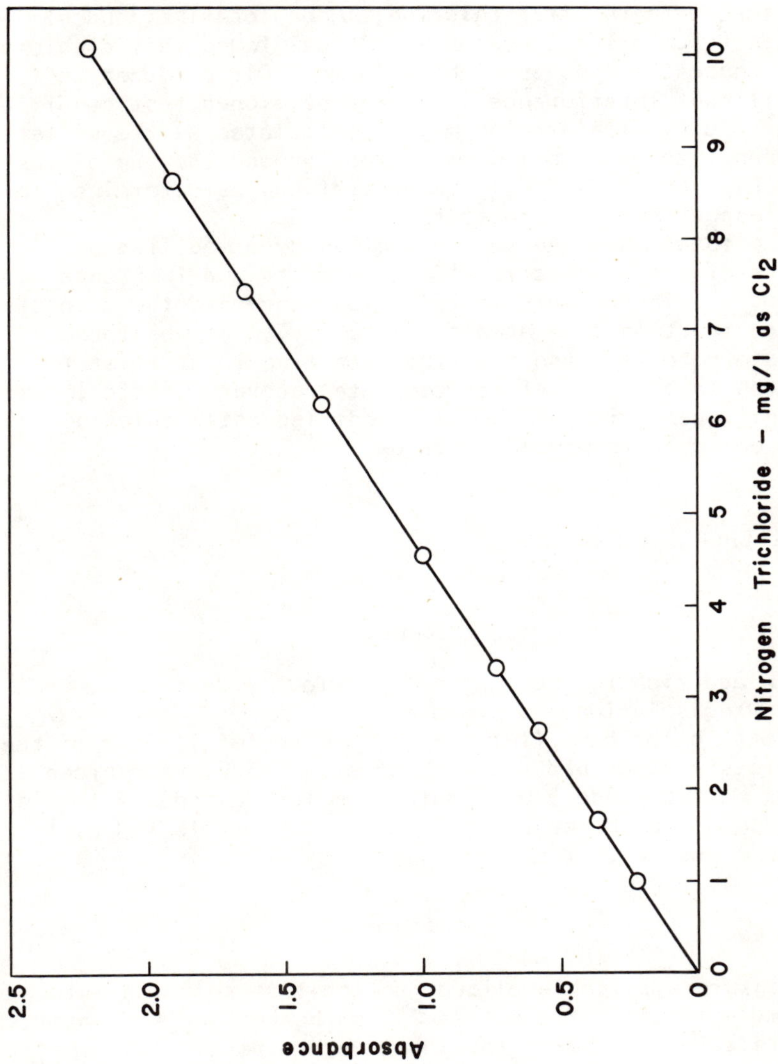

Figure 4.3. Standard absorbance curve for nitrogen trichloride.

DICHLORAMINE DETERMINATION

The dichloramine stock solution described previously was tested in the free chlorine, monochloramine, and nitrogen trichloride procedures. It was found that dichloramine concentrations as high as 10 mg/l Cl_2 produced no significant interference under the developed test conditions. As a result, dichloramine may be calculated as the difference between a total chlorine determination and the sum of the free Cl_2, NH_2Cl, and NCl_3 concentrations as determined in the respective test procedures.

The total chlorine was determined by a modified procedure of the leuco crystal method contained in Standard Methods.[4] The primary modification concerned the potassium iodide reaction time at a sample pH 3.7. It was found necessary to increase this reaction time to at least five minutes in order to effect complete recovery of dichloramine and nitrogen trichloride. The modified total chlorine test employed is presented below.

TOTAL CHLORINE PROCEDURE

Reagents

In addition to the reagents previously described, the following solution was prepared.

Total chlorine buffer, pH 3.7: Transfer 1 liter of the previously described acetate buffer, pH 3.7, to a brown glass bottle. Add 3.0 of potassium iodide and mix to dissolve. Store the solution in the brown bottle and avoid undue exposure to the air.

Procedure

Measure a suitable aliquot of the test solution into a 100-ml volumetric flask or a 250-ml beaker and add water to obtain a 50- or 100-ml total volume, respectively. The following quantities of reagents are added based on a 50-ml total sample volume:

1. Add 0.5 ml of total chlorine buffer, pH 3.7, mix and allow a reaction period of at least 5 minutes.
2. Add 1.0 ml of the mixed indicator solution, mix, dilute to mark and measure absorbance at 592 nm.

Discussion of Results

The total chlorine procedure was applied to solutions of free Cl_2, NH_2Cl, $NHCl_2$, and NCl_3 and the absorbance values for each chlorine form were plotted against concentration as shown in Figure 4.4. Within experimental error, all points fell on the same absorbance curve. The results of a regression analysis for each chlorine form are presented in Table 4.II.

DETERMINATION OF ALL ACTIVE FORMS OF CHLORINE

For the analysis of the four active forms of chlorine, the following sequence is performed on a given sample:

1. Determine free chlorine.
2. On a separate sample, determine the sum of free chlorine plus monochloramine and, by difference, obtain monochloramine.
3. On a separate sample determine the sum of free chlorine, monochloramine, and nitrogen trichloride and, by difference, obtain nitrogen trichloride.
4. Determine total chlorine and subtract the summation of chlorine forms in 3. above to obtain dichloramine.

By a judicious use of time, one person can determine all four forms in approximately 15 minutes. Previously established calibration curves may be used directly to obtain the concentration values. Similarly, the absorptivities for each chlorine form may be obtained from the slopes of the respective calibration curves and employed in a simultaneous solution of the absorbance, A, equations presented below.

$$A_{f+m} = \varepsilon_f C_f + \varepsilon_m C_m \tag{1}$$

$$A_{f+m+t} = \varepsilon_f C_f + \varepsilon_m C_m + \varepsilon_t C_t \tag{2}$$

$$A_T = \varepsilon_T C_f + \varepsilon_T C_m + \varepsilon_T C_t + \varepsilon_T C_d \tag{3}$$

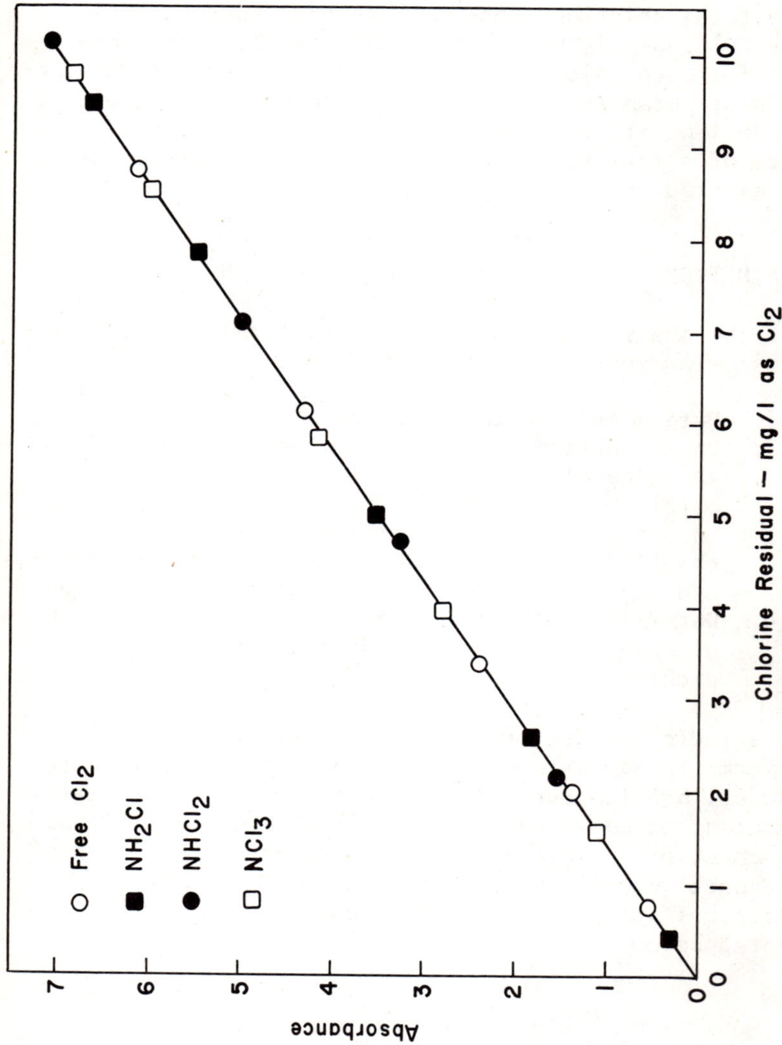

Figure 4.4. Standard absorbance curve for total chlorine.

Table 4.II

Characteristics of Absorbance Curves for the Various Chlorine Forms in the Total Chlorine Procedure

Chlorine Form-Total Chlorine Curve	Correlation Coefficient	Molar Absorptivity l, cm^{-1}, $mole^{-1}$, (Cl_2)	Per Cent Error
Free Chlorine	0.9998	5.05×10^4	±4.50
Monochloramine	0.9997	4.85×10^4	±0.28
Dichloramine	0.9997	4.92×10^4	±1.16
Trichloramine	0.9999	4.90×10^4	±0.70

where: ε_f, ε_m, ε_t are the absorptivities of free Cl_2, NH_2Cl, and NCl_3, respectively, in both the monochloramine and nitrogen trichloride procedures; ε_T is the absorptivity of all chlorine forms in the total chlorine procedure: and C_f, C_m, C_d, C_t are the concentrations of free Cl_2, NH_2Cl, $NHCl_2$, and NCl_3, respectively. In Equation 1 the concentration of free chlorine, C_f, is determined independently in the free chlorine procedure.

INTERFERENCE STUDIES

The methods were evaluated in regard to possible interference from constituents normally found in water and wastewaters. Each constituent in varying concentrations was added individually to samples and the effect on the various chlorine analyses was observed. Finally, all constituents evaluated were added to composite samples which were then subjected to the various developed chlorine test procedures. The data presented in Table 4.III indicate the approximate maximum concentrations of constituents which did not produce significant interference in the chlorine methods. As shown in Table 4.III, the most significant interferences were those produced by Mn^{4+} in all methods and chlorides in excess of 100 mg/l in the total method.

Interference from oxidized forms of manganese remains a serious problem in chlorine analysis. Results from a recent study[14] indicate suitable methods for correction for or elimination of manganese interference may be developed in future work.

The extreme sensitivity of the total chlorine method indicates that a dilution of the sample to reduce chloride concentrations of 100 mg/l or less would be an appropriate technique to eliminate chloride ion interference.

The effect of temperature on chlorine determination was studied for each proposed method by analyzing samples maintained at 4°, 20°, and 35° C. The effect of temperature within the range studied was negligible on the recovery of chlorine and the stability of the developed color. In all test methods, the final maximum color developed in less than one minute and remained stable for more than thirty minutes.

The effect of dissolved oxygen was studied by reducing the oxygen content of water samples by nitrogen gas displacement. The effect of a varying dissolved oxygen concentration within the range of 0.65 to 7.60 mg/l of O_2 was negligible in all proposed chlorine test methods.

Table 4.III

Maximum Constituent Concentration Producing No Interference in Proposed Chlorine Methods

	Chlorine Method Constituent Concentration in mg/l			
Constituent	*Free Cl_2*	*Monochloramine*	*Nitrogen Trichloride*	*Total Cl_2**
Alkalinity, as $CaCO_3$	300.	300.	300.0	300.0
Hardness, as $CaCO_3$	485.	485.	485.	485.
Sulfate, $SO_4^{=}$	550.	550.	550.	550.
Chloride, Cl^-	1000.	1000.	1000.	1000.
Iron, Fe^{3+}	5.0	5.0	5.0	5.0
Copper, Cu^{2+}	30.	30.	30.	30.
Nitrate, NO_3^-	88.	88.	88.	88.
Nitrite, NO_2^-**	0.40	0.40	0.40	0.40
Manganese				
Mn^{2+}**	0.55	0.55	0.55	0.55
Mn^{4+}	0.00	0.00	0.00	0.00
Color, p.c.u.	120.	120.	120.	120.

* Includes determination of dichloramine
** Maximum concentration not established

PERFORMANCE STUDIES

In an effort to evaluate the developed methods in the determination of all forms of active chlorine, breakpoint studies were conducted at pH 7.0 and 5.0.

Procedure

For the pH 7.0 study, a series of buffered tap water samples at room temperature was treated with an ammonium sulfate solution to contain 1.3 mg/l of ammonia nitrogen. The samples were treated successively with increasing initial free chlorine concentrations in the range of 2.0 to 14.0 mg/l of Cl_2 and allowed to react for four to six hours. Samples were taken and analyzed by all four of the developed methods.

A similar study was conducted at pH 5.0 with the exception that predetermined amounts of dilute sulfuric acid were added after the addition of the ammonium sulfate and the appropriate chlorine dosage.

Discussion of Results

The results of the studies at sample pH 7.0 and 5.0 are shown in Figures 4.5 and 4.6, respectively.

Figure 4.5 illustrates typical breakpoint relationships among the chlorine forms that occur at pH 7.0. At low added chlorine to nitrogen ratios, monochloramine is the principal form of active chlorine present. As the ratio of added chlorine to nitrogen increases, dichloramine is formed probably as a result of disproportionation of monochloramine. Correspondingly, the monochloramine concentration decreases. Finally, both mono- and dichloramine decrease to the breakpoint at an added chlorine to nitrogen weight ratio of approximately 8.5:1. This breakpoint ratio at pH 7.0 is significantly less than the 10:1 ratio reported by Griffin and Chamberlin,[15] but is in good agreement with the 8.25:1 ratio reported by Palin.[16]

Amperometric titration results obtained on several of the samples during this study were, for the most part, in good agreement with the colorimetric test results. Some difficulty was encountered in amperometric titration by a drifting of the meter indicator when monochloramine and

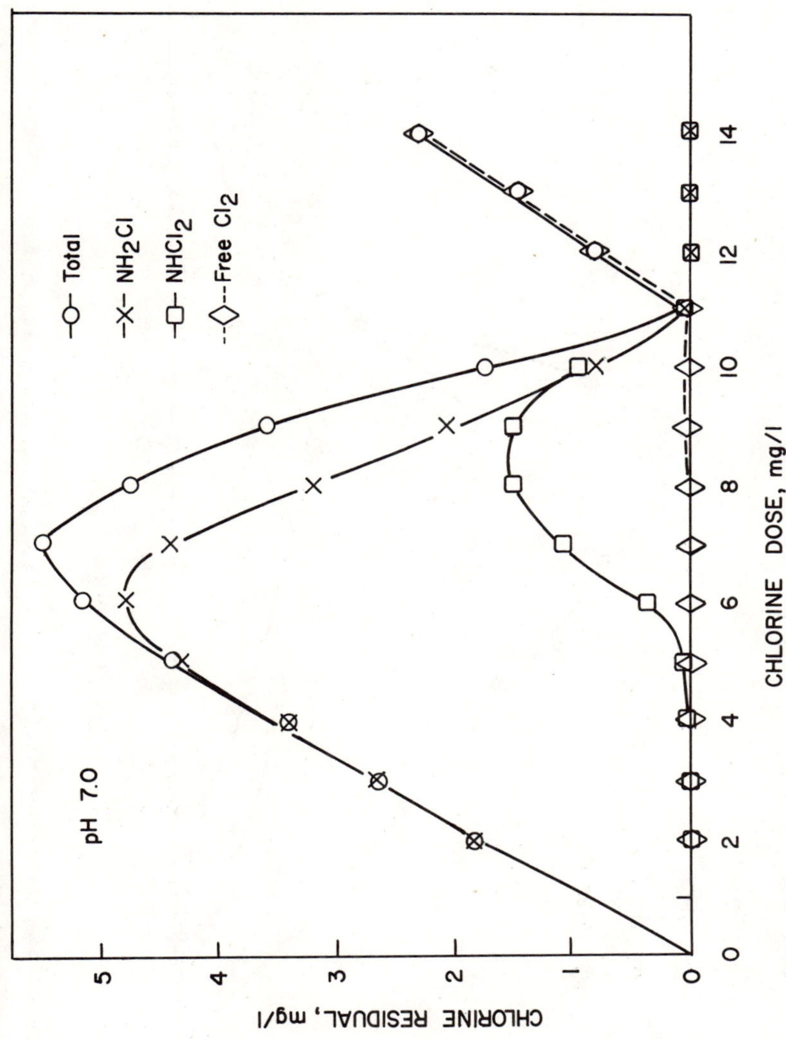

Figure 4.5. Chlorine residuals in breakpoint study at pH 7.0.

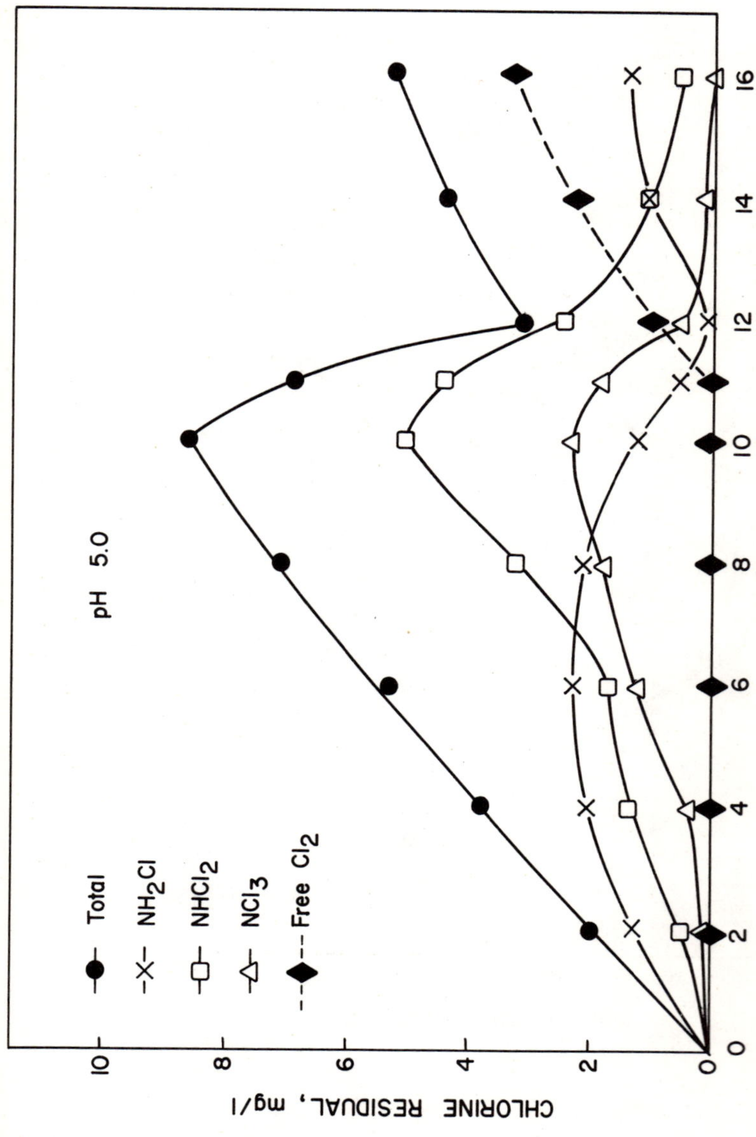

Figure 4.6. Chlorine residuals in breakpoint study at pH 5.0.

dichloramine were present together. In addition, it was difficult to distinguish between the two forms when there was a large disparity in concentration between the two forms.

While no nitrogen trichloride was found in the study conducted at pH 7.0, significant concentrations of NCl_3 were observed at solution pH 5.0. Figure 4.6 shows the early appearance of NCl_3 together with NH_2Cl and $NHCl_2$ at low added chlorine to nitrogen ratios. A true "breakpoint" was not achieved but an apparent breakpoint occurred at an added chlorine to nitrogen weight ratio of 9.25:1 which is near the 9.5:1 ratio at pH 6.0 reported by Palin.[16]

While no attempt will be made here to explain the complex phenomena exhibited in Figure 4.6, it is interesting to note that the disappearance, *i.e.* destruction, of NCl_3 appears to be a function of chlorine concentration. Also, the reappearance of monochloramine along with free chlorine after the apparent breakpoint is an unusual and unanticipated condition.

The objective of these breakpoint studies was not principally to verify the accuracy of the developed methods, but to determine if the methods gave reasonably self-consistent results in the analysis of all active chlorine forms. It is believed that this objective has been achieved in that predictable results and trends were obtained in the case of the breakpoint study at pH 7.0. Similarly, though not predictable, the results obtained at pH 5.0 were self-consistent in regard to changes in concentration relationships among the four active forms of chlorine. In this regard, it is significant to note that while dichloramine was always determined by difference, the sum of the free Cl_2, NH_2Cl, and NCl_3 in all cases never exceeded the total determination.

SUMMARY

New or modified methods have been developed for the determination of each of the active forms of chlorine normally encountered in water and wastewater treatment practice. The methods are capable of determining concentrations of free Cl_2, NH_2Cl, $NHCl_2$, and NCl_3 as high as 10 mg/l as Cl_2. It is believed these new methods will prove useful in research studies to elucidate the chemistry of chlorine reactions and disinfection processes and in application to water and wastewater chlorination practice.

REFERENCES

1. White, G. C. *Handbook of Chlorination*. Van Nostrand, Reinhold, New York, New York, 1972.
2. Lishka, R. J., E. F. McFarren, and J. H. Parker. "Water Chlorine (Residual) No. 2," Study No. 35, Analytical Reference Service, U.S. Department Health, Education and Welfare, 1969.
3. Lishka, R. J., and E. F. McFarren. "Water Chlorine (Residual) No. 2," Report No. 40, Analytical Reference Service, Environmental Protection Agency, 1971.
4. *Standard Methods for the Examination of Water and Wastewater*, 13th ed., American Public Health Association, New York, New York, 1971.
5. "The Evaluation of Existing Field Test Kits for Determining Free Chlorine Residuals in Aqueous Solution," Report No. 73-03, U.S. Army Medical Environmental Engineering Research Unit, Edgewood Arsenal, Maryland, 1972.
6. Bauer, R., and C. Rupe. *Anal. Chem., 43,* 421 (1971).
7. Black, A. P., and G. P. Whittle. *J. Amer. Water Works Assoc., 59,* 607 (1967).
8. Olivieri, V. P., T. K. Donovan, and K. Kawata. *J. San. Eng. Div., ASCE, 97,* 661 (1971).
9. Stringer, R., and C. W. Kruse. *J. San. Eng. Div., ASCE, 97,* 801 (1971).
10. Kruse, C. W., Y. C. Hsu, A. C. Griffiths, and R. Stringer. "Halogen Action on Bacteria, Viruses and Protozoa," Proc. National Specialty Conference on Disinfection, University of Massachusetts, Amherst, Massachusetts, 1970.
11. Kothandaraman, V., and R. L. Evans. *J. Water Pollut. Control Fed., 44,* 625 (1972).
12. Lapteff, A. "Improved Colorimetric Methods for the Determination of Chlorine Residuals in Aqueous Solution," Ph.D. Thesis, University of Alabama, 1973.
13. Metcalf, W. S. *J. Chem. Soc., 51,* 2113 (1929).
14. Andrews, E. R. "Determination of Total and Divalent Manganese in Water by a New Colorimetric Procedure," M.S. Thesis, University of Alabama, 1971.
15. Griffin, A. E., and N. S. Chamberlin. *J. New England Water Works Assoc., 55,* 371 (1941).
16. Palin, A. T. *Water and Water Engineering,* October 1950, November 1950, December 1970.

CHAPTER 5

REMOVAL OF CADMIUM FROM WATERS AND WASTES BY SORPTION ON HYDROUS METAL OXIDES FOR WATER TREATMENT

Hans S. Posselt and Walter J. Weber, Jr.*

INTRODUCTION

Treatment of water supplies for removal of traces of cadmium and certain other heavy metals is becoming increasingly necessary for production of potable water. Shortages of water resources and increased emphasis on re-use may also render such treatment essential for many industrial wastes. In accord with U.S.P.H.S. (now USEPA) recommendations, a water supply containing more than 0.01 mg/l of cadmium ($\sim 10^{-7}$ *M*) should not be considered as a source for drinking water. In view of the cumulative toxicity of cadmium, and the fact that drinking water recommendations serve merely as guidelines, it is of interest to examine the feasibility of methods for removal of traces ($< 10^{-6}$ *M*) of cadmium.

Ion exchange has been successfully employed for the removal and subsequent reclamation of metals in process waters. However, the economic feasibility of this method for treatment of large water volumes is questionable considering the large concentrations of ordinary competing cations (*e.g.*, Ca^{2+}, Mg^{2+}, Na^{+}) relative to trace amounts of heavy metals in most water sources.

---

*H. S. Posselt, Project Leader, Environmental Control Technology, Monroe County Community College, Monroe, Michigan. W. J. Weber, Jr., Professor of Environmental and Water Resources Engineering, The University of Michigan, Ann Arbor, Michigan.

An alternative approach to selective removal of certain heavy metals is that of sorption on suitable solid materials.[1-5] It has been proposed that naturally occurring oxides of manganese (IV) and iron (III) control the fixation of Mn, Fe, Co, Ni, Cu, and Zn in soils and in the aqueous environment.[2] High sorptive affinities of these substances for heavy metal ions have been observed in a number of investigations. A similar behavior should be expected with hydrolyzed products of aluminum (III) and probably some other solids which exhibit similar surface-chemical properties. Hydrous oxides of Al(III), Fe(III) and Mn(IV) are commonly employed or encountered in water treatment operations. Their general sorptive properties for cations, low price, and established use in water treatment technology suggest certain promise for cadmium removal via sorption.

Hydrous Manganese Dioxide

A rather amorphous but very active form of the hydrous oxide of Mn(IV), δ-MnO_2, is formed by oxidation of divalent manganese and/or by reduction of Mn(VII) in permanganate treated waters.[6] Hydrous manganese dioxide prepared *in situ* exhibits a very high surface area (*ca.* 300 m^2/gr) and a varied chemical composition ($MnO_{1.1}$ - $MnO_{1.95}$), depending on the conditions and mode of its formation. Under certain conditions, hydrous manganese dioxide may occur in a colloidal state of relatively high and long-term stability. The protonic acidity of MnO_2 has been sufficiently evidenced,[1] but there is less certainty with respect to the value for the pH of zero point of charge (ZPC). Recent reports give ZPC values from pH 1.5 for δ-MnO_2, pH 1.8 for Mn(II) manganite, pH 4.5 for α-MnO_2, pH 5.5 for γ-MnO_2, to pH 7.3 for β-MnO_2.[7] *In situ* preparations of hydrous and low ordered MnO_2 show ZPC values close to 2 and a higher sorptive affinity compared with well ordered oxides of Mn(IV).[5,8,9] Sorptive uptake of cations by hydrous MnO_2 is accompanied by proton release and generally increases in magnitude with increasing pH in the range pH > pH_{ZPC}.[5,8] Consistent with this, the colloid stability of hydrous MnO_2 increases over the same pH range, but is reduced concurrent with an increase in cation concentration in solution.[9]

The sorptive affinity of hydrous MnO_2 for cations is in fair agreement with Hofmeister's rule,[5] *i.e.*, the affinity tends to increase with increasing positive charge and decreasing ionic radius of metal cations.

Hydrolysis of Fe(III) and Al(III)

Ferric and aluminum salts are the most commonly used primary coagulants in water treatment. The main objective of coagulation in these applications is the removal of colloids (including suspended matter), colored compounds and similar impurities by agglomeration and subsequent sedimentation.

The trivalent ions Fe^{3+} and Al^{3+}, or their respective aquo-complexes, are unstable under pH conditions (pH 5-11) representative of water treatment operations, A number of hydrolysis species are formed, as illustrated for iron:

$$Fe(H_2O)_6^{3+} \rightleftharpoons [Fe(H_2O)_5OH]^{2+} + H^+ \rightleftharpoons$$
$$[Fe(H_2O)_4(OH)_2]^+ + 2H^+ \rightleftharpoons$$
$$[Fe(H_2O)_3(OH)_3]_{(s)} + 3H^+ \rightleftharpoons$$
$$[Fe(H_2O)_2(OH)_4]^- + 4H^+ \qquad (1)$$

The neutral insoluble species $[Fe(OH)_3(H_2O)_3]$ predominates in the neutral pH range, but there is also evidence for the existence of polynuclear complexes, for example:

$$2[Fe(H_2O)_5(OH)]^{2+} \rightleftharpoons$$
$$[(H_2O)_4Fe\langle{}^{OH}_{OH}\rangle Fe(H_2O)_4]^{4+} + 2H_2O \qquad (2)$$

and other species capable of colloid formation. There is little consistency with respect to ZPC values for Fe(III) and Al(III) sols, probably due to the aging and polymerization phenomena often observed with these compounds. Coagulation studies and electrophoretic measurements show that the ZPC values for both compounds fall in the range of pH 5-8.5.[9,10]

Cation sorption of Fe(III) and Al(III) hydroxides has been observed but capacities are less than those reported for hydrous MnO_2 under similar pH conditions.[1,3,4]

EXPERIMENTAL

Chemicals

All chemicals used in the course of the sorption experiments were of standard reagent quality. The preparation of stock solutions of stable and radioactive cadmium was described in a previous paper.[11]

Preparation of Manganese Dioxide

Formation of hydrous MnO_2 by oxygenation of divalent manganese is known to yield products of varied degrees of oxidation ($MnO_{1.1}$ - $MnO_{1.95}$) depending largely on experimental conditions.[8] This result is not surprising in view of the high sorption capacity of MnO_2 towards Mn^{2+} ions. A similar behavior, though much less in degree, must be expected under the conditions of the Guyard reaction:[5]

$$3\ Mn^{2+} + 2\ MnO_4^- + 2\ H_2O \longrightarrow 5\ MnO_2 + 4\ H^+ \qquad (3)$$

Sorption concurrent with oxidation can generally be minimized by maintaining an excess of the oxidant over the entire period of the reaction. This is achieved by observing the sequence of reactant additions. Thus, the reductant should be slowly added to a slight excess of the oxidant.

The problem of sorptive interference can be avoided by choosing a more suitable reducing agent, such as hydrogen peroxide. In moderate alkaline solution (pH 8-10) the reaction

$$3\ H_2O_2 + 2\ MnO_4^- \longrightarrow 2\ H_2O + 3\ O_2 + 2\ MnO_2 + 2\ OH^- \qquad (4)$$

is fairly rapid and quantitative. Slow addition of H_2O_2 to a slight excess of permanganate in dilute solution and at pH > 10 shows a stepwise reduction of permanganate to MnO_2. The green manganate (VI), formed as an intermediate product, undergoes disproportionation according to:

$$3MnO_4^{2-} + 2H_2O \longrightarrow 2MnO_4^- + MnO_2 + 4OH^- \qquad (5)$$

yielding permanganate and manganese dioxide. Sols of 2×10^{-4} *M* hydrous MnO_2 were prepared in accordance with these reactions (Equations 4 and 5) as described below.

A 20-ml portion of a standardized solution of 0.02 *M* $KMnO_4$ was pipetted into a beaker containing 900 ml of distilled water (pH $\approx$ 8). Approximately 98% of the stoichiometrically required hydrogen peroxide was mixed with the permanganate by dropwise addition from an approximately 0.03 *M* solution of H_2O_2. The resulting solution (pH $\approx$ 10) was kept for about 24 hours. It was then adjusted to the desired pH level by addition of nitric acid and a volumetric flask was filled to a volume of 2000 ml.

Residual permanganate in these solutions was determined by the following method. One ml of a solution of approximately 2 *M* $Ca(NO_3)_2$ was added to a portion of 50 ml of the MnO_2 soln. Coagulation of MnO_2 was observed after 15 minutes of rapid mixing. The coagulated MnO_2 was separated by filtration through a 0.1 μ membrane filter. The concentration of permanganate in the filtrate was obtained by spectrophotometric analysis at 526 mμ and comparison with a known standard. These analyses showed residues of less than 1% of the original permanganate used for the formation of the MnO_2 soln.

Iron and Aluminum Solutions

The aforementioned hydrolysis reactions of Fe(III) and Al(III) give rise to changes in the composition of solutions containing these salts unless they have been stabilized by acidification. Stock solutions were made approximately 10^{-2} *M* in $HClO_4$ and 10^{-1} *M* in $Al(ClO_4)_3$ or $Fe(ClO_4)_3$. Standardization of the reagent grade salts $Fe(ClO_4)_3 \cdot 6H_2O$ and $Al(ClO_4)_3 \cdot 9H_2O$ was made by alkalimetric titration.

The stock solutions exhibited good stability with respect to precipitation over the entire period of their usage. Secondary stock solutions, usually of 2×10^{-4} *M*, were made by dilution of the stock solutions with distilled water prior to the sorption experiments.

Equipment

The reactants were mixed in teflon beakers with a magnetic stirrer. Capped polyethylene bottles (125-ml capacity) were used for equilibration. A mechanical laboratory shaker was used to provide for agitation. The shaker was operated at approximately 400 displacements per min with a horizontal displacement of about 1 inch. A description of the instrumentation for pH measurement, radioassay and the filtration apparatus was previously reported.[11] Standard laboratory glassware was employed except for polyethylene pipettes, which were used for transfer of the radioisotope and of dilute solutions of cadmium.

Procedures

Predetermined volumes of the standard reagent and adsorbent stock solutions (except for cadmium) were mixed with distilled water in a teflon beaker to yield (100 - x) ml of the buffered and pH-adjusted solution of the absorbent. A known quantity of the radioisotope and stable cadmium were mixed with distilled water. Of this, a predetermined quantity was added, under rapid mixing, to the adsorbent solution to yield 100 ml of the experimental medium, specifically defined in terms of pH, ionic strength and concentration of all reagents. Since the concentration of the adsorbent was varied while that of the sorbate remained constant, the pH variation within the determination of one particular isotherm was usually less than 0.05 pH units. This has been verified repeatedly prior to and after filtration of sorption systems.

The well-mixed solutions were poured into polyethylene bottles, capped and mechanically shaken for approximately 60 minutes (except for sorption rate studies) at ambient temperature.

The solutions were buffered with 0.001 *M* sodium acetate between pH 5-7, and with 0.001 *M* sodium tetraborate at higher pH conditions. The usefulness of both buffer systems was established by comparisons with unbuffered experiments.

The samples were filtered according to a technique described previously for precipitation studies,[11] using 0.1 μ membrane filters. Filters were cleaned with acid and washed with distilled water after each individual filtration. For MnO_2 experiments, a mixture of dilute H_2O_2 and H_2SO_4 was used to dissolve the filter cake, while 3 *M* HCl was found more suitable for dissolving precipitated Al(III) and Fe(III).

Equilibrium concentrations of the sorbate were determined by radioassay of the acidified filtrates.[11] The small excess of permanganate applied in the MnO_2 experiments was rarely noticeable after final filtration of the samples. This is very likely due to decomposition which is catalyzed by MnO_2. Filtrates discolored by residual permanganate were decolorized by small additions of acidic hydrogen peroxide after weighing of the vials.

Treatment of Data

The sorptive uptake of cations on hydrous oxides of manganese(IV) and iron(III) has usually been presented in terms of the Langmuir equation for adsorption.[5,8]

$$X = \frac{X_m \cdot b \cdot C_{eq}}{1 + b \cdot C_{eq}} \tag{6}$$

In this equation, X represents the quantity sorbed per unit amount of sorbent, X_m the limiting value for X ("mono layer" capacity), C_{eq} the equilibrium concentration of the sorbate in solution, and b is a constant related to the energy of sorption. By rearrangement, Equation 6 yields a linear form of Langmuir's equation

$$\frac{1}{X} = \frac{1}{X_m} + \frac{1}{X_m \cdot b \cdot C_{eq}} \tag{7}$$

which is more convenient for the evaluation of sorption data and calculation of the constants X_m and b. In accordance with Equation 7, a plot of $1/X$ versus $1/C_{eq}$ should give a linear trace with a slope of $1/X_m \cdot b$ and an intercept of $1/X_m$ as exemplified in Figure 5.1.

RESULTS AND DISCUSSION

It has been previously observed that rates of attainment of sorption equilibria for cation uptake on freshly prepared hydrous oxides of Mn(IV) and Fe(III) are very rapid.[5,8] The kinetics of sorption with these oxides have been followed by the disappearance of the sorbate in solution,[5,8] as well as by the concurrent release of protons[1] which are displaced by cations of sufficient sorptive, or ion exchange, affinity. In these cases, equilibrium was usually established within 10 minutes or less. It is noteworthy to mention that sorption rates with well ordered (most often aged) crystals may be much slower than freshly precipitated or highly dispersed forms of these oxides. For example, contact times of 15 hours were required for equilibrium uptake of zinc on γ-MnO_2 at pH 7 and an ionic strength of 2.0 (2 *M* NH_4Cl).[12]

Apart from the possible influence of crystal habit and modification on sorption rates, the contribution of particle size and its effect on the transport of the solute to the

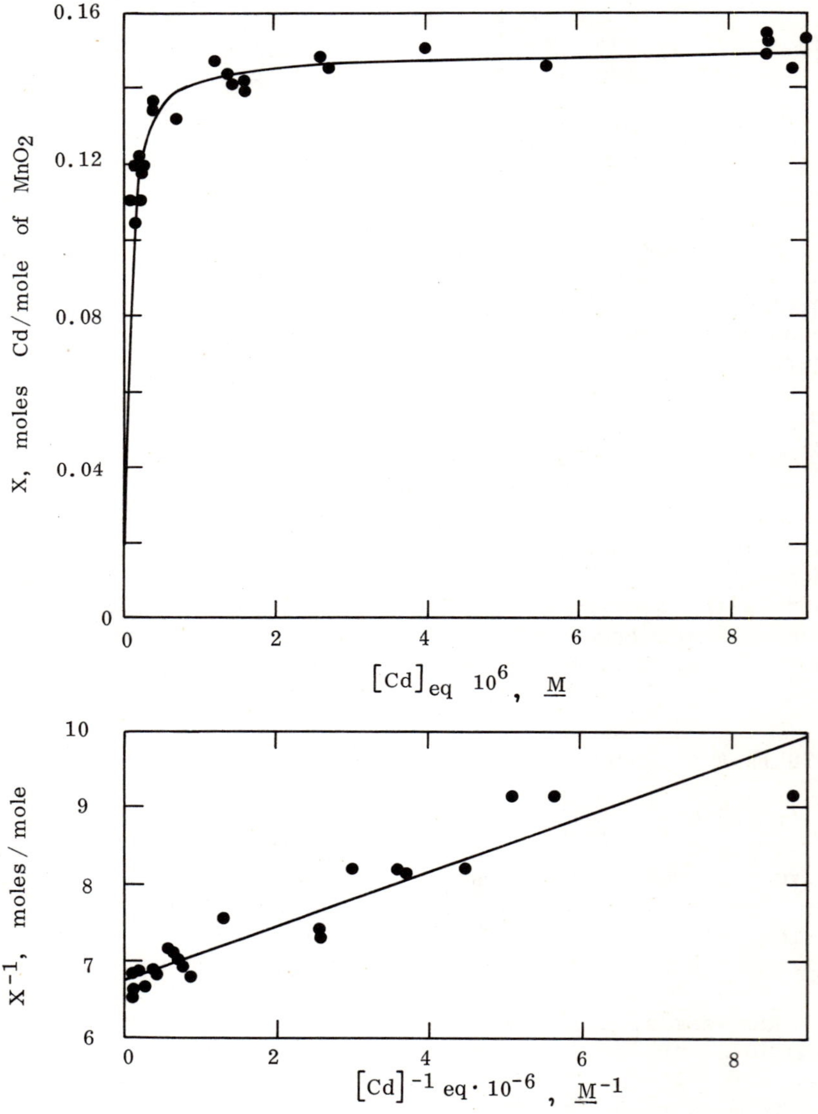

Figure 5.1. Sorption of cadmium on hydrous manganese dioxide at pH 5. Ionic strength 0.01, pH 5, ambient temperature. Lower graph shows Langmuir plot fitted by method of least squares. Curve in upper graph is calculated Langmuir isotherm.

solid surface site must be considered. Indications for this are given by the observation that the rate of Mn^{2+} uptake on hydrous manganese dioxide tends to decrease at higher ionic strengths at which particle aggregation of the sorbent is predominant.[8]

Rate studies within the context of this work were performed with the aim of establishing certainty of sorption equilibrium within the experimental contact times. This requirement seems to be sufficiently met by a contact period of 60 minutes as indicated by the rate studies for the oxides used and by similar results of other relevant investigations.[5,8]

Effect of pH on Cadmium Sorption

In most water treatment operations including sorption and colloid destabilization, pH is an important variable. In the process of sorptive uptake of cations by hydrous metal oxides, pH exerts direct influence on the chemical, and often physical, properties of both the sorbent and sorbate species.

The distribution of cadmium species as a function of pH has been defined in a previous study,[11] but there is much less certainty with respect to the influence of pH on the structural chemistry and physical morphology of hydrous oxides of Mn(IV). Manganese dioxide is generally regarded as insoluble under the pH conditions of natural waters.[1] The possibility of structural changes (*e.g.* particle size, polymerization, etc.) of MnO_2 is indicated by the observation of reversible color changes induced by variation of pH of hydrous manganese dioxide sols. Furthermore, as discussed in previous sections, the charge of colloidal hydrous manganese dioxide becomes increasingly negative as pH increases beyond the pH of zero point of charge. The simultaneous increase of the cation exchange capacity within the same pH range points toward a structurally related linkage between the colloid-chemical and sorptive behavior of hydrous MnO_2.[5,8] An analogous situation is expected to exist with other hydrous heavy metal oxides that exhibit similar structural and chemical properties.

While other mechanistic interpretations for cation sorption on hydrous MnO_2 cannot be ruled out, present concepts of the chemical behavior of highly dispersed forms of the oxide in the presence of relatively low concentrations of cations suggest a mechanism of exchange sorption. Proton release concurs with sorptive uptake of cadmium upon mixing of isoprotic solutions of hydrous MnO_2 and cadmium. Thus, the same qualitative interpretation may be extended to this study.

Individual sorption isotherms for cadmium were determined for a pH range of 5-8.3 and at an ionic strength of 0.05 and ambient temperature. A plot of the limiting sorption capacities and Langmuir constants against pH is shown in Figure 5.2. The increase of X_m values with increasing pH is qualitatively in agreement with sorption data for other cations on hydrous MnO_2. However, the quantitative effect of pH on the extent of cadmium sorption is less pronounced than has been reported for Mn(II).[8] This difference may be explained in part by a high specific affinity of hydrous MnO_2 for manganese(II) compared with other cations, but may also derive from differences in the distribution of aquo and hydroxy-species of the two metals.

An experimental investigation of the sorption of cadmium at a range higher than pH 8.3 is of questionable value, since extremely low concentration ranges are obscured by wall adsorption, and the limited range of sorbate concentration is insufficient to determine isotherm parameters with any degree of accuracy.

Since ionic equilibria of the type

$$Cd^{2+} + OH^- \rightleftharpoons CdOH^+ \tag{8}$$

are considered very rapid, an assessment with respect to which species is preferentially adsorbed cannot be easily made on the basis of sorption data as a function of pH. In view of the small competitive effect of monovalent cations[5] (see Figure 5.3), it could be assumed that $CdOH^+$ plays a lesser role than the more highly charged Cd^{2+} ion. On the other hand, the $CdOH^+$, which is less hydrated, may approach and penetrate surface sites more readily than the bulkier Cd^{2+} species. Thus, the process shown in Equation 8 may proceed from right to left in the course of exchange sorption.

While X_m values for cadmium compare well with those obtained for other divalent cations,[5,8] the b values found for cadmium are considerably higher. An explanation for the latter result may be sought in specific properties of the sorbate species or in possible differences of the sorbent structure.

The ion Cd^{2+} differs from the corresponding ions of the group IIa metals (Mg^{2+}, Ca^{2+}, Sr^{2+}, Ba^{2+}, and Ra^{2+}) in that

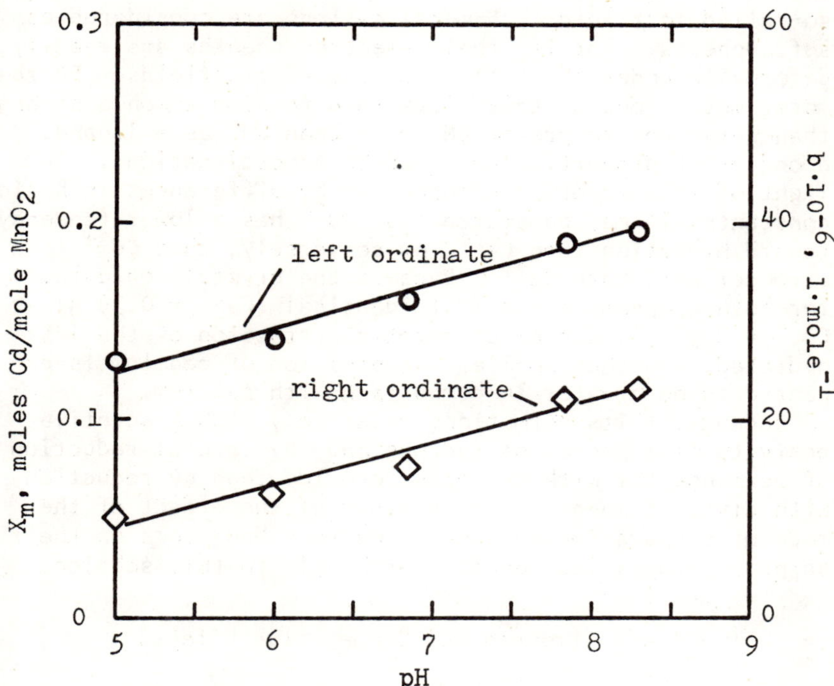

Figure 5.2. Effect of pH on sorption of cadmium on hydrous manganese dioxide. Ionic strength 0.05, ambient temperature.
<u>pH 5-7:</u> $[Cd]_o = 10^{-5}$ M, $[MnO_2] = (10\text{-}140) \times 10^{-6}$ M, buffer: 10^{-3} M sodium acetate.
<u>pH 7-8.3:</u> $[Cd]_o = 8 \cdot 10^{-7}$ M, $[MnO_2] = (1\text{-}7) \times 10^{-7}$ M, buffer: 10^{-3} M sodium tetraborate.
Circles represent plot of limiting sorption capacity (left-hand scale), diamonds represent plot of Langmuir constant b (right-hand scale) versus pH.

it possesses 10 electrons in the outer shell and thus does not have an inert gas configuration as is exhibited by the alkaline earth metal cations. By this definition, cadmium has been classified as a B-metal cation or a so-called soft acid.[1] B-metal cations are considered as soft spheres, that is, their electron sheaths are readily deformable under the influence of electric fields. Furthermore, metal ions of this class tend to bind ammonia stronger than water, or to prefer CN^- more than OH^- as a ligand, among other properties compared to A-metal cations. In light of this theory and supported by differences in basicity constants, it may be stated that Cd^{2+} has a lower tendency toward hydration than Ca^{2+}, or conversely, that Cd^{2+} is a stronger acid than Ca^{2+}. Because the crystal ionic radii for both elements are almost identical (Ca^{2+} = 0.99 Å; Cd^{2+} = 0.97 Å), the electrostatic attraction of the less hydrated, and thus smaller, aqueous ion of cadmium is expected to be greater in comparison with calcium.

Because of its formation, relatively higher sorptive activity is expected of hydrous MnO_2 by careful reduction of permanganate with hydrogen peroxide than by reduction with divalent manganese.[5] A study of the effect of the mode of preparation and age of hydrous MnO_2 sols on their sorptive properties appears worthwhile in this section.

Ionic Strength and Competitive Effects

In terms of electrostatic theory, ionic strength effects on sorption equilibria can be interpreted in a simplified manner by way of cation competition. In accordance with Hofmeister's rule a cation will be held more strongly by a negatively charged surface site the higher its charge and the smaller its size.[13] Thus, the effect of sodium ion concentration on the sorption of, or cation exchange capacity for, divalent ions is not expected to be very pronounced.

Figure 5.3 shows the effect of ionic strength ($NaClO_4$) on the Langmuir constants X_m and b for cadmium on hydrous MnO_2 at pH 5 and room temperature. Considering the low initial concentration of cadmium (10^{-5} *M*) in contrast to that of sodium ions, it is apparent that the influence of monovalent cations is rather weak. Greater ionic strength effect has been previously reported for the sorption of calcium on hydrous MnO_2 at the same pH.[5] This fact is quite consistent with the high sorptive affinity of hydrous MnO_2 for cadmium and the relatively small competitive effect

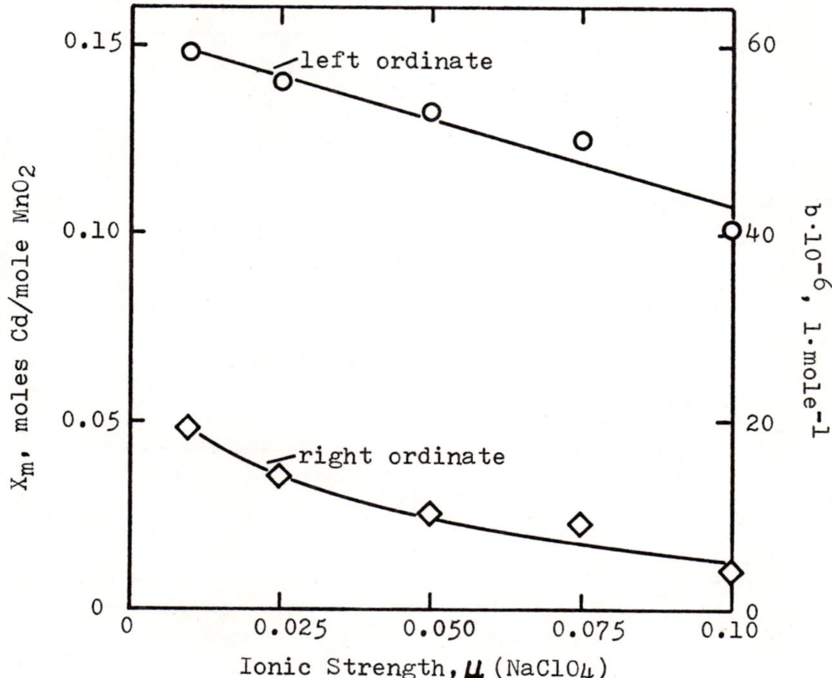

Figure 5.3. Effect of ionic strength on sorption of cadmium on hydrous manganese dioxide at pH 5. Ambient temperature, $[Cd]_o = 10^{-5}$ M, $[MnO_2] = (4\text{-}160) \times 10^{-6}$ M, buffer: 10^{-3} M sodium acetate. Circles represent plot of limiting sorption capacity (left-hand scale), diamonds represent plot of Langmuir constant b (right-hand scale) versus ionic strength.

of calcium ion on the molar sorption ratio Cd-MnO_2 as shown in Figure 5.4.

A loss of only about 20% of the original sorption capacity (at zero calcium) is encountered by addition of 2×10^{-3} *M* calcium nitrate. Relative to the initial concentration of cadmium, this corresponds to a molar $[Ca^{2+}]/[Cd^{2+}]$ ratio of 200. While the formation of specific surface complexes between the sorbate ion and functional groups cannot be excluded in ion exchange phenomena, the weak competition between calcium and cadmium is reasonably well interpreted in terms of differences in acidity and outer shell configuration, as discussed in the preceding section.

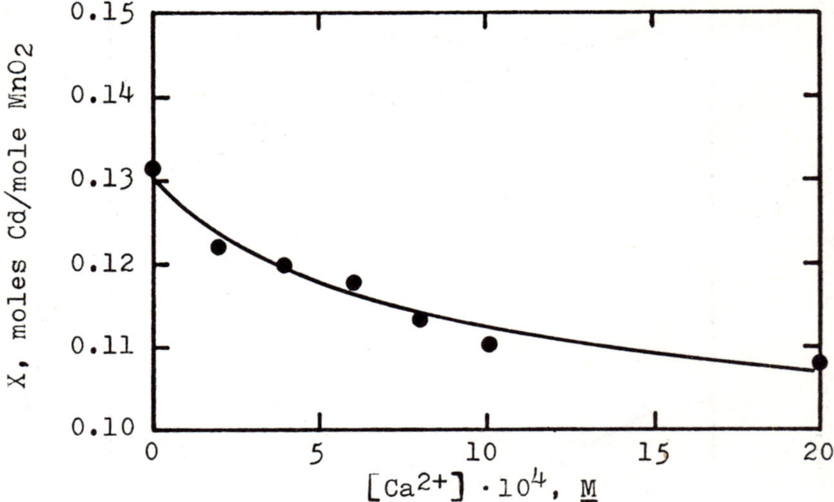

Figure 5.4. Effect of calcium on sorption of cadmium on hydrous manganese dioxide. pH 5, ionic strength 0.05, ambient temperature, $[Cd]_o$ = 10^{-5} M, $[MnO_2]$ = 4 x 10^{-6} M, buffer: 10^{-3} M sodium acetate.

Complexation of cadmium by chloride ion as a competing process for sorption is similarly weak in magnitude. This is shown by a plot of the chloride ion concentration versus the molar sorption ratio X in Figure 5.5. Distribution diagrams for chloride species of cadmium[11] reveal that $CdCl^+$ occurs as the only competitor for Cd^{2+} in this concentration range and at pH 5. While the observed effect of chloride and hydroxide on cadmium uptake by hydrous MnO_2 appears insignificant in view of the concentration and pH conditions of water treatment operations, considerable interference with sorption processes may occur in the presence of ligands with stronger complexing capability. This complication is expected due to complexation with certain organic compounds present in many natural and wastewaters. Decomplexation via oxidative degradation of organic ligands prior to sorption appears to be a feasible process to minimize this interference.[6]

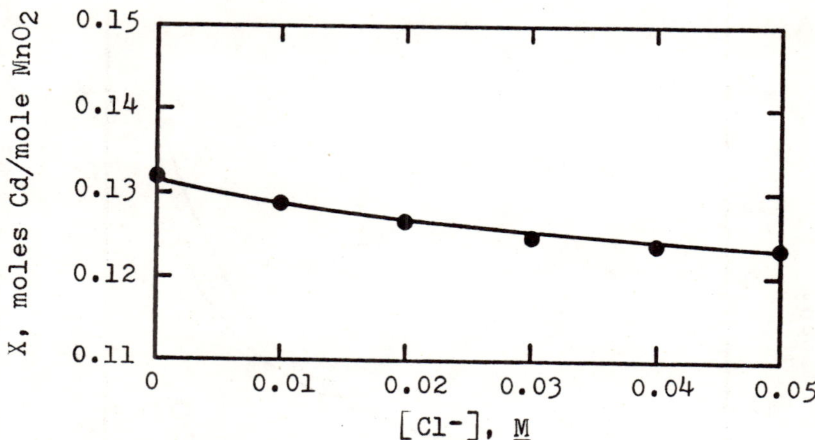

Figure 5.5. Effect of chloride on sorption of cadmium on hydrous manganese dioxide. pH 5, ionic strength 0.05, ambient temperature, $[Cd]_o$ = 10^{-5} M, $[MnO_2]$ = 4 x 10^{-6} M, buffer: 10^{-3} M sodium acetate.

Sorption of Fe(III) and Al(III)

Several accounts in the literature document sorptive properties of highly dispersed insoluble, largely X-ray amorphous products of the hydrolysis of Fe(III) towards cations.[1,2,4,8] From the general chemical resemblance of Fe(III) with Al(III) this behavior is, at least qualitatively, also expected for insoluble aluminum hydrolysis products.

Apart from quantitative differences in hydrolysis equilibria, it must be realized that profound qualitative differences in the type of species, degree of polymerization crystal modifications, etc. Exist between hydroxides and hydrous oxides of Mn(IV), Fe(III) and Al(III).

On a quantitive basis, it is important to consider the fact that hydrous oxides of Al(III) and Fe(III) exhibit ZPC values considerably higher than hydrous forms of MnO_2. As a consequence, the operational range of cation sorption by insoluble Al(III) and Fe(III) is much smaller compared to MnO_2. As shown in Figure 5.6, sorptive uptake of cadmium begins, approximately, at pH > 6 for Fe(III) and pH > 6.75 for Al(III), and continues to rise with increasing pH. Assuming Cd^{2+} and $CdOH^+$ as species capable of being

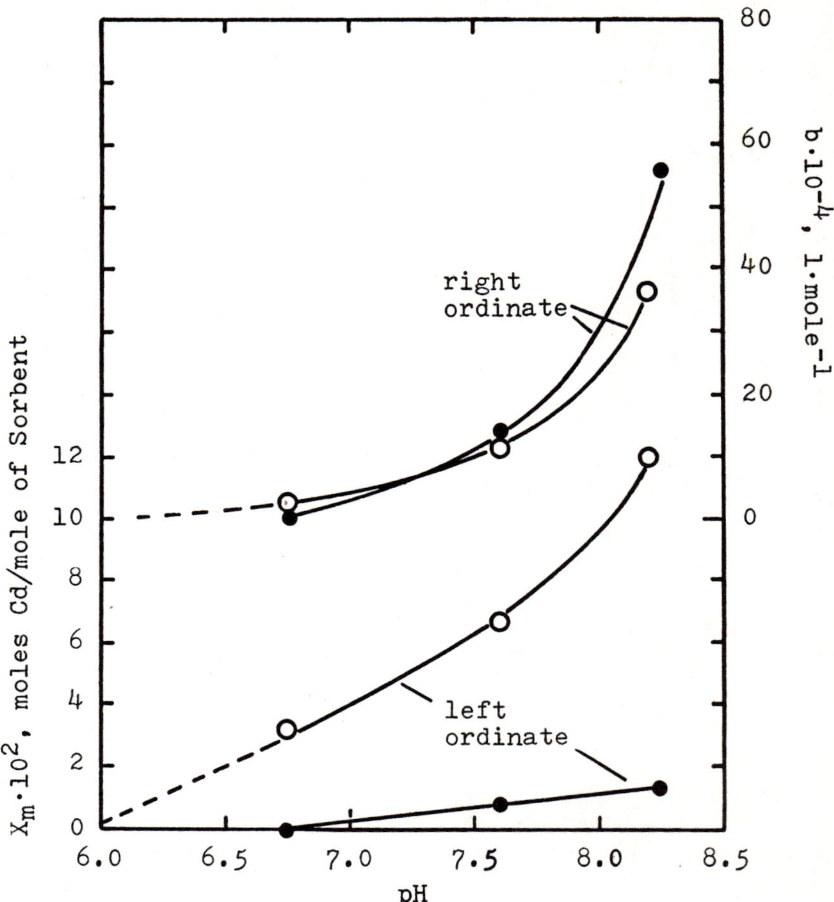

Figure 5.6. *Effect of pH on sorption of cadmium on hydrous oxides of Fe(III) and Al(III). Ionic strength 0.05, ambient temperature,*

pH 6.75: 10^{-3} *M sodium acetate,* $[Cd]_o = 10^{-5}$ *M,* $[Fe(III)] = (5-50) \times 10^{-5}$ *M;*

pH > 6.75: 10^{-3} *M sodium tetraborate,* $[Cd]_o = 8 \cdot 10^{-7}$ *M,* $[Fe(III)] = (4-140) \times 10^{-6}$ *M,* $[Al(III)] = (1-50) \times 10^{-5}$ *M.*

Open and solid circles stand for Fe(III) and Al(III), respectively. Upper two curves (right-hand scale) show plot of Langmuir constant b, lower two curves (left-hand scale) show plot of limiting sorption capacity as function of pH.

sorbed, and considering the increasing negative charge of the sorbent with pH, a further increase of sorption capacity for the approximate range pH 8.3-10 could be predicted. Opposing this trend, however, is the increased tendency of these sorbents to form soluble anionic complexes by combination with hydroxide ion.[10] Formation of ferrate(III) appears insignificant in the practical range of pH and concentration conditions, but solubilization of $Al(OH)_3$ becomes of serious consequence at pH > 8 and $[Al(III)] < 10^{-5}$ *M*. The very low sorptive capacity of hydrous Al(III) oxide may, at least in part, be attributed to this behavior. In light of these facts and considerations, Al(III) salts cannot be considered feasible for sorptive removal of either cadmium or, very likely, other cations.

At comparable pH conditions, hydrous oxides of Fe(III) show a somewhat lower sorption affinity toward cadmium than does hydrous manganese dioxide. This tendency is reflected in the values for limiting sorption capacities as well as in the Langmuir constants. The importance of the magnitude of the b value is often overlooked; in fact, some investigators have failed to report these at all. The following calculation will serve to demonstrate the practical significance of Langmuir b constants at comparable values for X_m, the limiting capacity.

Assume the cadmium content of a given water is 5×10^{-7} *M* (56 ppb) and is to be reduced to 10% of its original level (5.6 ppb) by sorption on hydrous oxides of Fe(III) or Mn(IV) at a pH of 8.25 and otherwise identical conditions ($\mu = 0.05$). The molar concentration of the sorbate is calculated by

$$m = \frac{(C_o - C_{eq})\ (1 + b \cdot C_{eq})}{X_m \cdot b \cdot C_{eq}} \tag{9}$$

and from the respective constants X_m and b. These are:

Fe(III): $X_m = 0.12$ moles/mole
$b = 3.7 \times 10^5$ l/mole

MnO_2: $X_m = 0.195$ moles/mole
$b = 23 \times 10^6$ l/mole.

The calculation of m yields 4.4×10^{-6} *M* (∿0.4 ppm as MnO_2) for Mn(IV) and 2×10^{-4} *M* (∿113 ppm as $Fe_2(SO_4)_3 \cdot 9H_2O$) in the case of Fe(III) as sorbent.

Cadmium sorption on hydrous oxides of Fe(III) exhibits a very similar response to chloride complexation and calcium competition as shown previously for hydrous manganese dioxide. On a quantitative basis, the effects of both Cl^- and Ca^{2+} ions were found to be relatively stronger when compared with MnO_2. However, accounting for the differences in the initial concentrations of cadmium tends to compensate for this apparent behavior.

SUMMARY AND CONCLUSIONS

Removal of trace amounts of cadmium ($< 10^{-6}$ *M*) from drinking water supplies is highly desirable in view of the low levels at which toxic effects are manifested by this metal.[14] Sorptive uptake of cadmium on suitable solid materials would constitute a method easily adaptable to present water treatment technology. Accordingly, the sorption behavior of hydrous oxides of Al(III), Fe(III), and Mn(IV) towards cadmium was examined over a wide range of variable conditions.

The equilibrium uptake of cadmium on the sorbents studied can be described well by the Langmuir equation for sorption. Sorption rates observed with highly dispersed hydrous oxides are extremely rapid, with equilibrium usually being attained in less than 10 minutes.

Both the limiting sorption capacity (X_m) and adsorption affinity (b) for cadmium tend to increase over the pH range beyond the pH of ZPC for all three oxides studied. Hydrous manganese dioxide, prepared by careful reduction of permanganate, was found superior on the basis of Langmuir parameters and by virtue of its low ZPC value. The practical range for Al(III) salts is rather limited due to the formation of aluminate ion and a relatively high ZPC for insoluble Al(III) hydrolysis products.

The sorption of cadmium exhibits a small dependence on ionic strength. Competitive effects from uptake of calcium or from chloride complexation are similarly weak in magnitude. These results suggest that sorption on hydrous oxides of Mn(IV) and Fe(III) is a feasible approach for removal of cadmium in water and wastewater treatment.

This work and related studies[5,7,8] indicate that the sorptive behavior of hydrous oxides is greatly influenced by the age and mode of preparation of these materials. Investigation of these aspects may prove very fruitful from the standpoint of optimizing the efficiency of sorptive processes on a separate basis or in combination with coagulation processes.

Furthermore, the chemical and structural resemblance between hydrous manganese dioxide and "silica" suggests similar sorptive properties for the latter material. Since the use of silica a coagulant aid in water treatment has already been established, an examination of its role as a potential sorbent for heavy metal ions appears worthwhile.

REFERENCES

1. Stumm, W., and J. J. Morgan. *Aquatic Chemistry*, Wiley-Interscience, New York, New York, 1970.
2. Jenne, E. A. In *Trace Inorganics in Water*, Advances in Chemistry Series 73, American Chemical Society, Washington, D.C., 1968, Chapter 21.
3. Weber, W. J., Jr., and E. Matijevic, editors. *Adsorption from Aqueous Solutions*. Advances in Chemistry Series 79, American Chemical Society, Washington, D.C., 1968.
4. Faust, S. D., and J. V. Hunter, editors. *Principles and Applications of Water Chemistry*. Wiley & Sons, New York, New York, 1967.
5. Posselt, H. S., J. A. Anderson, and W. J. Weber, Jr. *Environ. Sci. Technol., 2,* 1087 (1968).
6. Posselt, H. S., and W. J. Weber, Jr. In *Physicochemical Processes for Water Quality Control*. Wiley-Interscience, New York, New York, 1972, Chapter 8.
7. Murray, D. J., T. W. Healy, and D. W. Fuerstenau. In *Adsorption from Aqueous Solution*. Advances in Chemistry Series 79, American Chemical Society, Washington, D.C., 1968, Chapter 7.
8. Morgan, J. J. and W. Stumm. *J. Colloid Sci., 19,* 347 (1964).
9. Posselt, H. S., A. H. Reidies, and W. J. Weber, Jr. *J. Amer. Water Works Assoc., 60,* 48 and *60,* 1366 (1968).
10. Black, A. P. In *Principles and Applications of Water Chemistry,* Wiley-Interscience, New York, New York, 1967.
11. Posselt, H. S., and W. J. Weber, Jr. "Equilibrium Models and Precipitation Reactions for Cadmium in the Aqueous Environment," In press.
12. Gabano, J. P., P. Etienne, and J. F. Laurent. *Electrochem. Acta, 10,* 947 (1965).

13. Adamson, A. W. *Physical Chemistry of Surfaces*. Wiley-Interscience, New York, New York, 1967.
14. Schroeder, H. A. *N. Eng. J. Med., 280,* 836 (1969).

CHAPTER 6

CHEMISTRY OF SULFUR SPECIES AND THEIR REMOVAL FROM WATER SUPPLY

Kenneth Y. Chen*

INTRODUCTION

The study of the chemistry of sulfur species and their removal is tremendously significant in water supplies. Several of the thermodynamically stable species are environmental pollutants; the more noted ones are the sulfides and, to a lesser degree, sulfur and sulfate. Sulfides are given a special role here, mainly due to their undesirable properties; most serious are their obnoxious odor and toxicity, together with their corrosive ability. Their removal can readily be effected through aeration and chemical oxidation. Sulfates are usually found in dissolved form in water. An excess amount of sulfate causes physiological effects on man, reducing the utility of water for domestic and industrial use. Elemental sulfur, although nearly insoluble in water may appear in colloidal form as a white, milky-colored dispersion. It is also subject to disproportionation to sulfide and thiosulfate in alkaline solution.

Treatment processes for the removal of sulfide have been successfully practiced over the years. However, few details have been known as to their chemistry, reaction mechanisms, and kinetics. This paper presents an overall review of the chemistry of sulfur species in dilute aqueous

---

*K. Y. Chen, Environmental Engineering Programs, University of Southern California, Los Angeles, California.

solution as related to water treatment practices, with special emphasis on the removal of sulfides. The chemical oxidants discussed are oxygen, chlorine, ozone, permanganate, and chlorine dioxide.

EQUILIBRIUM CHEMISTRY OF SULFUR SPECIES IN AQUEOUS SOLUTION

Out of the thirty-plus ionic and molecular sulfur species that possibly exist, only five are thermodynamically stable in aqueous solution at room temperature and one atmosphere pressure, namely, HSO_4^-, $SO_4^=$, $S°$, H_2S, and HS^-.[1] Other species such as thiosulfate, polysulfides, and polythionates also appear in the natural environment. However, they are considered to be thermodynamically unstable. A pε-pH predominance diagram of sulfur species in aqueous solution is shown in Figure 6.1. Free energy data are obtained from stability constants.[2] The methodology for the construction of such a diagram is detailed elsewhere.[3]

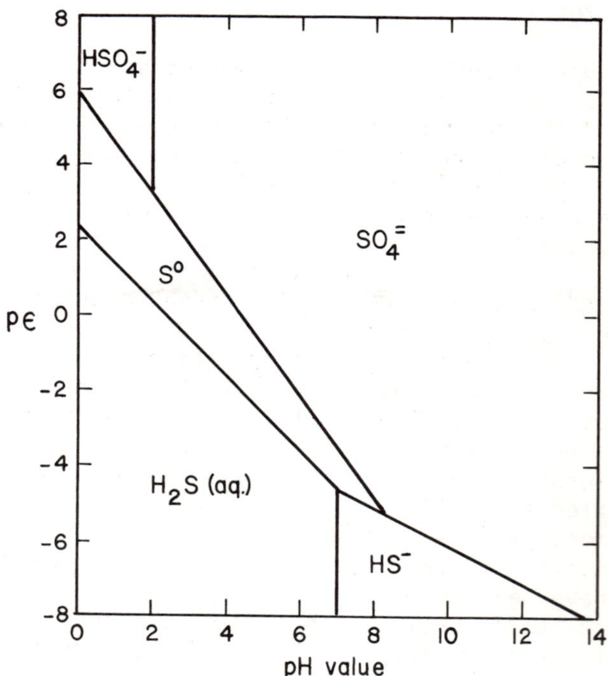

Figure 6.1. pH-pε predominance diagram of sulfur species.

The major features of the equilibrium chemistry of sulfur species related to their removal in water supply treatment are discussed in the following sections.

Hydrogen Sulfide

H_2S dissociates in water according to equations:

$$H_2S \rightleftharpoons H^+ + HS^-;\ K_1 = 1.0 \times 10^{-7} \quad (1)$$

$$HS^- \rightleftharpoons H^+ + S^{2-};\ K_2 = 1.0 \times 10^{-14} \quad (2)$$

Hydrogen sulfide has a pronounced, unpleasant sulfuretted odor (rotten egg) discernable even in amounts as low as a few tenths of one part per million. It is frequently found in ground waters, arising from the decomposition of organic matter. Sulfides do not accumulate in nature except under anaerobic conditions. On formation they may be removed from the system through one of the following methods: (1) biological oxidation, (2) metal sulfide precipitation, (3) volatilization, or (4) chemical oxidation.

The presence of small quantities of reduced sulfur species (H_2S, HS^-, and S^{2-}) can limit the concentration of Fe(II) and Mn(II) in aqueous solutions. Most of the metal sulfides are insoluble in water, except those belonging to Groups I, II, and III.

Elemental Sulfur

Elemental sulfur exists in nine crystalline forms which are all nearly insoluble in water. In the insoluble form, it can be removed with the aid of filters. However, sulfur may also appear in colloidal form, as a milky-white dispersion in water, which may not be amenable to filtration.

The mechanism of sulfur formation and growth is comprised of three parts:

1. The homogenous reaction, where molecularly dispersed sulfur forms immediately upon mixing and ends with the supersaturation of the sulfur solution.
2. The condensation stage, where the supersaturated solution begins to nucleate and starts a slow growth of sulfur micelles.
3. The heterogenous stage, where diffusion of the dissolved sulfur to the surface continually increases particle size.

As soon as colloidal sulfur is formed, a "Tyndall beam effect" can be detected spectrophotometrically at a wavelength of 3000Å. The solubility in water at this point is about 3.2×10^{-6} g-atom/l. Particle size may reach a maximum of 0.5 μm.[4]

In alkaline and neutral solutions, sulfur is unstable with respect to the formation of sulfide and thiosulfate (or sulfate). Solid elemental sulfur is stable only in acidic solution. The following equations describe this fact which is especially significant at high pH.

$$4S + 4\,OH^- = 2HS^- + S_2O_3^= + H_2O; \quad K = 10^{17.9} \qquad (3)$$

$$4S + 5\,OH^- = SO_4^= + 3HS^- + H_2O; \quad K = 10^{27.2} \qquad (4)$$

Simple calculations indicate that formation of sulfur is not likely at pH greater than 8 unless excessive amounts of thiosulfate and sulfide are present. Aside from the oxidation of sulfide by oxygen in both acidic and neutral solutions, elemental sulfur may also be formed from acidification of thiosulfate, or even interactions of sulfur oxyanions and hydrogen sulfide in acidic solution.[5] When sulfur is formed in the presence of sulfide, polysulfides can be formed immediately.[6]

Figure 6.2 shows the relative electron activity of sulfur species as a function of redox potential, with the activity of S(s) as reference. The dependence of log $(SO_4^=/S)$ and log (HS^-/S) is expressed as function of pε for pH 6-9. The relative activity of elemental solid sulfur as seen in the diagrams diminishes with increasing pH. At pH 9, the solid sulfur cannot exist in thermodynamic equilibrium.

Polysulfides

The formation of polysulfides may be the result of: (1) the interaction of sulfur with an aqueous solution of sulfide, or (2) aging or oxygenation of sulfide and hydrosulfide. This can be demonstrated by the fact that when crystalline sodium sulfide is added to oxygen-free water at room temperature a clear solution is formed at any pH value. If, however, the water contains dissolved oxygen and has a pH between 6 and 9, a straw to greenish color characteristic of polysulfides soon develops, the intensity of which depends on the initial concentration of sulfide. Upon acidification a whitish colloidal sulfur

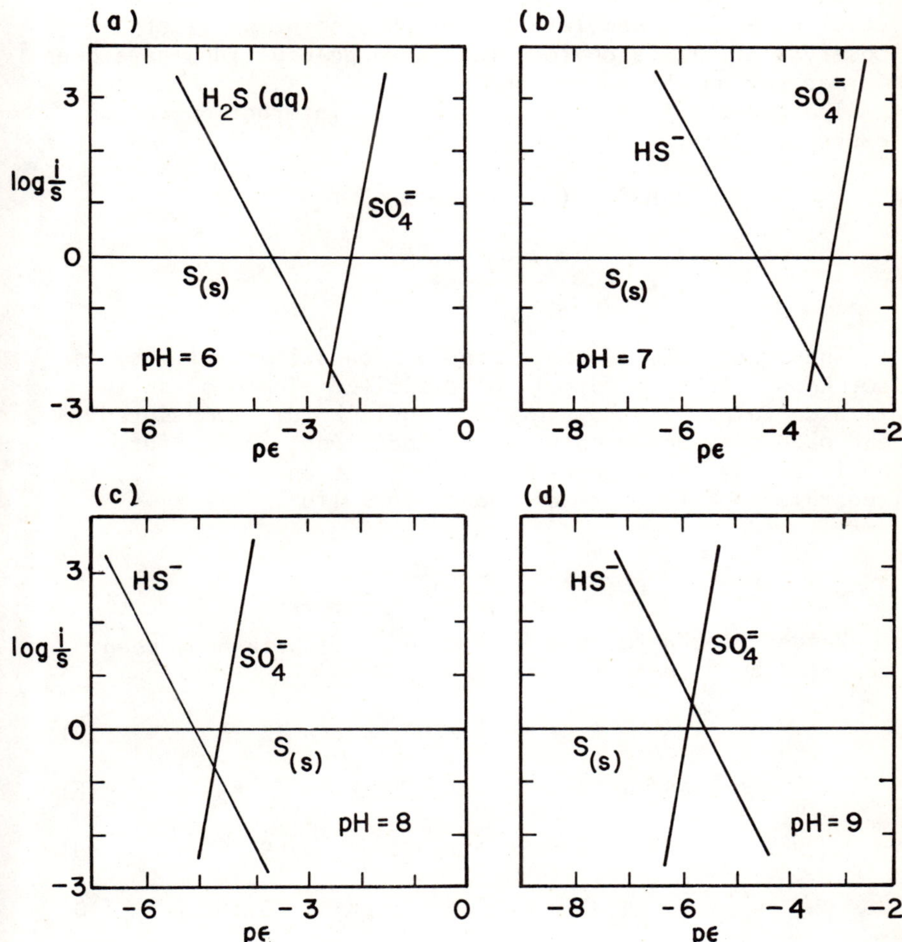

Figure 6.2. *Relative activity of sulfur species as a function of redox potentials.*[6]

suspension forms and shows a Tyndall beam effect. The colored solution exhibits an absorption band in the UV region at 285-290 nm. When the colored solution is diluted to an initial sulfide concentration of 2×10^{-4} *M*, the color disappears completely; the UV absorption is still observable. It is obvious that some polysulfides have been formed during the oxidation.[6]

The general form of sulfide-sulfur interaction can be expressed as:

$$HS^- + (x - 1)S = S_x^{2-} + H^+;$$

$$K_x = \frac{(H^+)(S_x^{2-})}{(HS^-)(S)^{x-}} \tag{5}$$

The exact form of the dependence on sulfur activity is not known, for the formula of dissolved elemental sulfur is uncertain. If, however, the equilibrium is measured and used only under saturation conditions, the sulfur activity may be taken as unity, so that one may write, regardless of the form of the solid sulfur,

$$K_x = \frac{(H^+)(S_x^{2-})}{(HS^-)} \tag{6}$$

Values of pK_x for the following equilibria have been reported by Maronny.[8]

$$HS^- + S = S_2^{2-} + H^+; \quad pK_2 = 12.16 \tag{7}$$

$$HS^- + 2S = S_3^{2-} + H^+; \quad pK_3 = 10.85 \tag{8}$$

$$HS^- + 3S = S_4^{2-} + H^+; \quad pK_4 = 9.86 \tag{9}$$

$$HS^- + 4S = S_5^{2-} + H^+; \quad pK_5 = 9.18 \tag{10}$$

Many investigators (References 9-14) have confirmed that the polysulfuration index, x, is between 2 and 5 in aqueous solution. Schwarzenbach and Fisher[14] have also shown that only S_4^{2-} and S_5^{2-} can be measured in aqueous polysulfide solution; S_2^{2-} and S_3^{2-} are either unstable or at such low concentrations that they cannot be detected.

It is noted that K_2 values of the acid dissociations of H_2S_4 and H_2S_5 are $10^{-6.3}$ and $10^{-5.7}$, respectively.[14] Over the entire pH range from 6 up, S_4^{2-} and S_5^{2-} are the dominant polysulfide species present and comprise more than

90% of the polysulfide species at pH greater than 7.3. On acidification, solutions of polysulfides form H_2S, free sulfur (milk of sulfur) in a white colloidal form, small amounts of H_2S_2, H_2S_3, and possibly higher hydrogen polysulfides.[15]

Thiosulfate

Thiosulfates may be formed by the reaction of sulfites with sulfur in alkaline solution:

$$SO_3^{2-} + S \longrightarrow S_2O_3^{2-} \tag{11}$$

or by the oxygenation of HS^-. When slightly acidified, dilute solutions decompose to sulfite, free sulfur, and polythionate.[16] The sulfur formation starts with a clear solution for several minutes, followed by turbidity and final precipitation into milky-white colloidal sulfur.

From the electronic formula, the oxidation state of sulfur in thiosulfate is +2, but may be considered the average between +6 for the central sulfur atom and -2 for the coordinated sulfur atom.[15] Halogens oxidize thiosulfate to tetrathionate. In alkaline solutions, thiosulfate reduces permanganate to MnO_2 .

Sulfur Cycle in Impounded Water

The transformation of sulfur species in aqueous media found in nature is largely due to microbial mediation. In the sulfur cycle in aquatic systems, the interrelationships of the various sulfur species depend on the redox potential of the environment. The biological oxidation of H_2S to $SO_4^=$ passes through the elemental sulfur stage resulting from the activity of photosynthetic purple and green sulfur bacteria or aerobic colorless sulfur bacteria. The direct reduction of sulfate into sulfide is mediated by the obligate anaerobic bacteria of *genus Desulfovibrio*. Oxidation of sulfide to sulfate may take place with aerobic *Thiobacillus* bacteria, an example being the formation of mine acid from pyrite-containing ores. A small portion of the sulfide becomes sequestered in the insoluble form, the most common of which is the black-colored iron sulfide found in sediments. Oxidation and reduction of organic sulfur may be mediated by microorganisms converting the sulfur to sulfate or sulfide, respectively.

The chemistry of the sulfur species in the impounded water is likely to be predominantly biological in nature and may involve numerous inorganic and organic sulfur compounds.

CHEMISTRY OF OXYGENATION OF SULFIDE

Extensive studies on the kinetics and mechanism of the oxygenation of sulfide have been made in recent years (References 6,7,17,18,19,20). The extreme complexity of the reaction is evidenced by the following observations.

1. Variation in the induction periods.
2. Complex dependence of reaction rates on oxygen and sulfide concentrations.
3. Sensitivity of reaction to impurities.
4. The number of intermediates and their related products.
5. Great dependence of specific rate on pH.

Based on the observations of reaction intermediates and products, the change in reaction rate and stoichiometry with reagent proportions, and the study of catalysis and inhibition, Chen and Morris[20] proposed a reaction pathway for the neutral and slightly acid or alkaline solutions.

$$HS^- + O_2 \longrightarrow HS\cdot + O_2^- \quad (12)$$

$$HS\cdot + O_2 \longrightarrow HO_2 + S \quad (13)$$

$$HS\cdot + O_2^- \longrightarrow S + HO_2^- \quad (14)$$

$$HS^- + (x - 1)S \longrightarrow H^+ + S_x^{2-},\ x \text{ can be } 2\text{-}5 \quad (15)$$

The reaction scheme following the polysulfide formation was presented as follows:

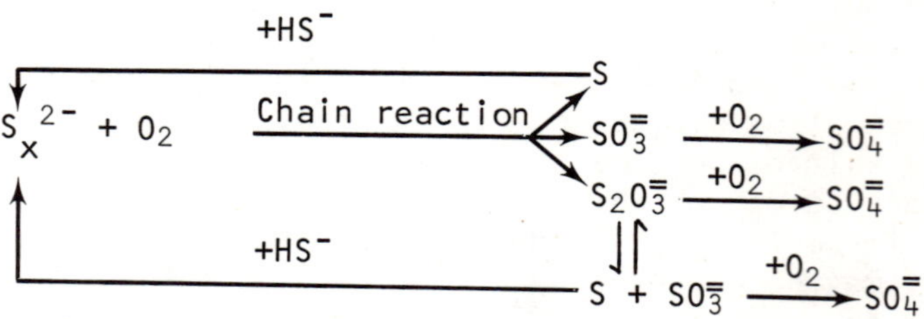

The reaction mechanism of oxygenation through polysulfide formation was further studied by Chen and Gupta.[7] Figure 6.3 shows the effect of pH on polysulfide formation, while

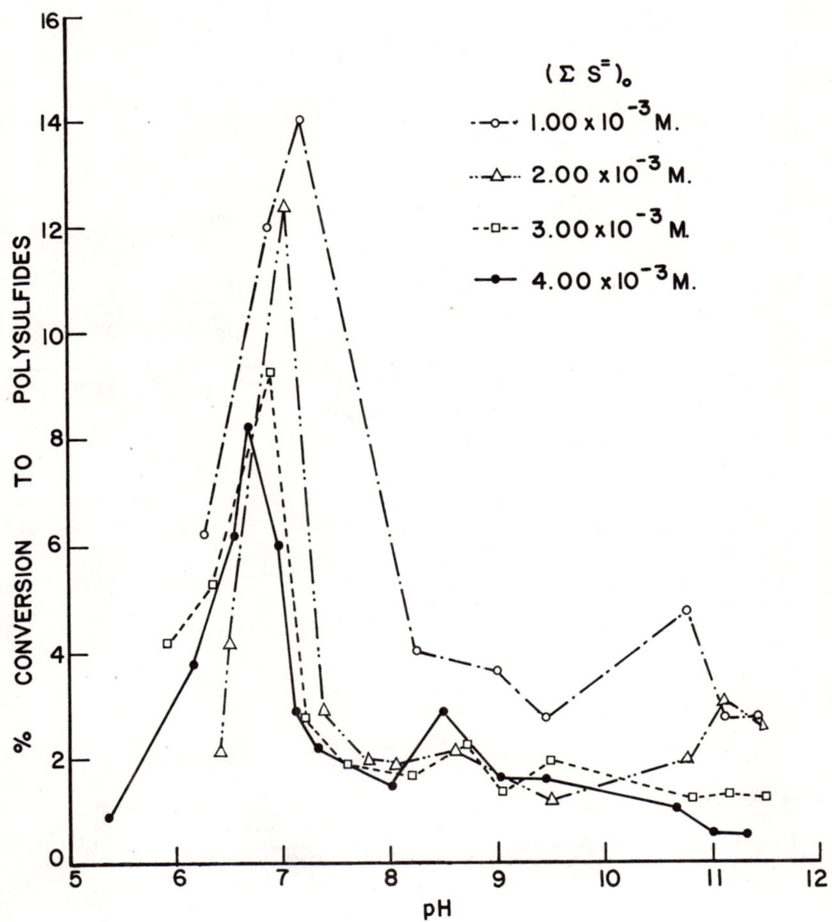

Figure 6.3. Effect of pH on polysulfides formation, P_{O_2} *= 1 atm.*[7]

Figure 6.4 shows the time dependence of polysulfide formation in dilute aqueous solution. From observation of the reaction intermediates and products, the cycle of sulfide, sulfur, polysulfides, and oxyanions is well demonstrated.

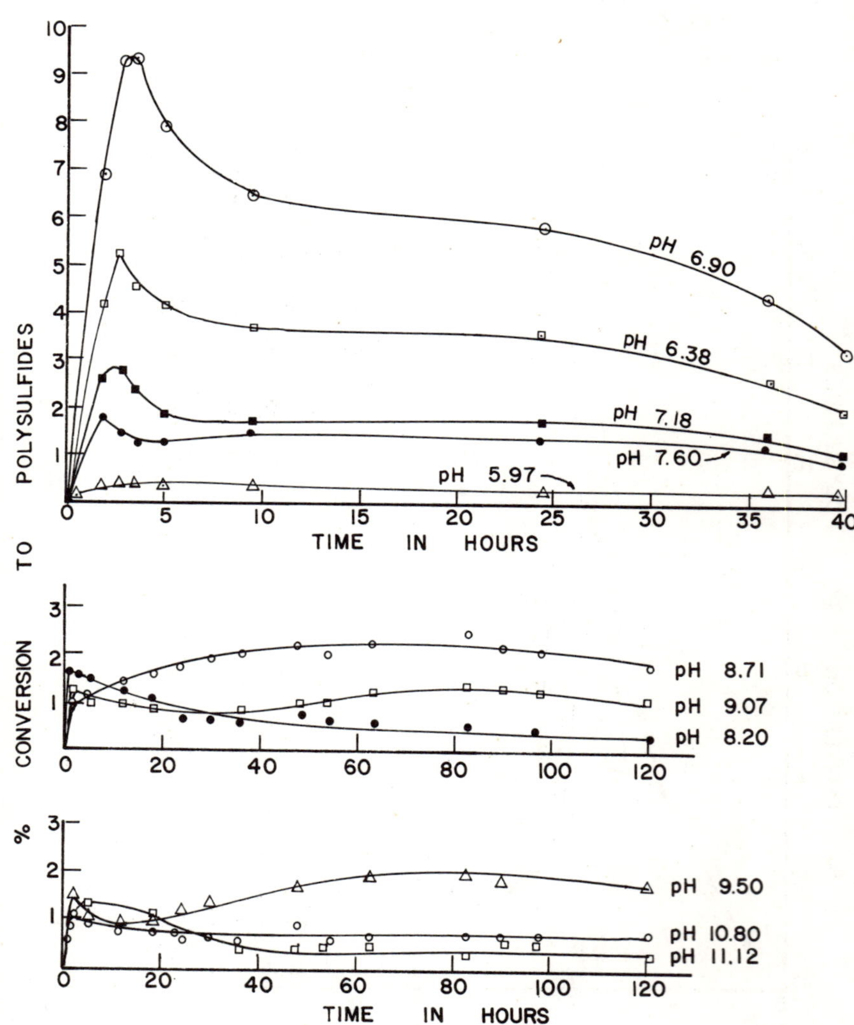

Figure 6.4. *Formation of polysulfides, $(\Sigma S^{=})_o = 3.0 \times 10^{-3}$ M; $P_{O_2} = 1$ atm.*

Chen and Morris[20] in another investigation found wide variation in oxygen uptake in the presence of catalysts or inhibitors. Most notable among the catalysts were the transition metals in mildly alkaline solutions, where an increase in rate of oxygenation of 10 to 100 fold may occur. Organic substances of phenolic nature (*e.g.*, hydroquinone and phenol) accelerate oxygen-uptake, but to a lesser degree. EDTA and nitriloacetic acid are examples of substances that act as chain-breakers and may inhibit reactions to an extent ranging from 33-90%. Conclusions are that sulfides may coexist with oxygen in water for great lengths of time; however, sulfide persistence is made unpredictable with the presence of metal ion catalysts.

REMOVAL OF SULFIDES FROM WATER SUPPLIES

Sulfides exhibit many undesirable properties; hence, the study of their oxidation is given special attention in this paper. They are most often found in ground water and bottom water of impoundments where anaerobic conditions prevail. In the presence of reduced forms of iron and manganese, the concentration of sulfide is generally low. In this study, the discussion is limited to dilute aqueous solutions under conditions generally encountered in water supply.

The common oxidizing agents most used in water treatment are oxygen, chlorine, hypochlorite, chlorine dioxide, ozone, and potassium permanganate. They are generally known to be effective in eliminating taste and odor associated with sulfides. Unfortunately, with a few exceptions, little is known about the mechanisms involved. The lack of information may be attributed to the difficulty in the segregation of end products with their subsequent quantification and the establishment of rate and reaction parameters, as well as complications posed by catalysts and inhibitors in the minute quantities found in water. The principles and applications of these treatment processes are discussed in the following sections.

Aeration

The earliest, simplest and least costly form of oxidation, using atmospheric oxygen, is by aeration. Two general types are in common use, sprays and diffused-air aeration (where small bubbles are blown in the water). The overall direct oxidation rate is slow due to the low solubility and diffusion of oxygen in water as well as slow reaction rate between oxygen and sulfide. In addition to oxidation, aeration can also liberate H_2S, CO_2 and many other volatile organic substances from water under favorable pH conditions. Beneficial effects resulting from aeration are as follows: (1) it eliminates odors and taste, (2) it decreases the corrosive effects of H_2S on metals, cement and concrete, and (3) it reduces interference during chlorination. Blower power requirements average about 1.0 Kw/MGD capacity.

In the study on the kinetics of oxidation of aqueous sulfide by oxygen, Chen and Morris[6] have demonstrated that oxygenation of sulfide is extremely slow in the absence of catalyst. Therefore, the air oxidation process is not viewed favorably as a treatment process in water works. In addition, the release of sulfide from the aqueous phase into the gas phase will not be acceptable with respect to air pollution standards. However, oxygenation does play an important role in the removal or prevention of sulfide formation in impounded water, for example, in reservoirs or lakes which are the sources of water supplies.

Typical applications where sulfides are successfully controlled by aeration are illustrated in the two cases which follow.

Ground waters from wells at Herscher, Illinois, had an intense H_2S smell which was readily removed by aeration; from an original H_2S concentration of 60 ppm, a 66% reduction was effected. It was also found that by lowering the hydraulic loading on the aerators, better stripping efficiency (92-97%) could be realized. Liberated hydrogen sulfide was collected and destroyed by ozonation.[21]

Ridley and Symons[22] summarized the efforts of natural and artificial destratification of impoundments both in the United Kingdom and the United States and found that aeration plays an important role in the reduction of sulfide concentration in impounded water. For example, normal sulfide concentration during midsummer in the bottom layer of a stratified eutrophic reservoir may reach a level of 10 ppm. As a result of periodic mixing of bottom waters by pumping to the surface during the 1965 test a low level of sulfide was attained. In tests performed during 1966 using diffused-air pumping the concentration of sulfide was reduced to tolerable levels. Figures 6.5 and 6.6 illustrate the effectiveness of aeration in the prevention of sulfide build up in lakes during the 1965 and 1966 tests.

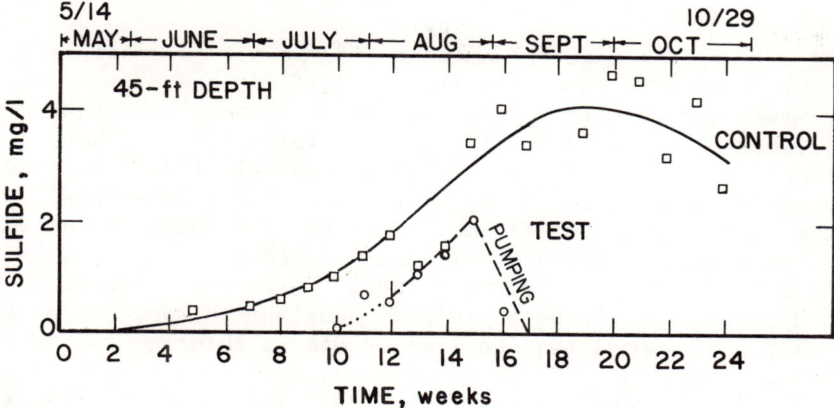

Figure 6.5. *Comparison of changes in sulfide concentration with time at 45-foot depth in test and control lakes.*[22]

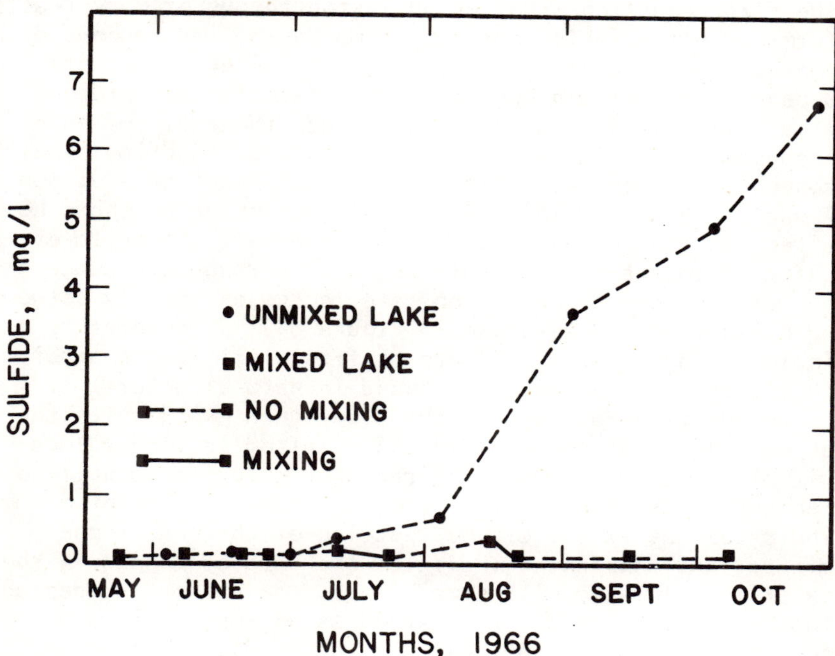

Figure 6.6. *Influence of diffused-air pumping on sulfide.*[22]

Chlorination of Sulfide

Chlorine, a powerful oxidant, is the most commonly used disinfectant in water treatment. Sulfides are effectively removed and converted into less objectionable substances (elemental sulfur and sulfate) by chlorination . The reaction occurs very rapidly and, theoretically, 8.5 parts of chlorine are needed for every part of H_2S to assure complete chemical oxidation. However, the dosage required is generally less than that necessary to give a residual. To achieve efficient removal, factors like pH, temperature and dosage have to be optimized. Optimum pH ranges are 6.5-8.5.[23] Among the chemical oxidants, chlorine is the least expensive.

The stoichiometric relationship in sulfide oxidation by chlorine may be represented by two equations, according to Black and Goodson:[23]

$$Cl_2 + S^{2-} \longrightarrow 2\,Cl^- + S \qquad (16)$$

$$4\,Cl_2 + S^{2-} + 4\,H_2O \longrightarrow 8\,HCl + SO_4^{=} \qquad (17)$$

The first equation involves an instantaneous primary reaction forming sulfur, while the second may be a simultaneous or a consecutive oxidation of a portion of sulfur to sulfate. Depending on certain conditions the reaction may proceed at a slower pace. Factors that affect the ratio of end products (S to $SO_4^{=}$) are: pH, temperature, reaction time, concentration of reactants, and rate of reactant addition. A variation of the initial sulfide concentration shows no effect on the extent of oxidation; however, the ratio of chlorine to the reacted sulfide shows strong dependence on pH. The secondary oxidation rate increases sharply between pH 6.5 and pH 7.3 as shown in Figure 6.7. Temperature and ionic strength variations are of less significance under conditions similar to those found in natural waters.

In the time-rate study, the equation $R = at^b$ suits the experimental ratios obtained and resulted in linear logarithmic plots where R is the ratio of chlorine to sulfide reacted, t is the reaction time in minutes, constant a indicates the extent of the reaction at the end of one minute, and constant b indicates the rate of change in the experimental ratio with time. Thus, the authors suggested that constant b might be regarded as an empirical rate constant for the oxidation under the conditions of the experiment.

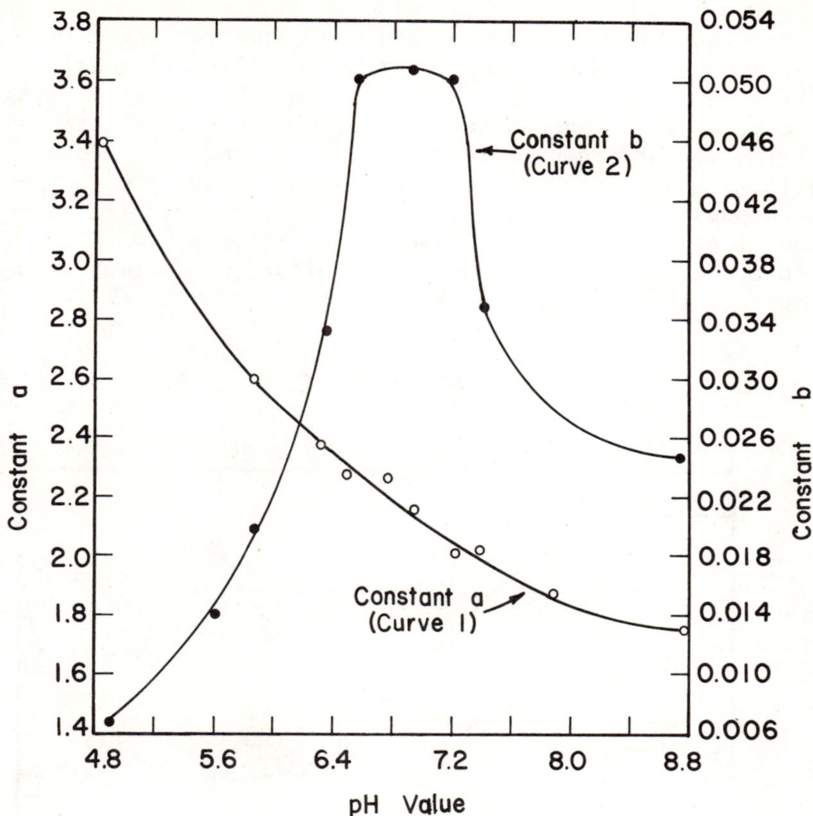

Figure 6.7. *Effect of pH on the constants of the empirical equation $R = at^b$. Four moles Cl_2 added per moles ΣS^{2-}.*[23]

Sodium hypochlorite is another common disinfecting compound which gives 95% available chlorine by weight when compared with Cl_2. It is generally used in smaller size water treatment facilities. It reacts rapidly with sulfides to form sulfur and sulfate. As in the other methods, pH, temperature, and quality of influent water are factors to be considered to get maximum removal of sulfide at least cost. The cost of sodium hypochlorite is several times more expensive than chlorine in gas form.

Previous study by Choppin and Faulkenberry[24] revealed that hypochlorite does not react with sulfide in equal equivalents. This again indicates that the reaction may proceed in simultaneous parallel or consecutive reactions forming sulfur and sulfate (Equations 16 and 17). Four variables influence the ratio of sulfur to sulfate in the end product, namely: (1) temperature, (2) pH, (3) reaction time, and (4) rate of addition. Figure 6.8 illustrates the marked changes in sulfate production as a result of temperature and pH variation.

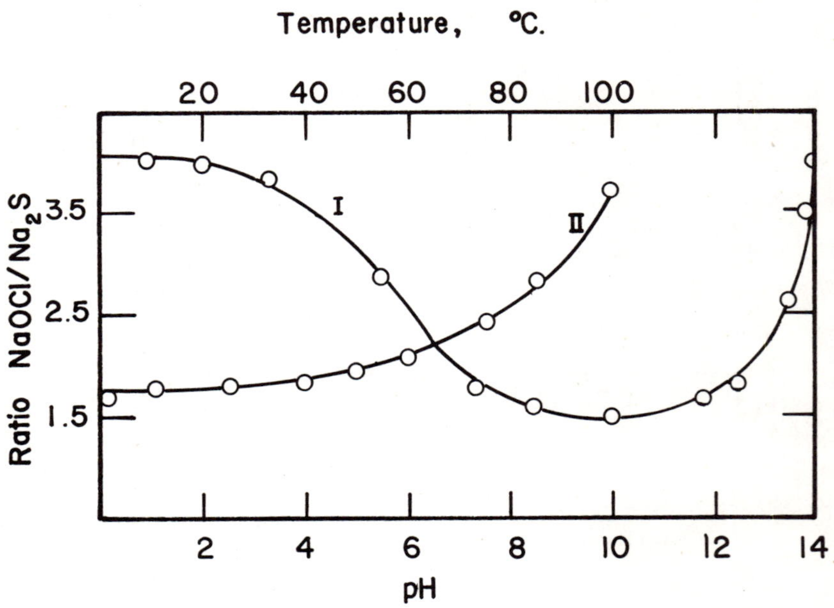

Figure 6.8. Effect on sulfate production:
I. pH ratio II. temperature ratio.[24]

The high ratios of hypochlorite consumed per mole of initial sulfide at both very high and very low pH probably reflect the fact that at high pH soluble sulfur from the primary oxidation of sulfide by hypochlorite can be easily disproportionated to sulfide and thiosulfate, both of which are subjected to further oxidation. While at very low pH, the presence of chlorine monoxide (Cl_2O) can carry the oxidation much further, because Cl_2O is a stronger oxidizing agent than HOCl. It seems that once precipitated sulfur is formed further oxidation by chlorine is slow. A concurrent increase in $NaOCl/Na_2S$ ratio and sulfate production occurs at high temperature, and is probably related to the higher solubility of sulfur at increased temperature. Changes in reaction time and rate of addition affects sulfate production also, possibly as a result of re-solution of sulfur and its subsequent oxidation by excess hypochlorite during the secondary reactions.

Ozonation of Aqueous Sulfide

Ozone, a strong oxidizing agent, is extensively used in Europe, whereas applications in the United States are very limited. Used primarily for sterilization of water, it also effectively removes color, odor, taste, iron and manganese as well. Due to its high oxidative capacity, ozone converts sulfide into sulfur and sulfate almost instantaneously.[25] It shows definite advantages over chlorine; since no residual products are left it leaves no taste and odor problem and renders the water more palatable.

Ozone in aqueous solution is very unstable,[26] and will readily decompose to molecular oxygen. At room temperature, it has an effective life measured in minutes. The decomposition depends strongly on pH and trace amounts of catalyst present in water.

Ozone must be generated on the site due to its rapid decomposition. To achieve optimum results at least cost, rigid control of variables like pH, temperature, bubble size, dosage, and influent oxidizing capacity are necessary. Contact time for ozone is only a fraction of that needed for chlorine. Water with high threshold odor is best treated by ozonation. Dosage sufficient to disinfect water is usually adequate to remove sulfide. Typical ozone requirements are shown in Table 6.I.

Table 6.I

Ozone Dosages for Various Water Qualities[27]

Water Quality	*Approximate dosage*
Clear, pure ground water	1/2 - 1 ppm
Treated (filtered) surface water	2 - 3 ppm
Same as above, but heavily polluted	2.5 - 5 ppm
Clear, colored water	3 - 6 ppm

Operating costs are low, about 3-4¢/1000 gal, but comparatively more expensive than chlorine. Likewise, capital costs for ozone units are many times higher than those for chlorination equipment. Energy required for every pound of ozone generated and applied is placed at 10 kw-hr.

Although ozone has been generally regarded as very effective in the removal of odor and taste, a literature search revealed no paper covering ozonation of aqueous sulfide. In an effort to establish some knowledge on its chemistry and mechanism, preliminary investigations were conducted. The study was limited to pH between 6.0 and 10.0 at room temperature, which was about 20°C. Buffering with 0.2 *M* $NaOH-KH_2PO_4$ was used for pH 6-7, and 0.2 *M* $NaOH-H_3BO_3$ was used for pH 8-10.

Two hundred fifty ml of buffered solution was bubbled for 10 minutes using an ozone-oxygen mixture at a flow rate of 0.568 l/min, then immediately transferred to the appropriate vessel for determination of either ozone content or remaining ozone after reaction with sulfide. Results of the experiments can be summarized as follows:

1. The saturated values of ozone in buffered solutions, determined after a fixed time of 15 seconds by iodometric methods for residual ozone, were found to be pH dependent. Experimental values are plotted in Figure 6.9, with the higher saturation concentrations found in mildly acidic conditions decreasing with higher pH. Close control of the time interval prior to analysis was essential to get consistent and meaningful values as ozone decomposes rapidly in aqueous solution. With a longer lapsed time, lower ozone saturation values were noted. The rate of decay was relatively slower at lower pH. At pH 10.0 no ozone was left after several minutes. Similar observations were also noted using UV absorbance readings at 260 nm, at which level ozone exhibits maximum value. This finding agrees with Stumm[26] on the rapid decomposition of ozone with time and at an increased rate in basic solutions.

2. Kinetic studies could not be undertaken at present without using special fast-rate determining equipment. Reaction was observed in the spectrophotometer to be instantaneous, within a fraction of a minute following the addition of ozone-saturated buffered solution to

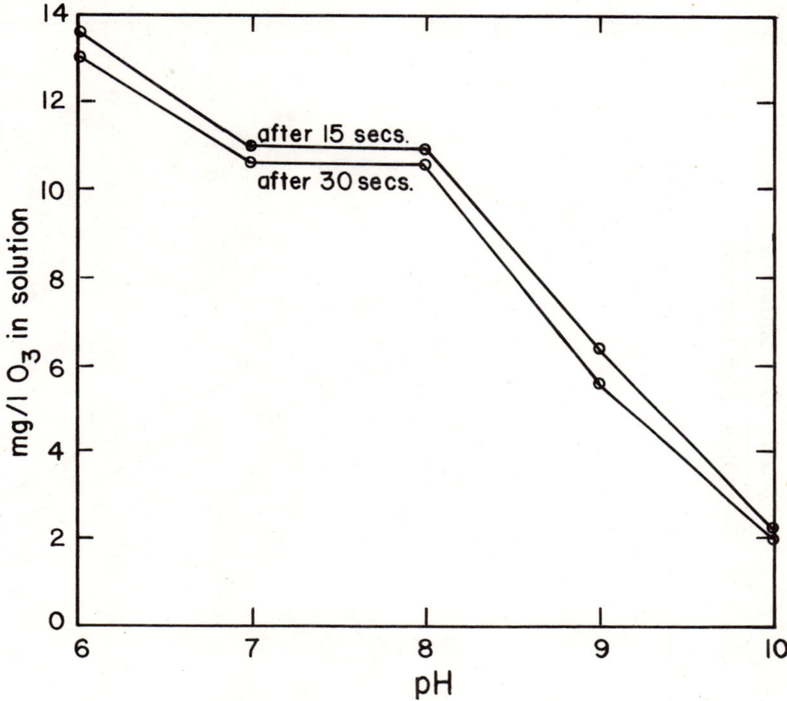

Figure 6.9. Effect of pH on ozone conc. in solution.

a predetermined amount of Na_2S standard, no trace of sulfide could be detected by the addition of strong acid after 30 seconds.

3. Ozone demand by sulfide cannot be conveniently determined by direct bubbling of ozone into the sulfide solution without incurring losses due to the stripping action. The method chosen consisted of transferring the ozonated buffered solution into a reaction bottle containing the sulfide, with subsequent analysis by iodometric method. Ozone reaction was quenched by the addition of KI.

4. The ozone demand by sulfides using the same amounts in each test were found to be strongly influenced by pH. As shown in Figure 6.10, the ratio of ozone consumed per mole of reacted sulfide diminishes with increasing pH; the lower concentrations of available ozone at higher pH suggest it is a limiting factor. At lower

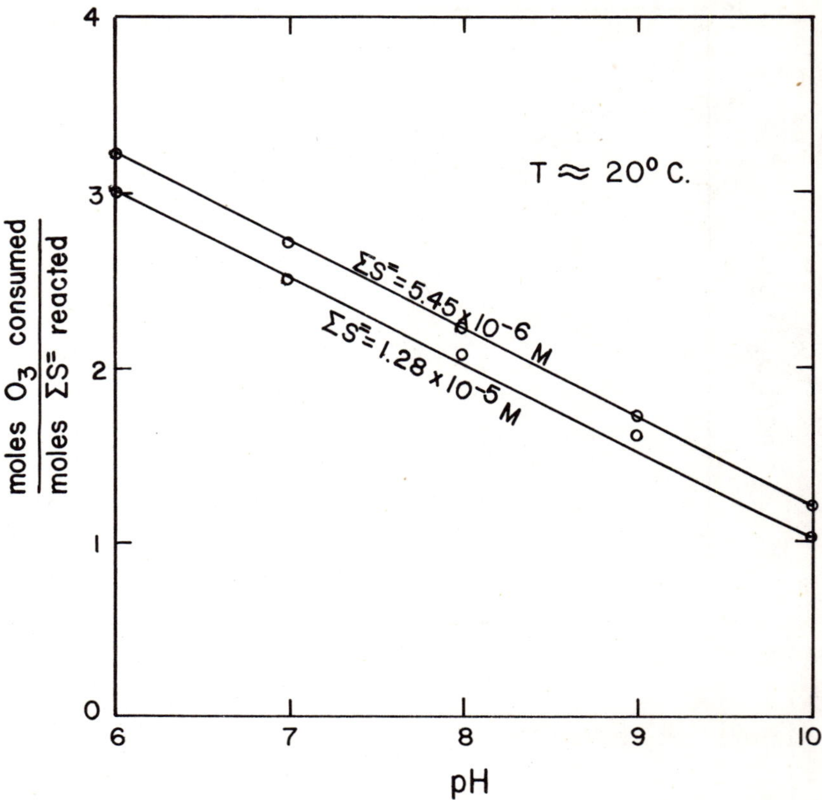

Figure 6.10. Effect of pH on ozone demand of sulfide solutions.

pH, the experimental ratio was approximately 3.0-3.5, while at higher pH values the ratios are close to equivalence (evidenced by less excess ozone residual), and a value close to unity was obtained. Based on the above observations, the reaction seems to proceed according to the equations:

$$S^{=} + O_3 + H_2O \longrightarrow S + 2\ OH^- + O_2 \qquad (18)$$

$$S^{=} + 4O_3 \longrightarrow SO_4^{=} + 4\ O_2 \qquad (19)$$

The theoretical molar ratios (O_3 to Na_2S) are 1:1 for sulfur and 4:1 for sulfate. Experimental ratios ranged within these limits.

5. Gravimetric analysis of end products for sulfate (using similar amounts of O_3 and Na_2S in each test) gave relatively lesser amounts of sulfate with increasing pH. This further supports the suggested reactions; where available ozone was limited at higher pH, the oxidation tends to follow Equation 18.

6. When the ozone to sulfide ratio was higher than one, an end product exhibiting a peak at 216 nm could be observed in the UV spectrophotometer. This suggests that aside from sulfur and sulfate, other intermediate or end products such as thiosulfate or even polythionates can be formed.

Removal of Sulfide With $KMnO_4$

Potassium permanganate finds successful application in the removal of H_2S from water. It is generally used for its capability to remove certain tastes and odors rather than for water disinfection. Available in solid or liquid forms it is easy to handle and dispense. Normal dosages vary from 1-5 ppm and proper dosage is readily established by the pinkish tint it imparts with excess.

Potassium permanganate is a powerful oxidant and reaction with it is strongly influenced by pH. In waters with normal pH values ranging from 6.0-10.0, the reduction proceeds through a 3-electron transfer with formation of MnO_2 as product. The equations under the different conditions may be written:

basic solution

$$MnO_4^- + 2\ H_2O + 3\ e^- \longrightarrow 4\ OH^- + MnO_2 \qquad (20)$$

slightly acid to neutral

$$MnO_4^- + 4\ H^+ + 3\ e^- \longrightarrow 2\ H_2O + MnO_2 \qquad (21)$$

Stewart[28] reported a similar reaction in basic solutions to yield sulfate products according to the equation:

$$8\ MnO_4^- + 3\ S^{2-} + 4\ H_2O \longrightarrow 8\ MnO_2 + 3\ SO_4^= + 8\ OH^- \qquad (22)$$

In either neutral or milkly acid medium, the products are sulfur and tetrathionate and the reaction tends to be incomplete.[28]

Wiley, *et al.*[29] indicated that at neutral pH conditions 6.2 parts of $KMnO_4$ were needed for every part of H_2S reacted (molar ratio 4:3). Similar ratios were obtained from laboratory titrations of various well waters. Analysis showed that two-thirds of the sulfide was oxidized to sulfate. Based on the experimental data gathered, the stoichiometry of the reaction assumes the equation:

$$4\ MnO_4^- + 3\ H_2S \longrightarrow 2\ SO_4^= + S + 3\ MnO + MnO_2 + 3\ H_2O \qquad (23)$$

The inconsistent stoichiometric relationships gathered by many investigators prompted some experiments on a limited scale to study the chemistry of sulfide oxidation with $KMnO_4$. These met with little success. In this study initial attempts to determine the kinetics of reaction were dropped because the reaction proceeds at a rapid pace, making measurements difficult without special equipment, and because serious color interferences as well as MnO_2 rendered spectrophotometric and titration methods unreliable. Some of the observations are noted as follows:

1. The reaction between $KMnO_4$ and Na_2S goes to completion within 5 seconds, as exhibited by the rapid loss of the purple $KMnO_4$ color.
2. The $KMnO_4$ demand of sulfide using oxygen-saturated permanganate solution was found to be one-half that of deaerated permanganate solution as shown in Figure 6.11. Back titration with thiosulfate using excess KI was the method used for determining the $KMnO_4$ equivalent. Actual $KMnO_4$ used was based on the difference between thiosulfate in blank and sample. The importance of the level of dissolved oxygen is shown and can be a factor in estimating dosage requirements.
3. When Na_2S was added to the buffered solution no endpoint was recognizable, and the color changed from purple to yellow. Visible MnO_2 was noted at higher pH.

Factors that are important in $KMnO_4$ applications are retention time, filters, and pH. Filters are essential to remove the insoluble manganese dioxide. In alkaline conditions, excessive consumption of $KMnO_4$ may result. $KMnO_4$ may be added to any point, although the mixing and sequence of addition of other chemicals can improve the efficiency to some extent.

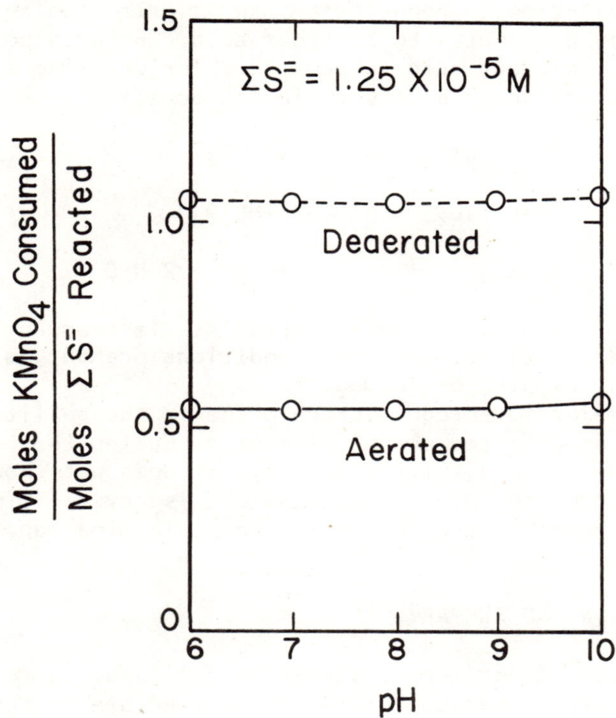

Figure 6.11. Effect of pH on permanganate demand of sulfide solutions.

Oxidation of Sulfide by Chlorine Dioxide

Closely related to chlorine, chlorine dioxide is a more chemically active oxidant with an irritating and unpleasant odor. On site generation through the reaction between chlorine and sodium chlorite in acid solution is necessary due to its instability. The action of ClO_2 is very rapid; though its capacity for removing hydrogen sulfide is demonstrable, there is no obvious advantage to its use.

In water, ClO_2 reduces into several forms due to the numerous oxidation stages of chlorine, depending on conditions in the system and the reactants. In acid conditions, chloride is formed through a five-electron transfer:

$$ClO_2 + 4\ H^+ + 5\ e^- \longrightarrow Cl^- + 2\ H_2O \tag{24}$$

In basic conditions, ClO_2 reduction proceeds under a single-electron exchange to produce chlorite, which subsequently protonates to chlorous acid. Being a powerful oxidant, chlorous acid reduces to chloride. The steps involved are shown by the following equations.

$$ClO_2 + e^- \longrightarrow ClO_2^- \quad (25)$$

$$ClO_2^- + H^+ \longrightarrow HClO_2 \quad (26)$$

$$HClO_2 + 3\ H^+ + 4\ e^- \longrightarrow Cl^- + 2\ H_2O \quad (27)$$

Disproportionation of the intermediate is frequently experienced. Depending on the conditions present, alternate products may also be formed.[30]

Its major advantage over chlorine is the ability to oxidize ammonia and phenol without producing the odorous end products. On sulfide removal, it does not show much benefit over the other oxidants. It is several times more costly than chlorine for equivalent oxidation capacity.

DISCUSSION AND SUMMARY

From the experimental studies and literature review it is obvious that sulfur is an intermediate in the oxidation of sulfide under conditions normally encountered in water supply. Once sulfur is formed, four competing reactions may occur:

1. Further oxidation of sulfur by the oxidant.
2. Formation of polysulfide through the interaction of sulfur and sulfide. This will lead to further oxidation.
3. Nucleation and eventual precipitation of sulfur from solution. This is favored in the neutral pH region.
4. Disproportionation of sulfur to sulfide and thiosulfate. This is significant in highly alkaline solutions.

The end products remaining will depend to a great extent on the rate of each competitive reaction as well as the relative concentrations of reactants under different solution conditions. For example, if the polymerization and nucleation of sulfur are faster than the alternative reactions, elemental sulfur would be the major end product. Such would likely be the case in the reaction between sulfide and a less powerful oxidizing agent in the neutral pH region if the remaining sulfide were not sufficient for the formation of polysulfide.

Despite the fact that there are still many unknown factors in dealing with sulfur species in water supplies, there is no doubt that most strong oxidizing agents are capable of removing undesirable inorganic sulfur species from solution. Therefore, the final choice will be dependent on the economy of the treatment and ease of operation. Due to slow kinetics, the aeration process is suitable only for the prevention of sulfide formation and its removal from impounded waters. Chlorine, ozone, permanganate, and chlorine dioxide are excellent oxidizing agents for the removal of sulfide. However, as long as chlorine remains the predominant choice as disinfectant in the water industry, the removal of undesirable sulfur species will be most economically accomplished by chlorination. On the other hand, processes such as combined usage of air and permanganate could be potentially useful if further developed. Such processes may be especially attractive for the treatment of sulfide-bearing ground water in small to medium installations. The formation of manganese dioxide from the permanganate treatment could also enhance the removal of trace amounts of iron and manganese from water.

ACKNOWLEDGEMENT

Acknowledgement is made to the donors of the petroleum research fund, administered by the American Chemical Society, for support of this research. The assistance of Mr. Amancio Sycip in this study is greatly appreciated.

REFERENCES

1. Garrels, R. M., and C. R. Naeser. *Geochim. Cosmochim. Acta, 15,* 113 (1958).
2. Sillen, L. G. *Stability Constants of Metal-Ion Complexes*. The Chemical Society, London, 1971.
3. Stumm, W., and J. J. Morgan. *Aquatic Chemistry*. Wiley-Interscience, New York, 1970.
4. Zaiser, E. M., and V. K. La Mer. *J. Colloid Sci., 3,* 571 (1948).

5. Pollard, F. H., and D. J. Jones. Special Publication No. 12, The Chemical Society, London, 1959.
6. Chen, K. Y., and J. C. Morris. *Environ. Sci. Technol., 6*, 529 (1972).
7. Chen, K. Y., and S. K. Gupta. *Environ. Letters, 4*, 187 (1973).
8. Maronny, G. *Electrochim. Acta, 1*, 58 (1959).
9. Peschanski, D., and G. Valensi. *J. Chim. Phys., 46*, 602 (1949).
10. Maronny, G. and G. Valensi. *Proc. Comite. Int., Thermodynamique Cinetique Electro-chimiques, 6*, 180 (1954).
11. Maronny, G., and G. Valensi. *Proc. Comite. Int., Thermodynamique Cinetique Electro-chimiques, 7*, 266 (1955).
12. Maronny, G., and G. Valensi. *Proc. Comite. Int., Thermodynamique Cinetique Electro-chimiques, 8*, 204 (1956).
13. Maronny, G., and G. Valensi. *Proc. Comite. Int., Thermodynamique Cinetique Electro-chimiques, 9*, 155 (1957).
14. Schwarzenback, G., and A. Fisher. *Helv. Chem. Acta, 169*, 1365 (1960).
15. Sisler, H. H., C. A. Vanderwurf, and A. W. Davidson. *General Chemistry*. The Macmillan Co., New York, New York, 1950.
16. Karchmer, J. H. *The Analytical Chemistry of Sulfur and Its Compounds*. Wiley-Interscience, New York, New York, 1970.
17. Avrahami, M., and R. M. Golding. *J. Chem. Soc., 1968*, 647A (1968).
18. Cline, J. D., and F. A. Richards. *Environ. Sci. Technol., 3*, 838 (1969).
19. Chen, K. Y., and J. C. Morris. *Adv. Water Pollution Research, 3*, 32/1 (1972).
20. Chen, K. Y., and J. C. Morris. *J. San. Engr. Div., ASCE, 98*, SA1, 215 (1972).
21. O'Connor, J. T. "Hydrogen Sulfide Removal at Herscher," Final Report to the Village Board of Herscher, Illinois, 1969.
22. Ridley, J. E., and J. M. Symons. *Water Pollution Microbiology*, R. Mitchell, Ed., Wiley-Interscience, New York, New York, 1971.
23. Black, A. P., and J. B. Goodson, Jr. *J. Amer. Water Works Assoc., 44*, 309 (1952).
24. Choppin, A. R., and L. C. Faulkenberry. *J. Am. Chem. Soc., 59*, 2203 (1937).

25. Kirk-Othmer. *Encyclopedia of Chemical Technology.* Vol. 19, Interscience, New York, New York, 1969.
26. Stumm, W. *J. Boston Soc. Civil Engrs., 45,* 68 (1958).
27. Hann, V. A. *J. Amer. Water Works Assoc., 48,* 1317 (1956).
28. Stewart, R. *Oxidation in Organic Chemistry, Part A.* K. B. Wibert, Ed., Academic Press, New York, New York, 1965.
29. Wiley, B. F., H. Jennings, and F. Muroshi. *J. Amer. Water Works Assoc., 56,* 475 (1964).

CHAPTER 7

CHEMICAL FORMS OF ARSENIC IN WATER SUPPLIES AND THEIR REMOVAL

John F. Ferguson and Marc A. Anderson*

INTRODUCTION

Concern about toxic elements in water supplies and receiving waters seems to have abated slightly from the virtual hysteria that was recently focused on mercury and cadmium, two elements which exert toxic effects on man through the aquatic food chain rather than through his drinking water. Though arsenic is accumulated by plants and aquatic organisms, it is not known to have caused toxic effects on man as a result of consumption of foods that have high concentrations accumulated from polluted waters. Rather, acute and chronic arsenic poisoning is due to ingestion of polluted drinking water. The potential problems of arsenic toxicity are water to man interactions with organisms altering the chemical state of arsenic but not accumulating toxic quantities through a food chain.

Arsenic in water supply systems is not a current problem in the United States. McCabe *et al.*[1] found arsenic exceeding the U.S.P.H.S. recommended limiting concentration (0.01 mg/l) in 0.5 per cent of the treated water samples, and exceeding the maximum permissible limit (0.05 mg/l) in 0.2 per cent of the samples. The U.S.P.H.S. standards are stringent, and water purveyors are doing an excellent job of providing water free of this and other toxic elements. However, arsenic is found in fresh surface waters of the

*J. F. Ferguson and M. A. Anderson, Department of Geography and Environmental Engineering, The Johns Hopkins University, Baltimore, Maryland.

U. S. frequently enough to merit interest in arsenic removal in natural systems and in water treatment processes. In analyses of fresh surface waters in the U. S., arsenic was found in about 7 per cent of the 1500 samples from 150 rivers,[2] and in 21 per cent of 727 samples from rivers and lakes.[3] In almost every case in which arsenic was detected, the concentration exceeded the recommended limiting concentration of 0.01 mg/l. In the first survey, the mean of arsenic values in violation of the U.S.P.H.S. permissible limit was 0.1 mg/l, a factor of 10 above the permissible limit and 2 above the maximum limit. Arsenic concentrations exceeded the U.S.P.H.S. standards far more frequently than any other element potentially hazardous to human health. It is not known whether arsenic is effectively removed by any water treatment system in the U. S. Conventional wisdom has conservatively assumed ineffective removal of toxic elements, including arsenic,[4] although studies have appeared from time to time demonstrating rather high removal of arsenic to residual levels within the U.S.P.H.S. standards. The U.S.P.H.S. standards are customarily applied to the raw water source with no assumed removal of the element in the water treatment process. It is very likely that water supply systems have used alternative sources when high levels of arsenic have been encountered, rather than relying upon removal in processing. In future decades, as demand for water continues to increase, it is highly likely that some localities will prefer to process water to remove arsenic in order to meet standards for drinking water rather than find an alternative source.

Treatment technology for removing arsenic from water supplies has been reviewed recently by Patterson and Minear.[5] A bewildering variety of chemical processes have been successfully used, as summarized in Table 7.I. Either arsenic can be readily removed by almost any process, or arsenic removal processing is subject to few generalizations. Review of the studies cited in Table 7.I discounts the former possibility. Processes that were effective in some instances were often used unsuccessfully in others.

The purpose of this paper is to review the chemical speciations of arsenic in natural waters, the possible conversions between species, and the significance of this speciation for removal technology. Included are a discussion of the removal of different forms of arsenic and a description of the results of experimental work on adsorption of arsenic.

Table 7.I

Arsenic Treatment Methods and Removals Achieved. Adapted from Patterson and Minear.[5]

Treatment	*Initial Arsenic mg/l*	*Final Arsenic mg/l*	*Per Cent Removal*	*Reference*
Lime Softening	0.2	0.03	85	5
Precipitation with Lime (plus Iron)	---	0.05 or less	--	5
Charcoal Packed Bed	0.2	0.06	70	5
Ferrous Sulfide Filter Bed	0.8	0.05	94	7
Coagulation with Ferric Sulfate	25.0	5 or less	80 or more	26
Coagulation with Ferric Chloride	3.0	0.05	98	27
Precipitation with Ferric Hydroxide	---	0.6	--	5
Precipitation with Ferric Hydroxide	272.0	11-15	94-96	5
Precipitation with Sulfide Salt	---	0.05	--	5
Alumina Packed Bed	0.06	0.006	90	21

SPECIATION AND MECHANISMS OF CONVERSION

The equilibrium relations among arsenic species can be computed using thermodynamic data. An Eh-pH diagram for the arsenic system (10^{-5} *M* arsenic), including 10^{-3} *M* sulfur summarizes information on predominant soluble species and solids.[6]

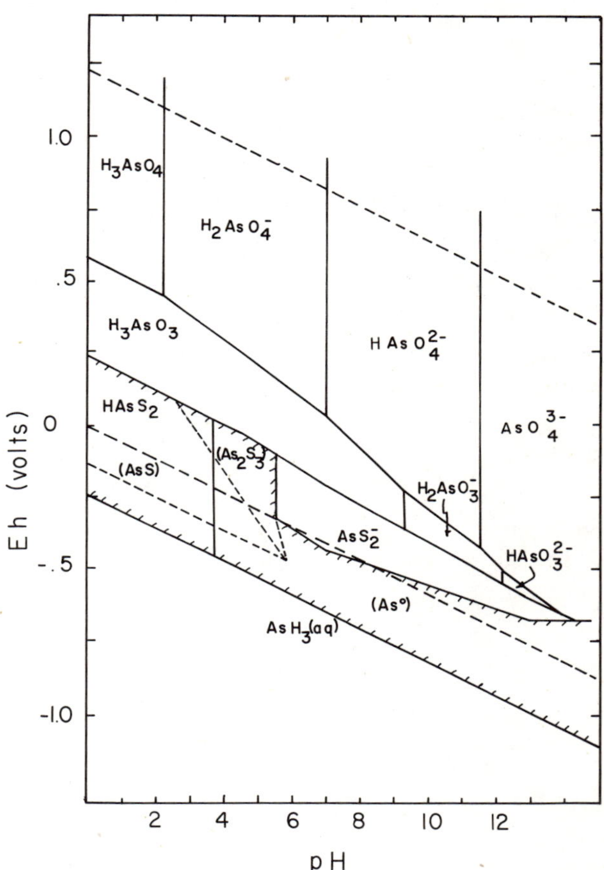

Figure 7.1. *The Eh-pH diagram for As at 25°C and one atmosphere with total arsenic 10^{-5} M and total sulfur 10^{-3} M. Solid species are enclosed in parentheses with dashed boundary line. Cross-hatched area indicates solubility less than $10^{-5.3}$ M. Stability of water is indicated by dashed lines. Adapted from Ferguson and Gavis.[6]*

In oxygenated waters, arsenic acid species (H_3AsO_4, $H_2AsO_4^-$, $HAsO_4^{2-}$ and AsO_4^{3-}) are stable. Under mildly reducing conditions, arsenious acid species (H_3AsO_3, $H_2AsO_3^-$ and $HAsO_3^{2-}$) become stable. The arsenic oxides, As_2O_5 and As_2O_3, are too soluble to appear in the diagram. Under conditions where sulfide is stable, realgar (AsS) and orpiment (As_2S_3) occur as stable solids at pH values below approximately 5.5 and Eh values of approximately 0 volts. The predominant species at low pH in the presence of sulfide is $HAsS_2$(aq). It has maximum solubility of $10^{-6.5}$ *M* (0.025 mg/l As). AsS_2^- predominates at pH greater than 3.7; above pH 5.5 its concentration is 10^{-5} *M*. At still lower Eh values elemental arsenic is thermodynamically stable. Although As° solubility in water has not been reported, it is undoubtedly low. At very low Eh values, arsine (AsH_3) may be formed. Arsine is only slightly soluble; the calculated line for a concentration of $10^{-5.3}$ *M* in solution corresponds to a partial pressure of about 1 atm. If iron were included, ferric arsenate (pK_{sp} 20.24) would have a small region of stability at pH below 2.3 and Eh above +0.74 V. There are no other arsenic-containing solids with measured solubilities low enough to result in arsenic concentrations less than 0.05 mg/l ($10^{-6.2}$ *M*).

The available thermodynamic data indicate that low Eh conditions in the presence of sulfide should result in effective removal of As(III). Packed beds of ferrous sulfide and the addition of hydrogen sulfide have been used in removal processes.[7] Precipitation of ferrous sulfide in a solution originally containing arsenate adsorbed on ferric hydroxide did not cause removal of arsenic, probably because the arsenate was not reduced to arsenic(III).[8] Thermodynamic data provide few other clues about possible removal processes. There are several stable, soluble arsenic species, but few sufficiently insoluble arsenic solids. The data, of course, provide no information about rates of any of the possible reactions.

It is necessary to look elsewhere in order to understand arsenic concentrations in nature and the removal of these concentrations in water treatment processing. In Figure 7.2, a schematic cycle of reactions is shown for arsenic in a stratified lake.[6] The diagram indicates that arsenic may be oxidized from 0 to (III) to (V), that ligand exchange reactions with OH^-, H_2O, and S^{2-} may occur, and that removal to sediments may occur through precipitation as well as sorption and biological uptake. The cycle is hypothetical

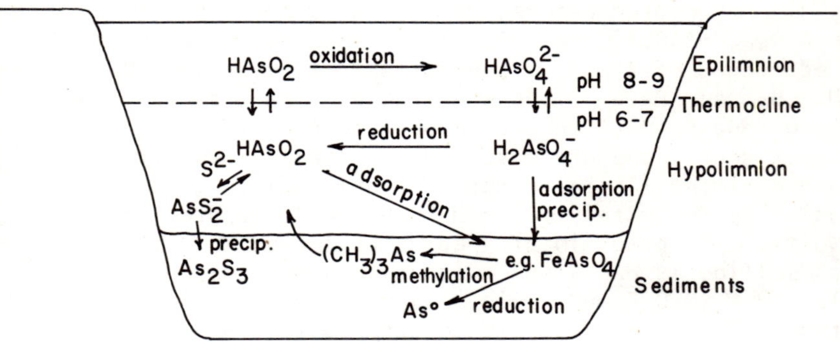

Figure 7.2. Cycle proposed for arsenic in a stratified lake. Adapted from Ferguson and Gavis.[6]

since some of the reactions shown have not been studied in an aquatic environment; however, all the reactions are consistent with experimental observations.

There seems to be a well-developed biogeochemical cycle for arsenic. Microorganisms enter the cycle by mediating reactions that are kinetically slow or in some cases thermodynamically unfavored. In any particular water, arsenic may be present in various chemical forms that will react differently in water treatment processes, as well as pose different hazards to consumers of the water.

The two parts of this cycle that will be discussed in some detail are the balance between arsenite and arsenate in aerobic waters and the possible methylation of arsenic.

Recent measurements by Johnson and Pilson[9] of arsenic (III) and (V) in the Sargasso Sea can be interpreted with care for implications for concentrations in freshwaters. These measurements confirmed earlier ones by Sugawara and Kanamori[10] that arsenic in seawater is present in significant amounts in the (III) oxidation state, as well as the stable (V) state. The persistence of 20 to 75% of the total arsenic as arsenite can be explained by postulating addition of mostly As(III) and a negligible rate of oxidation, or a steady state between the rate of oxidation of As(III) and the reduction of As(V). Johnson and Pilson[11] explored these possibilities for seawater and found the rate of arsenite oxidation to be very slow, but still much higher than needed to oxidize arsenite completely in the oceans, where the residence time for arsenic is estimated at 60,000 years. They also demonstrated that arsenate in non-sterile seawater enriched with glucose can be reduced to arsenite, and postulated that in the sea a steady state exists between the chemical oxidation and the bacterial reduction. The authors have not found measurable chemical oxidation of arsenite in fresh water systems over a four month period and believe chemical oxidation to be unimportant in freshwaters. Others have found rapid oxidation of arsenite solutions to be catalyzed by bacteria.[12,13]

Reduction of arsenic species to either dimethyl or trimethyl arsine has been identified in several systems but not yet in water.[14-17] The authors are not aware of any study of the rate of oxidation of methylarsines in water. It is possible that the reduction observed by Johnson and Pilson was to a methylarsine, which was subsequently oxidized to arsenite during their experiment or to arsenate during chemical analysis. There is no doubt that methylarsines occur in nature. The possible production of another methylated species in water, its importance in aquatic cycling of arsenic, its stability, accumulation, toxicity, and removal in processing all merit study, even though there is no reason to suspect toxic effects like those found for mercury.

In lakes or other fresh surface waters there is likely to be a partitioning of arsenic between the oxidation states. Depending on the arsenic sources, the relative rates of chemical and bacterial oxidation of As(III), the rate of bacterial reduction of As(V) and the presence or absence of an anaerobic hypolimnion, the As(III) content could be large or small. Though the concentrations of the species are important, little more can be reported at present. No measurements in freshwaters are reported, and information about the rates of the reactions is insufficient to make predictions.

In addition to the single sweep polarographic technique described by Whitnack and Brophy,[18] with modifications given in this paper, there are now colorimetric techniques for separately measuring arsenic(III) and (V) at environmental levels.[19,10] The measurement of the forms of arsenic should become routine in any case where levels approach the drinking water standard. Toxicity is associated primarily with As(III), and removability is affected by the chemical state. Future drinking water standards might well be set for both arsenic(III) and total arsenic, as is done for hexavalent and total chromium.

The oxidation state of arsenic is equally important in ground waters. It seems likely that the soluble species would be near equilibrium with their environment.

ARSENIC REMOVAL

In the face of uncertainty about species, the task of removing arsenic has been approached by trial and error. Plausible, though not necessarily accurate, explanations can be contrived for chemical conditions leading to success or failure of the processes listed in Table 7.I.

Removal in water treatment by precipitation processes with calcium, iron or aluminum salts may be accomplished by precipitation of an insoluble solid, for which thermodynamic information is not available, by occlusion in the precipitated solids, or by adsorption on their surfaces. The terms precipitation, occlusion and adsorption are used to distinguish loosely between the following:

1. Formation of a stoichiometric phase, which may be undetectable by conventional chemical or X-ray analysis and constitutes only a minor fraction of the precipitate;
2. Incorporation into the interior of a solid either in lattice sites or in interstices; or,
3. Association with the surface of the precipitate without specifying the kind of bonds causing the association.[20]

There are no data to indicate that any solids formed with arsenic or arsenious acid species are insoluble enough to cause removal to low levels in water treatment. However, there are no thermodynamic solubilities reported for arsenite solids and only a few reported arsenate solubilities. For arsenate, which is very similar to phosphate, it seems probable that the variety and complexity of the possible precipitated solids have not been appreciated. There may be highly insoluble solids that can be precipitated from solution.

The experimental studies described later in this paper are intended to determine the importance of adsorption of arsenite and arsenate onto amorphous ferric and aluminum hydroxides. The findings cast doubt on the importance of adsorption as a removal mechanism in water treatment precipitation processes. The remaining phenomenon, occlusion, seems the most likely mode of arsenic removal. Occlusion has been restricted to include only phenomena such as ion entrapment, lattice substitution and solid solution formation. Even so, occlusion is vague, hard to measure, and difficult to distinguish from adsorption in rapid precipitations.

Removal in water treatment by packed bed processes is evidently accomplished by adsorption in the case of charcoal beds and alumina beds. Epitaxial precipitation may be important in the case of ferrous sulfide beds, as arsenic readily forms strong, predominantly covalent bonds with both iron and sulfur. The tendency to bond with these elements causes arsenic to change its coordinative partners readily, especially if it is in the (III) oxidation state. In the presence of sulfide, arsenic(V) should be reduced; however, it is doubtful whether this reaction occurs rapidly.[8] In Taiwan,[7] removal of arsenic from ground water onto ferrous sulfide beds is consistent with the presence of As(III). The chronic arsenicism found in the area also implies that As(III) was present. In Fallon, Nevada, the removal onto both alumina columns and charcoal beds provides no clues to the oxidation state.[2] Arsenate species have a high affinity for aluminum, though perhaps no greater affinity than arsenite has.

The need to know the chemical state of arsenic in order to understand its removal is repeatedly apparent. This knowledge is certainly not vital as there is no immediate need to remove arsenic. Trial and error has served well when removal methods have been needed. However, arsenic, as a constituent of surface waters, is important enough to merit serious study of the removal of its various chemical forms. The tools are at hand to acquire a detailed understanding of arsenic removal. The experimental work described in the remainder of this paper is an attempt to begin this task.

Procedure

All glassware used in the experiments was soaked in 0.25 N HCL for 24 hours or more, washed, and then rinsed three times with distilled deionized water, which was also used in all sample and reagent preparations. Polarography cells were treated in similar fashion, but were rinsed five times.

Standard arsenate solutions were prepared by weighing Fisher reagent grade sodium arsenate ($Na_2HAsO_4 \cdot 7\ H_2O$) and diluting it to a working stock of 2.67×10^{-2} M As(V). Standard arsenite solutions were prepared with Baker "Dilut it" ($NaAsO_2$) vials and kept under nitrogen to prevent oxidation. A working stock solution was 2.51×10^{-2} M As(III). Arsenate must be reduced to arsenite to facilitate polarographic analysis of samples. The reducing agent used by Johnson and Pilson[19] was used in this instance. Fresh reagent was prepared daily and kept stoppered during use.

In order to obtain solids with reproducible characteristics, it was necessary to prepare aluminum and ferric hydroxide precipitates using the procedures which follow:

To prepare aluminum hydroxide, 33.32 g of Fisher reagent grade $Al_2(SO_4)_3 \cdot 18\ H_2O$ was weighed out and dissolved in 950 ml water. This was stirred for 1 hour and then titrated with 30 ml of 10 *N* NaOH at a dropping rate of 1 ml/30 sec. At the end of the titration, the pH, which was above 8.0, was adjusted to pH 7.0 by adding 0.1 *N* HCL. After aging the precipitate for another hour, the suspension was washed with 950 ml water over Whatman 42 filter paper. The aluminum hydroxide precipitate was then resuspended in another 950 ml water. The washed, resuspended precipitate was aged once more for 12 to 20 hours before use. The pH in each batch was monitored and was stable at the end of the aging. In various batches the final pH ranged from 6.2 to 6.9.

To prepare iron hydroxide, 48.20 g of reagent grade $Fe(NH_4)_2(SO_4)_2 \cdot 6H_2O$ was dissolved in 950 ml water. The solution was then treated using a similar procedure as was used for aluminum hydroxide. However, it should be noted that the final pH, after adding 30 ml of 10 *N* NaOH, was very close to 7.0; therefore, no attempt was made to adjust it further.

Measures of the characteristics of an amorphous precipitate that affect its adsorptive behavior might include surface charge and isoelectric point, the amount of hydrolysis, the surface area, and any structural features the solid may have. The pH change was used as a surrogate for a direct measure of the precipitate characteristics. The solids were aged until pH changes were negligibly slow as compared to the reaction times in the experiments. From X-ray analyses of the precipitates with time, the suspensions used at the time of the experiments were amorphous $Al(OH)_3(s)$ and $FeOOH(s)$. Only after one and one-half months did $FeOOH(s)$ show some poorly characterized ordering peaks of goethite. The $Al(OH)_3(s)$ showed no signs of ordering even after one and one-half months.

Surface areas were not measured in the experiments but can be expected to be of the order of 100 m^2/g for these precipitates. Isoelectric points from 6.6 to 7.3 for amorphous $Al(OH)_3(s)$ and from 6 to 8.6 for amorphous

FeOOH(s) have been reported.[22] The precipitates used in these experiments were prepared in a manner similar to procedures used by Mattson,[23] who found isoelectric points of 7.1 for FeOOH(s) and 7.5 for $Al(OH)_3(s)$. Because the isoelectric point of metal hydroxide precipitates is variable as a function of sample history, it is quite possible that in these tests the isoelectric point was as much as a pH unit away from reported values.

Ten-ml aliquots of stirred suspended precipitates were taken and filtered on pre-weighed Whatman 42 filter paper. These were dried for 4 hours at 104° C and weighed to establish the weight concentration of the suspension. Experimental samples were prepared in volumetric flasks by adding, in order, a small portion of water, arsenic stock, and an aliquot of suspension, and then filling the flask to mark. "Zero time" was considered to be after the prepared solution was shaken 20 times, at which time the samples were transferred to stoppered Erlenmeyer flasks. Kinetic experiments were run in a 25 ± 0.5° C temperature bath. Samples were stirred with a magnetic stirring bar. After drawing an aliquot from the flask for analysis, the flask was flushed with N_2 and resealed to prevent oxidation. The pH and adsorption isotherm samples were placed in a shaker in a thermostatically controlled room at 25 ± 0.5° C.

The duration of equilibrium adsorption experiments was between 20 to 24 hours after which aliquots were drawn for analysis. Arsenic(III) analysis was performed immediately. Ten-ml aliquots of samples to be assayed for As(V) were placed in 25-ml Erlenmeyer flasks, to which 0.4 ml of reducing agent was added. These samples were stoppered and analyzed the following day.

Arsenic analysis was done on a Model 1660 Davis Differential Cathode Ray Polarotrace. Traces were reproduced on a Honeywell strip chart recorder with a chart speed of 1 in./sec. Mercury dropping rate was set at 0.037 g/min. Bath temperature was 26.4° C. One ml of Hg was used in each polarography cell. Arsenic(III) was polarographed with a start potential of 0.8 volts; the sweep takes the applied potential through 1.3 volts. RC derivative proved useful to lessen slope complications. Three drops of concentrated H_2SO_4 were used as a carrier electrolyte in 3 ml of sample. Degassing with N_2 was done for two minutes without skirt and for one minute with skirt in place. A standard curve plotting log units for peak height against log concentration proved to be linear at concentrations from 1×10^{-3} to 5×10^{-5} molar As(III).

Arsenic(V) will not produce a polarotrace and must be reduced to arsenic(III). Prior to running the arsenic(V) samples, the Erlenmeyer flasks were unstoppered and aerated for 30 minutes to relieve the systems of dissolved SO_2. It was found that 0.4 ml would reduce up to 5.0×10^{-3} *M* As(V); however, at lower concentrations between 1×10^{-5} *M*, 0.2 ml would be more satisfactory. A concentration of 0.4 ml can be used for all samples, but lower ranges require greater amplification. The unused portion of the reducing agent causes a reduction wave which tends to reduce the size of the peaks. The scanning potential was the same as for As(III). Sample size was 4 ml, and no supporting electrolyte was needed as it was already present in the reducing agent itself. Degassing with N_2 was done for two minutes without skirt and for two minutes with skirt in place. Standard curves were prepared for As(III), as previously mentioned.

EXPERIMENTAL RESULTS

Adsorption Kinetics

In all cases adsorption was initially rapid with 50 to 95 per cent of the total removal taking place before the first sample was taken as shown in Figure 7.3. Removal continued

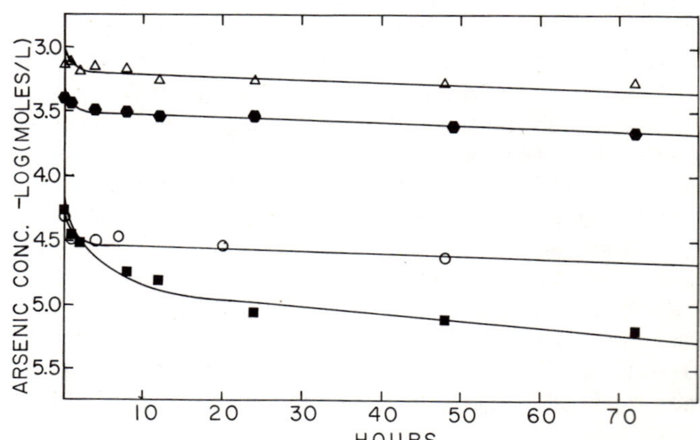

Figure 7.3. Arsenic adsorption as a function of time (initial As Conc. 1×10^{-3} M) ● As(V) on FeOOH (.766 g/l), ○ As(V) on $Al(OH)_3$(1.29 g/l), ■ As(III) on FeOOH (.685 g/l), △ As(III) on $Al(OH)_3$ (1.04 g/l).

slowly over the 72 hour length of the experiments. The extent of the reaction varied significantly with about 99.5% removal of As(III) on FeOOH(s), 97% removal of As(V) on $Al(OH)_3(s)$, 86% of As(V) of FeOOH(s), and 50% of As(III) on $Al(OH)_3(s)$. Adsorption varied from 0.435 mmol/g for As(III) on $Al(OH)_3(s)$ to 1.45 mmol/g for As(III) on FeOOH(s). These values are large and indicate that specific adsorption of the molecules was involved.

The pH values of the unbuffered suspensions were monitored throughout the test, and the values, shown in Figure 7.4,

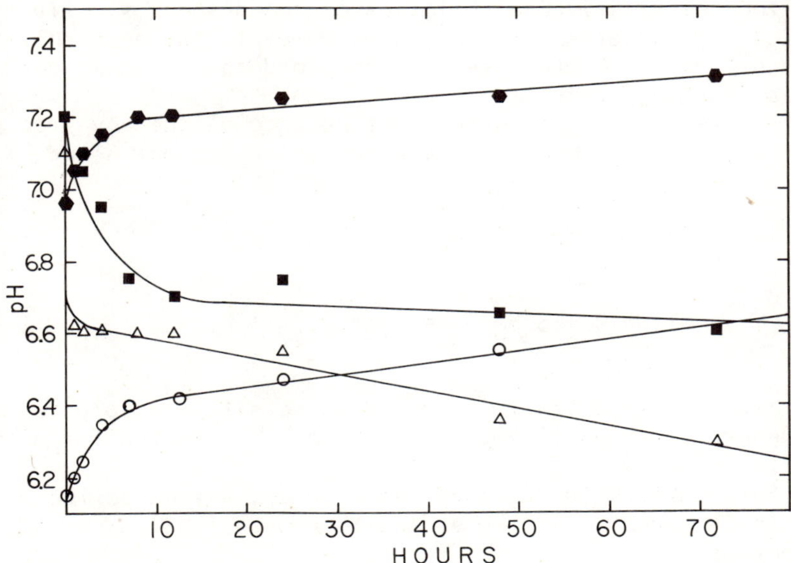

Figure 7.4. Arsenic adsorption pH changes as a function of time (initial arsenic Conc. 1 x 10^{-3} M) ● As(V) on FeOOH (.766 g/l), ○ As(V) on $Al(OH)_3$(1.29 g/l) ■ As(III) on FeOOH (.685 g/l), △ As(III) on $Al(OH)_3$ (1.04 g/l).

drifted slowly upward for As(V) adsorption and slowly downward for As(III) adsorption. Initially, the pH values changed more slowly than the As concentrations. The precipitates showed no pH drift in the absence of arsenic.

Using solubility data for ferric and aluminum arsenate and amorphous hydroxides, it was calculated that the initial arsenate solutions were undersaturated ($K/Q \simeq 10^{-1}$) with respect to conversion of the hydroxide to the arsenate (Equations 1 and 2).

$$FeOOH + H^+ + H_2AsO_4^- = FeAsO_4 + 2\ H_2O, \quad \log K = 7.8 \qquad (1)$$

$$Al(OH)_3 + H^+ + H_2AsO_4^- = AlAsO_4 + 3\ H_2O, \quad \log K = 8.0 \qquad (2)$$

At the end of the experiment, undersaturation was increased ($K/Q \simeq 10^{-3.1}$).

Arsenate was taken up only slightly by ferric hydroxide and much more extensively by aluminum hydroxide. Precipitation is evidently not involved, and adsorption does not follow the relative solubilities of the arsenates. Aluminum, which demonstrates more extensive removal, has a solubility product for arsenate more than four orders of magnitude greater than ferric iron.

The removal of arsenite might be accomplished by precipitation of aluminum or ferric arsenite (Equations 3 and 4).

$$FeOOH + 3\ HAsO_2 = Fe(AsO_2)_3 + 2\ H_2O \qquad (3)$$

$$Al(OH)_3 + 3\ HAsO_2 = Al(AsO_2)_3 + H_2O \qquad (4)$$

Ferric orthoarsenite is reported to be slightly soluble; however, no solubility products have been located for either solid.

For "equilibrium" adsorption studies, a time period of 20 to 24 hours was selected and then used in all subsequent experiments.

Effect of pH

Experiments were conducted in which the initial pH was varied by addition of acid or base so that the final pH would range between 6 and 8. The final pH values for arsenite adsorption were 0.3 to 0.5 units lower than the initial values, while the pH for arsenate adsorption drifted upward 0.2 to 0.3 pH units. The results of these experiments are shown in Figure 5. As(V) adsorption is at maximum or saturation values, but, as described in the following section, no saturation adsorption was found for As(III). There was considerable "scatter" in the data but little variation in removal with pH over the range tested.

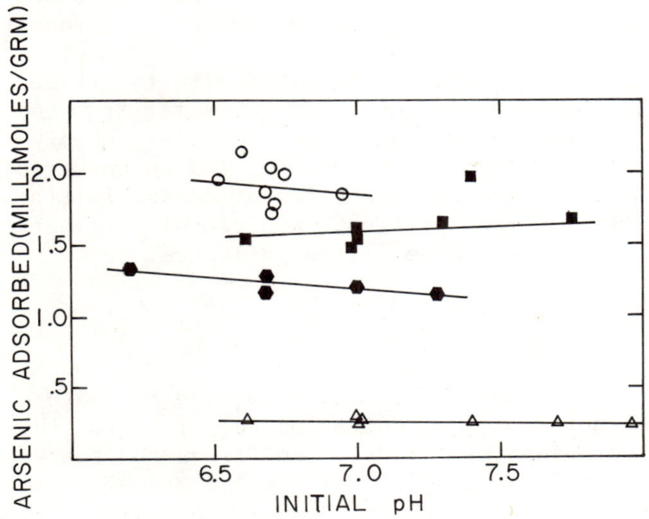

Figure 7.5. *pH effects on adsorption (initial As conc. 1 x 10^{-3} M) ● As(V) on FeOOH, ○ As(V) on $Al(OH)_3$, ■ As(III) on FeOOH, △ As(III) on $Al(OH)_3$.*

Reported zero point of charge (ZPC) values for amorphous $Al(OH)_3$(s) and amorphous FeOOH(s) are slightly basic. Negatively charged or uncharged molecules were removed onto an initially neutral or slightly positively charged surface. It was expected that if electrostatic interactions were important, adsorption of As(V) would decrease as the pH increased towards pH_{ZPC}. Alternatively, based on studies of anion adsorption onto goethite,[24] decreasing adsorption of arsenate and arsenite species with increasing pH might be expected, with inflections in the adsorption pH curve at pK values. No significant effects were observed.

For arsenate, the pH drifted upwards indicating an exchange, qualitatively similar to Equations 1 and 2, where an initially uncharged or positively charged surface hydroxyl group reacts with $H_2AsO_4^-$ and takes up protons during the process. The stoichiometry of the proton exchange has not been determined; however, it is believed that the adsorption process results in a net uptake of anions and, hence, a reduction of the surface charge (Stern Layer) of the particles and in a reduction in

pH_{ZPC}.[25] Further experiments are planned to determine the reaction stoichiometry and the effect on electrophoretic mobility.

For arsenite, H_3AsO_3 is adsorbed with the release of protons. The experimental pH range is far from the pK_1 of 9.2. Since the adsorption reaction is probably between an uncharged surface hydroxyl group and an uncharged molecule, the observed lack of an effect of initial pH on adsorption is not surprising. Additionally, electrostatic interactions should have little influence.

Adsorption Isotherms

A series of experiments was conducted to measure the adsorbent/adsorbate concentration relations for the two solutes onto the two solids. Conditions for the experiments are shown in Table 7.II. The results are given in Figure 7.6 and Figure 7.7, which show, respectively, the entire concentration range and an expanded diagram of the low concentration range. Adsorption of As(V) shows saturation at high concentrations. The lines are Langmuir isotherms fitted to the data. The data in a linearized form are shown in Figure 7.8, together with the selected isotherms. As(III) adsorption showed no saturation up to surface excesses of several mmol/g of adsorbate, which are extremely large values even for a high surface area material.

If the isotherms are extrapolated to the low concentrations of interest in water treatment, it can be calculated that for a dose of 100 mg/l of alum or 50 mg/l of ferric chloride, removal of an initial arsenic concentration of 0.1 mg/l would be 34% for As(V) on $Al(OH)_3(s)$, 3.8% for As(III) onto $Al(OH)_3(s)$, and 11% for As(III) or (V) onto FeOOH(s). These calculations are made assuming the same surface area to weight ratios as for the precipitates. Such an extrapolation to the origin is legitimate only to make the observation that adsorption on hydroxides does not appear promising for purposes of removal to meet drinking water standards. The literature provides examples of iron and alum precipitation that uphold these results,[26] as well as ones that do not.[27] It is believed that another mechanism, precipitation or occlusion, must be operating in the cases where high removals to low residuals have been observed.

Table 7.II

Adsorption Isotherm Experimental Conditions

Adsorption	*Precipitate Concentration g/l*	*pH Range*		*Final Concentration Range (mol/l)*
		Initial	*Final*	
As(V) on $Al(OH)_3$	0.29 - 1.40	6.9 - 8.0	7.2 - 8.3	5×10^{-6} - 5×10^{-3}
As(V) on FeOOH	0.29 - 2.90	5.4 - 6.4	5.6 - 6.6	5×10^{-6} - 5×10^{-3}
As(III) on $Al(OH)_3$	0.20 - 1.30	6.8 - 7.8	6.3 - 7.3	2×10^{-4} - 6×10^{-4}
As(III) on FeOOH	0.12 - 0.78	7.0 - 8.0	6.5 - 7.5	1×10^{-4} - 6×10^{-4}

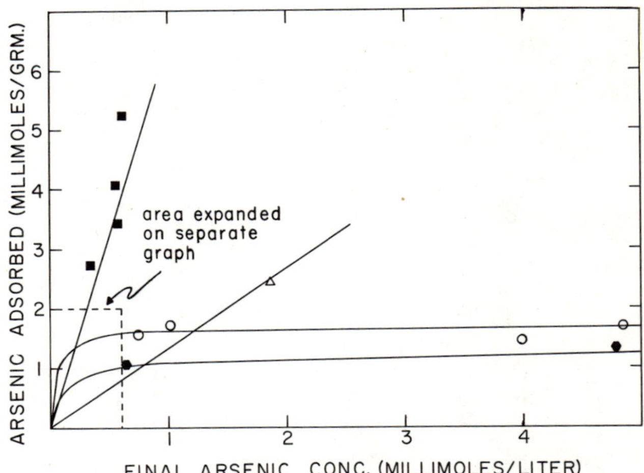

Figure 7.6. *Arsenic adsorption isotherms (not showing data points in dashed area)* ● *As(V) on FeOOH,* ○ *As(V) on Al(OH)* $_3$, ■ *As(III) on FeOOH,* △ *As(III) on Al(OH)* $_3$.

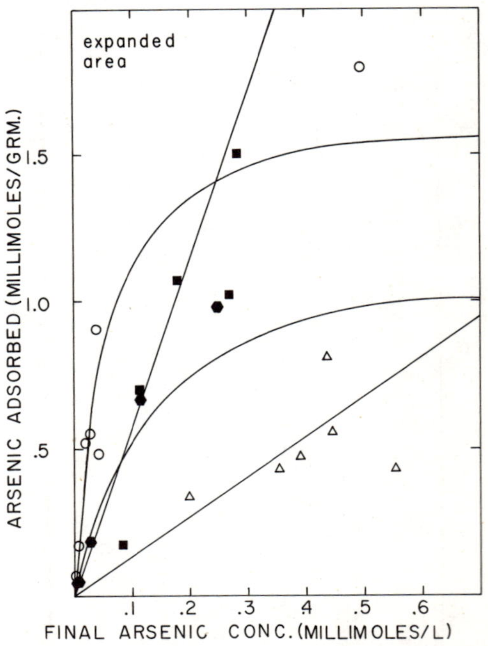

Figure 7.7. *Adsorption isotherms (low concentration range)* ● *As(V) on FeOOH,* ○ *As(V) on Al(OH)* $_3$, ■ *As(III) on FeOOH,* △ *As(III) on Al(OH)* $_3$.

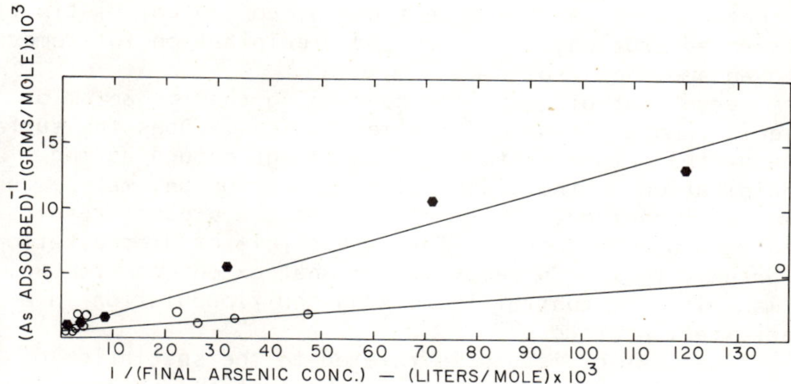

Figure 7.8. Linearized adsorption isotherms using Langmuir equation $\left(\frac{1}{\Gamma} = \frac{1}{\Gamma_o K_L S} + \frac{1}{\Gamma_o}\right)$

● *As(V) on FeOOH* (Γ_o = 1.25 m M/g, K_L = 7.30 x 10^3 l/mol),

○ *As(V) on* $Al(OH)_3$ (Γ_o = 1.70 m M/g, K_L = 19.7 x 10_3 l/mol).

SUMMARY AND CONCLUSIONS

It was found that arsenic species may be adsorbed in large quantities onto the surfaces of amorphous ferric and aluminum hydroxides. The adsorption varies greatly with the oxidation state of the arsenic, but not with the pH within the range of 5.5 to 7.5. This adsorption evidently does not account for the high removals that are sometimes found in arsenic removal studies. The adsorption of arsenic species has posed many interesting questions. These relate first to understanding interactions between anions or neutral molecules and hydroxide surfaces. It is not clear, for instance, whether arsenate adsorption can be explained solely in terms of the molecule-surface interaction or whether the charge, concentration and identity of the cation must also be considered. Also, the question remains as to why the same initial removal of both As(III) and (V) is found on FeOOH(s), and why saturation is not observed with As(III). At the highest surface excesses, on the order of mmol/g, the cation to arsenic ratio is between 10 and 2.

At these values can there be a useful conceptual distinction between adsorption, occlusion and precipitation for removal onto an amorphous solid?

A second set of questions concerning the mechanism of arsenic removal in water treatment arises. Does the surface area or the nature of the surface change enough during precipitation so that adsorption, or occlusion, may be the removal mechanism in contradiction to the present results with aged precipitates? What is the role of precipitation of arsenic solids in removal? To what extent can removal be maximized by controlling mixing and flocculation in a precipitation process?

A final set of issues is related to the speciation of arsenic in water sources. The concentrations of arsenite and arsenate can now be measured at environmental levels, and information about the cycling of arsenic species is rapidly accumulating. These advances can be utilized in routine measurement of arsenic species and the establishment of separate standards for arsenic(III) and (V). Relaxation of standards for arsenic(V) may allow use, without removal processing, of many of the surface waters that now equal or exceed the recommended limiting concentration of the U.S.P.H.S. Drinking Water Standards. In other cases, water sources that present a true hazard can be identified with certainty.

The relative abundance in the past of uncontaminated water sources near metropolitan areas no longer exists. Present understanding of arsenic removal mechanisms is still not advanced; however, studies to date indicate that removal processing seems to be approaching technological feasibility. Perhaps it is now reasonable to rely on processing to remove arsenic. In any case, the high levels of arsenic in surface waters and the demand for increased quantities of water for municipal supply indicate that careful consideration of standards and policies with respect to treatment is in order.

REFERENCES

1. McCabe, L. J., J. M. Symons, R. D. Lee, and G. G. Robeck. *J. Amer. Water Works Assoc., 62,* 670 (1970).
2. Kopp, J. R., and R. C. Kroner. "Trace Metals in the Waters of the United States," Federal Water Quality Administration, Cincinnati, Ohio, 1970.

3. Durum, W. H., J. D. Hem, and S. G. Heidel. "Reconnaissance of Selected Minor Elements in Surface Waters of the United States," October 1970, Geological Survey Circular 643, U. S. Dept. of the Interior, Washington, D.C., 1971.
4. "Water Quality Criteria," Federal Water Pollution Control Administration, U. S. Dept. of the Interior, Washington, D. C., 1968.
5. Patterson, J. W., and R. A. Minear. "Wastewater Treatment Technology," 2nd Ed., Illinois Institute for Environmental Quality Document 73-1, February 1973.
6. Ferguson, J. F., and J. Gavis. *Water Research, 6,* 1259 (1972).
7. Shen, Y. S., and C. S. Chen. "Relation Between Blackfoot Disease and the Pollution of Drinking Water by Arsenic in Taiwan," In *Advances in Water Pollution Research,* Vol. 1, O. Jaag, ed., Pergamon Press, Oxford (1964), pp. 173-185.
8. Tanaka, M. "Discussion," In *Advances in Water Pollution Research,* Vol. 1, O. Jaag, ed., Pergamon Press, Oxford (1964), pp. 118-123.
9. Johnson, D. L., and M. E. Q. Pilson. Presented at 35th Annual Meeting, American Society of Limnology and Oceanography, Tallahasee, Florida, March 1972.
10. Sugawara, K., and S. Kanamori. *Bull. Chem. Soc. (Japan), 37,* 1358 (1965).
11. Johnson, D. L., and M. E. Q. Pilson. Presented at 53rd Annual Meeting, American Geophysical Union, Washington, D.C., April 1972.
12. Green, H. H. *Rep. Vet. Res. S. Afr., 5/6,* 595 (1918).
13. Turner, A. W., and J. W. Legge. *Aust. J. Biol. Sci., 7,* 452, 496, 504 (1954).
14. Challenger, F., C. Higginbottom, and L. Ellis. *J. Chem. Soc., 1933,* 95 (1933).
15. McBride, B. C., and R. C. Wolfe. *Biochem. J., 10,* 4312 (1971).
16. Woolson, E. A., and P. C. Kearney. *Environ. Sci. Technol., 7,* 47 (1973).
17. Zussman, R. A., E. E. Vicher, and I. Lyon. *J. Bact., 81,* 157 (1961).
18. Whitnack, G. C., and R. G. Brophy. *Anal. Chim. Acta, 48,* 123 (1969).
19. Johnson, D. L., and M. E. Q. Pilson. *Anal. Chim. Acta, 58,* 289 (1972).
20. Laitenen, H. A. *Chemical Analysis.* McGraw Hill, New York, New York, 1960, pp. 165-175.

21. Bellack, E. *J. Amer. Water Works Assoc., 63,* 454 (1971).
22. Parks, G. A. *Chem. Reviews, 65,* 177 (1965).
23. Mattson, S. *Soil Science, 30,* 459 (1930).
24. Hingston, F. J., R. J. Atkinson, A. M. Posner, and J. P. Quirk. *Nature, 215,* 1459 (1967).
25. Hingston, F. J., A. M. Posner, and J. P. Quirk. "Adsorption of Selenite by Goethite," in *Adsorption from Aqueous Solution,* W. J. Weber and E. Matijevic, eds., Advances in Chemistry Series 79, American Chemical Society, Washington, D.C., 1968.
26. Buswell, A. M., R. C. Gore, H. F. Hudson, H. C. Wiese and T. E. Larson. *J. Amer. Water Works Assoc., 35,* 1303 (1943).
27. Irukayama, K. "Discussion," in *Advances in Water Pollution Research,* Vol. 1, O. Jaag, ed., Pergamon Press, Oxford, 1964, pp. 185-187.
28. Skripach, T., V. Kagan, M. Romanov, L. Kamen, and A. Semina. "Removal of Fluorine and Arsenic from the Wastewater of the Rare-earthy Industry," In *Advances in Water Pollution Research*, Vol. II, S. H. Jenkins, ed., Pergamon Press, Oxford, 1971.
29. "Drinking Water Standards," U. S. Public Health Service, U. S. Dept. Health, Education and Welfare, Washington, D. C., 1962.

CHAPTER 8

EFFECT OF ALUMINUM(III) HYDROLYSIS ON ALUM COAGULATION

Alan J. Rubin and Thomas W. Kovac*

INTRODUCTION

Aluminum sulfate (alum) is the most widely used chemical for the removal of colloidal substances and other finely suspended materials contributing to the turbidity of water supplies. In order to fully understand the chemical coagulation process it is imperative that the aqueous chemistry of aluminum be understood. This is especially true with regard to its hydrolytic reactions and interactions with colloidal particles.

The aqueous chemistry of iron(III), another hydrolyzing metal widely used in water clarification, has been firmly established. The addition of a ferric salt to water results in its dissolution and simultaneous hydration followed by lowering of the solution pH. This occurs because a series of stepwise hydrolysis reactions takes place in which iron acts as a Bronsted-Lowry acid. In these reactions hydrogen ions are released from the bound waters of hydration to form hydroxo species. The overall hydrolysis scheme for iron(III) is summarized in simplified form in Figure 8.1. As a consequence of hydrolysis amorphous ferric hydroxide is precipitated even in very acid solutions. For example, 2×10^{-4} *M* (11 mg/l) iron(III) will just begin to precipitate at pH 2.67 so that at pH 4, 99.85 per cent of the metal is insoluble.[1] Aging of these suspensions, which occurs fairly rapidly with iron, leads to the formation of even more insoluble crystalline phases. Raising the pH to alkaline levels results in the dissolution of the precipitate.

---

*A. J. Rubin, Associate Professor, and T. W. Kovac, Research Associate, Water Resources Center, Ohio State University, Columbus, Ohio.

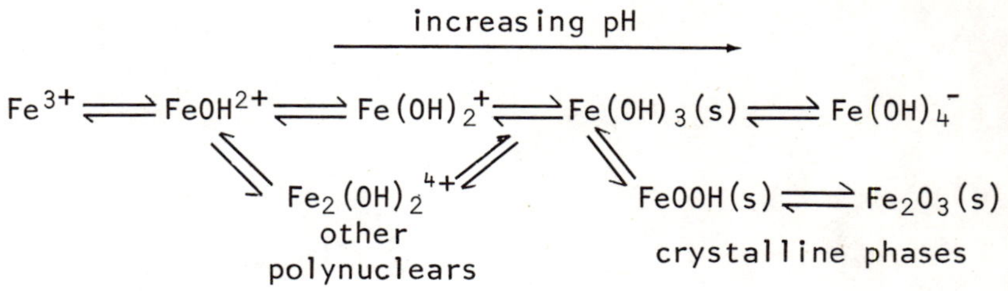

$$\xrightarrow{\text{increasing pH}}$$

$$Fe^{3+} \rightleftharpoons FeOH^{2+} \rightleftharpoons Fe(OH)_2^{+} \rightleftharpoons Fe(OH)_3(s) \rightleftharpoons Fe(OH)_4^{-}$$

$$FeOH^{2+} \rightleftharpoons Fe_2(OH)_2^{4+} \text{ (other polynuclears)} \rightleftharpoons Fe(OH)_3(s)$$

$$Fe(OH)_3(s) \rightleftharpoons FeOOH(s) \rightleftharpoons Fe_2O_3(s) \text{ (crystalline phases)}$$

Figure 8.1. Hydrolysis scheme for iron(III).

An important aspect of iron hydrolysis is the formation of the dimeric ion, $Fe_2(OH)_2^{4+}$. This has an olated structure in that the metal atoms are bound through hydroxo bridges rather than directly to one another. Its relative concentration is dependent on the applied concentration of iron as well as pH. The formation of this polynuclear species has been demonstrated conclusively.[2-5] It is to be emphasized that the presence of such species in iron and aluminum solutions cannot be ignored, if only because of their high charges which, according to the Schulze-Hardy rule,would be significant in classical coagulation.

Upon hydrolysis aluminum(III) undergoes a similar sequence of reactions. There are some differences, partly because aluminum is less acid than iron(III), and the exact nature of the hydrolyzed species is still being debated. Present indications are that the hydrolysis sequence for aluminum(III) can be accurately represented by the scheme shown in Figure 8.2.[6] The importance of these hydrolytic reaction products has been documented in the literature.[7-10]

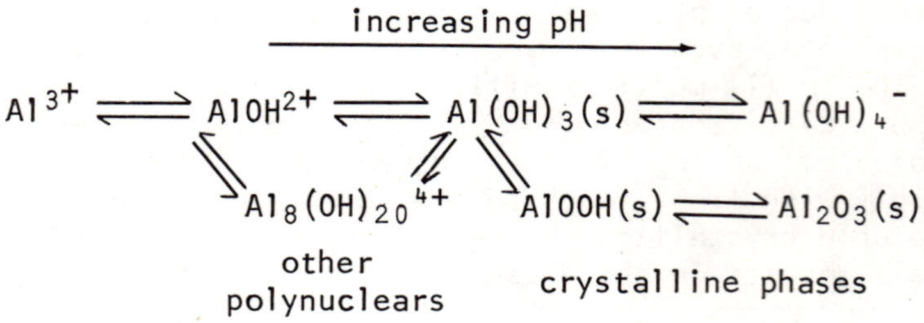

$$\xrightarrow{\text{increasing pH}}$$

$$Al^{3+} \rightleftharpoons AlOH^{2+} \rightleftharpoons Al(OH)_3(s) \rightleftharpoons Al(OH)_4^{-}$$

$$AlOH^{2+} \rightleftharpoons Al_8(OH)_{20}^{4+} \text{ (other polynuclears)} \rightleftharpoons Al(OH)_3(s)$$

$$Al(OH)_3(s) \rightleftharpoons AlOOH(s) \rightleftharpoons Al_2O_3(s) \text{ (crystalline phases)}$$

Figure 8.2. Hydrolysis scheme for aluminum(III).

The major purposes of the research described in this paper were to examine the aggregation and restabilization of three concentrations of colloidal titanium dioxide by freshly prepared aluminum sulfate solutions and to determine the relationship between "coagulant" concentration and pH for the sol. The experimental techniques used in similar studies with the bacterium *Escherichia coli* were employed.[9-11] Turbidity changes during settling, as estimated by absorbance measurements, served to indicate aggregation. Data extrapolated from the settling curves were used to establish the entire log aluminum sulfate concentration—pH stability limit diagram for titanium dioxide. Several workers using different experimental techniques have used this type of graphical procedure to summarize their results. Basically, the stability limit diagram is a sort of phase diagram showing the results, aggregation or stability, expected at any given pH and metal concentration. Rubin and Hanna,[9] using the technique of precipitate boundary analysis,[12,13] have demonstrated that its straight line portions can be related to the hydrolytic behavior of the coagulant metal. Using differences in settling rates, it was possible to distinguish between different modes of aggregation. At low pH the bacterial sol settled slowly whereas at higher pH in the presence of aluminum hydroxide very rapid clarification was obtained. The pH ranges over which the insoluble hydroxide was formed were designated at the "sweep zone" and enmeshing of the sol in the flocculating gelatinous precipitate, as described by Packham,[14,15] was proposed to explain the results. The solubility of aluminum hydroxide to form the negatively charged aluminate ion was shown to determine the sweep zone limits in alkaline solutions.

EXPERIMENTAL

Titanium dioxide suspensions were prepared with Zopaque RG (ASTM D 476-48, Class 1) obtained from the Gildden-Durkee Division of SCM Corporation, Baltimore, Maryland. This material is a water-dispersible powder of the anatase crystal type used in paper making. Zopaque RG has an extremely small particle size of 0.1 to 0.4 microns and forms a negatively charged sol in aqueous media. Because of its colloidal properties and high refractive index, about 2.55, TiO_2 forms very stable suspensions in dilute concentrations having an intense white to blue color.

All solutions and suspensions and their dilutions were prepared from carbonate-free distilled-deionized water. A Waring blender was used to disperse various weights of the dry powder to prepare the stock suspensions of TiO_2. Suspension age did not appear to have any effect on the stability of the sol except for a short period of time during which larger particles were removed by settling. The experimental suspensions at concentrations of 50.2, 41.9 and 3.2 mg/l are referred to in the text as the high, intermediate and low sol concentrations, respectively.

Fresh dilutions from a stock solution of 0.295 *M* aluminum (reagent grade aluminum sulfate Merck) were prepared just prior to experimentation. Solutions of reagent grade sodium hydroxide and nitric acid were used for pH adjustment. For each experimental run two separate series of solutions were prepared. One series, in small glass vials, consisted of base (if needed) and the exact amount of stock TiO_2 suspension needed to give the desired initial sample turbidity. The second series, in round 19-ml Coleman cuvettes, consisted of aluminum solution, acid (if needed), and distilled water. An experimental sample of 15-ml total volume was prepared by adding the contents of a vial to a cuvette. Both acid and base were never added to the same sample. The cuvette was shaken vigorously for about 10 seconds and then set aside without further mixing for the duration of the experiment. Turbidity and pH readings were made at specified time intervals depending upon the applied aluminum dose and sol concentration.

Absorbance measurements were taken at 550 mμ using a Coleman Model 14 spectrophotometer. Turbidity measurements for the low sol series were made by 90° light scattering using a Coleman Model 9 Nepho-Colorimeter. Sargent-Welch Model NX and Model DR pH meters fitted with Sargent combination electrodes were used for all pH measurements. Streaming current measurements to determine the sign of the charge on the sol were taken with a Water's Association Streaming Current Detector. The procedure has been described by Cardwell.[16]

In addition to the studies with aluminum sulfate, several experiments were run to determine the concentrations of nitric acid, sodium nitrate, and calcium nitrate required to coagulate TiO_2. Reagent grade chemicals were used with essentially the same procedure as described above (see reference 11).

RESULTS AND DISCUSSION

Hydrogen Ion Coagulation

The effect of pH on the stability of titanium dioxide without the addition of aluminum (*i.e.*, pH adjustment with nitric acid or sodium hydroxide alone) was determined at the three sol concentrations. These studies demonstrated that TiO_2 is slowly coagulated by very low concentrations of hydrogen ion. The rate of settling was highly dependent on sol concentration. For example, approximately 10 hours was required to observe an 80% reduction from the initial absorbance with the high sol concentration, whereas after 100 hours only 40% reduction was obtained with the low concentration. In general, the critical pH just below which settling occurred increased slightly with time to a limit where no further change was observed. At all three sol concentrations this critical pH ranged from pH 5.8 to 6.0. For example, representative turbidity-pH data after 24 hours settling at the intermediate sol concentration are shown in Figure 8.3. Critical pH values (pH_s) were

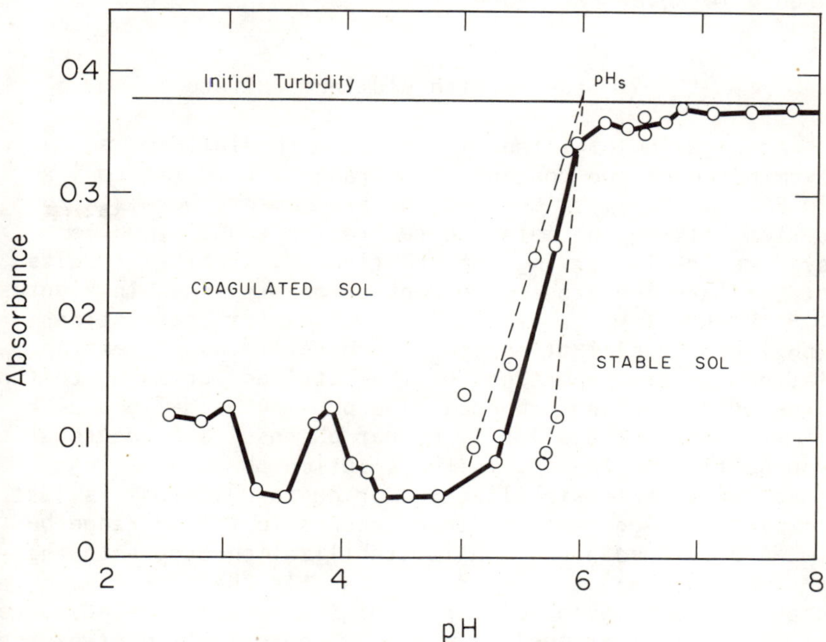

Figure 8.3. Effect of pH on the stability of titanium dioxide. No alum added. 24-hr. settling data for the intermediate sol concentration.

determined by extrapolating the steepest portions of the curve up to the average initial absorbance. As indicated by the figure, the data in the region of the critical value were rather erratic; however, the several different extrapolations possible all resulted in approximately the same pH_s of around 6.0 (streaming current measurements indicated an isoelectric point at 3.2). The hydrogen ion concentration of about 10^{-6} *M* at the pH_s is well below that predicted by the Schulze-Hardy rule for coagulation by simple single-plus counterions.[17] This observation coupled with the small size of the ion would tend to support the occurrence of an adsorptive coagulation mechanism although a greater dependence of pH_s on sol concentration would have been expected.

The pH for freshly prepared TiO_2 stock suspensions in distilled water is around 6.1. Within one month these dispersions experience a lowering in pH to 5.7 with significant decreases in turbidity. Protecting the suspensions from carbon dioxide absorption and maintaining the pH at 6 or above resulted in relatively stable suspensions which changed only slightly in absorbance. Suspensions initially adjusted to pH 7 and exposed to the atmosphere have remained stable for over two years.

Studies with Aluminum Sulfate

Aggregation of titanium dioxide with aluminum was examined over the concentration range 1×10^{-1} *M* to 1×10^{-8} *M* $Al(SO_4)_{3/2}$. Most of the experiments were run in a series holding the metal concentration constant while systematically varying the solution pH. Typical results at the intermediate sol concentration are shown in Figures 8.4 through 8.6. Critical pH values for coagulation (pH_c) and stabilization (pH_s) were determined by extrapolating the steep portions of the settling curves up to the line of initial absorbance. The pH_c may be defined as a limit such that a slightly higher pH results in aggregation and settling. The pH_s is the solution pH above which stabilization (restabilization or no coagulation) is just completed. Aggregation, then, occurs in the pH range between a pH_c and a pH_s, while stabilization occurs in the pH range between a pH_s and a pH_c and is characterized by a stable unsettleable sol. Each of these extrapolated values at its initial or applied aluminum concentration gives one point along a boundary of the aluminum sulfate concentration--pH stability limit diagram (domain) for titanium dioxide.

Figure 8.4 shows the data obtained with the samples at 1×10^{-3} *M* aluminum. Three distinct zones or regions can be distinguished. Between pH 4 and 9 is a region of rapid or "sweep" aggregation characterized by heavy floc formation and excellent clarification. This is the sweep zone in which the solution is oversaturated with respect to aluminum hydroxide. To its left is a region of slower clarification, presumably due to coagulation by hydrogen ion. On the right is a region of stability, that is, of no coagulation or aggregation.

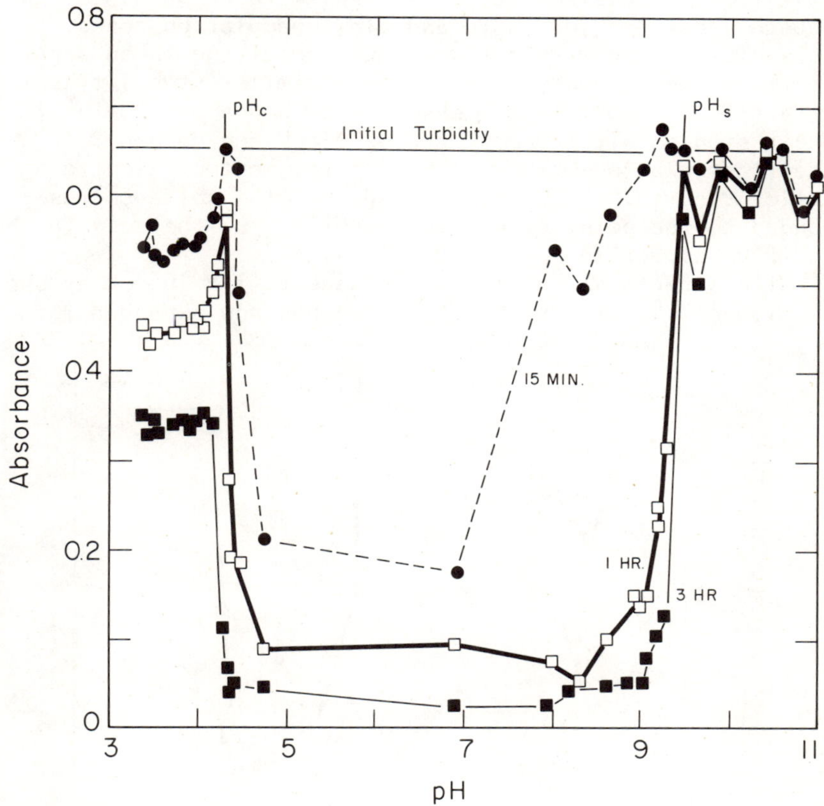

Figure 8.4. Aggregation of titanium dioxide with 1×10^{-3} M $Al(SO_4)_{3/2}$. Intermediate sol concentration. Settling times: ● 15 min., □ 1 hr., ■ 3 hr., ◇ 6 hr., ▲ 12 hr., ◆ 18 hr., ○ 24 hr., △ 40 hr.

It was observed that small volumes of highly concentrated sodium hydroxide would act as a coagulant. In these cases the right stability region would slowly settle behaving very much like the slow coagulation region on the left. Upon using larger quantities of more dilute base to achieve the same pH adjustment, stability in the right-hand zone was maintained. Consequently, dilute base solutions were used for increasing the pH in all subsequent studies. In general, the left boundary at the pH_c was relatively time independent compared to the boundary between the sweep zone and the stabilization region. An attempt was made to extrapolate the critical pH values after the time dependence period of the boundaries had passed.

An unusual feature of the turbidity—pH curve in Figure 8.4 is the transient stability "spike" near pH 4.3 at the boundary between the rapid and slow coagulation zones. This disappeared within three hours resulting in an apparent shift of the boundary with time. A similar stability spike with aluminum sulfate was observed by Hanna and Rubin[9] in their studies with *E. coli* and by Black and Hannah[18] in their coagulation work with kaolinite, montmorillonite and Fuller's earth. The latter found that the spike corresponded to the point of greatest mobility of the sols in the positive direction.

With a decrease in aluminum dosage to 1 x 10^{-4} *M* as shown in Figure 8.5, the stability spike has now expanded into a

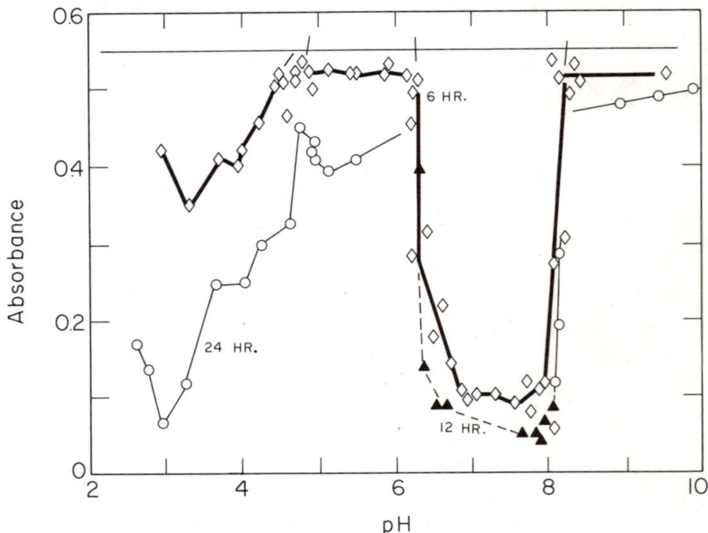

Figure 8.5. Aggregation of titanium dioxide with 1 x 10^{-4} M $Al(SO_4)_{3/2}$. Intermediate sol concentration Symbols as described in Figure 8.4.

fully developed region of stability between the slow and rapid settling zones. The turbidity of samples in this pH range decrease in time but at a considerably lower rate than in the slow coagulation region. The sweep zone has further narrowed, occupying the range between pH 6.2 and 8.2. After 12 hours of settling, the right boundary is comparatively time independent.

Figure 8.6, representing the addition of 3.16×10^{-5} *M* $Al(SO_4)_{3/2}$, shows a further broadening and a shifting of the central stability region to the basic side. The sweep

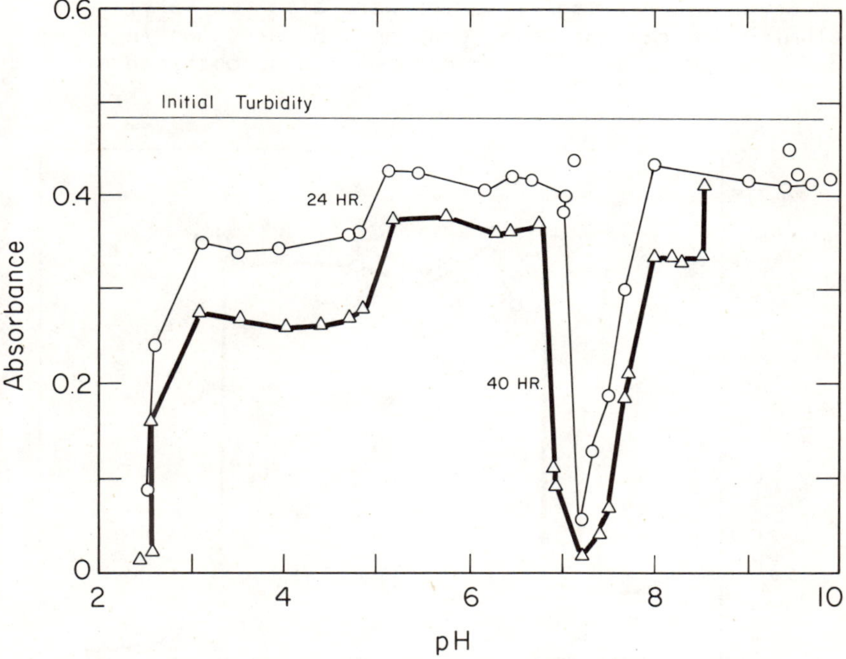

Figure 8.6. Aggregation of titanium dioxide with 3.16×10^{-5} M $Al(SO_4)_{3/2}$. Intermediate sol concentration. Symbols as described in Figure 8.4.

zone continued its narrowing trend with decreasing alum dose, and forty-hour readings were necessary to clearly differentiate between the various zones. It is also to be noticed that sample turbidities in the slow coagulation zone between pH 2.5 and 5.0 are not greatly different from those in the central stability region. Figure 8 .5 also indicates the formation of this new zone which is approximately equal in width to the central stability region.

Assuming that stability occurs because of restabilization and charge reversal, since hydrogen ion coagulation would be expected, the abrupt drop in turbidity at the left extremity (pH 2.5 in Figure 8.6) of this "extended" zone could be caused by coagulation with nitrate ions introduced as nitric acid for pH adjustment. Most likely the extended zone was not observed at higher aluminum sulfate concentrations because of coagulation by sulfate ions. The extended zone is outside the range of aluminum hydroxide formation, whereas the samples in the central stability region are saturated with the precipitate.[6]

Several experimental series were also run varying the aluminum concentration at constant pH, as shown in Figure 8.7 for pH 5.71. Similar data were also obtained at other

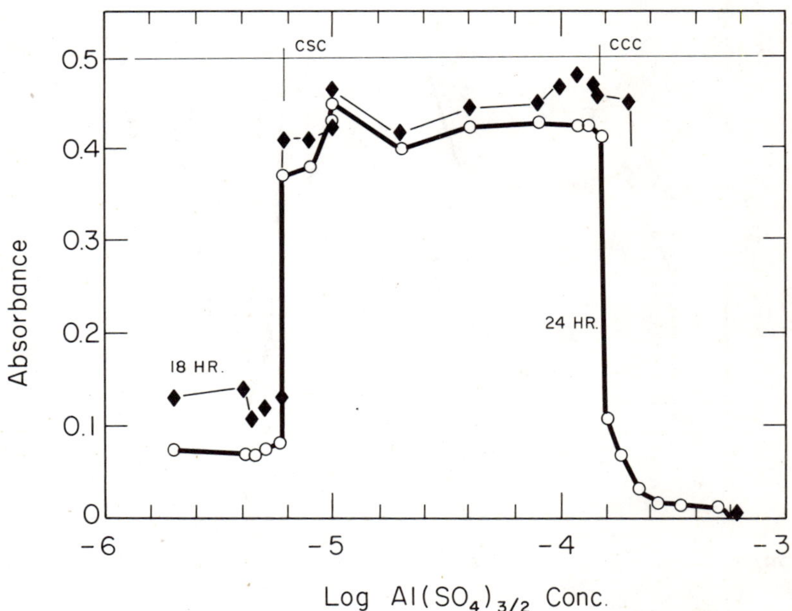

Figure 8.7. Aggregation of titanium dioxide with $Al(SO_4)_{3/2}$ *at pH 5.71. Intermediate sol concentration. Symbols as described in Figure 8.4.*

pH indicating three consecutive boundaries (*i.e.*, the "irregular series" to be mentioned in a subsequent section). Critical coagulation concentration (c.c.c.) and critical stabilization concentration (c.s.c.) values were extrapolated from these curves in the same fashion as described for obtaining the critical pH values. Again, the c.c.c. and c.s.c. at their corresponding pH are points along the boundaries of the stability limit diagram for titanium dioxide.

At pH 5.71 readings at 24 hours were required to obtain the time independent values of -3.82 for the log c.c.c. and -5.20 for the log c.s.c. at the intermediate sol concentration. The lowest log c.c.c. of -6.1 was found with both the high and intermediate sol concentrations.

Similar studies at very low aluminum concentrations, less than 1×10^{-6} *M*, gave results that were essentially a reproduction of the data shown in Figure 8.3. Results at other sol concentrations were almost identical with those just described except for the extent of the central stability region. These will also be discussed in the following sections.

Stability Limit Diagram

The critical pH and concentration values obtained from Figures 8.4 through 8.7 and similar data were plotted log $Al(SO_4)_{3/2}$ molar concentration against pH to establish the stability limit diagram as shown in Figure 8.8. The open symbols are critical data obtained for the intermediate concentration of TiO_2 while the blackened symbols represent the data for the low sol concentration. Circles are critical pH values and diamonds are critical concentration values. The data at the high sol concentration are represented by the dotted line and, except for the restabilization zone, were in perfect agreement with the results at the other concentrations. These data are not shown since they would only serve to clutter the figure. The slopes of the straight line boundaries of the figure and their intercepts were determined by least squares analysis of the data for all three titanium dioxide concentrations.

The stability limit diagram is characterized by four distinct zones. The central area is dominated by the sweep zone where heavy floc formation results in rapid clarification. This zone surrounds the central stability zone and narrows with decreasing concentration reaching a minimum at the c.c.c. of 8×10^{-7} *M* aluminum. This minimum is independent of sol concentration and considerably lower than has previously been observed with other sols. Although the area of the sweep zone to the right and beneath the central stability region is not characterized by extensive

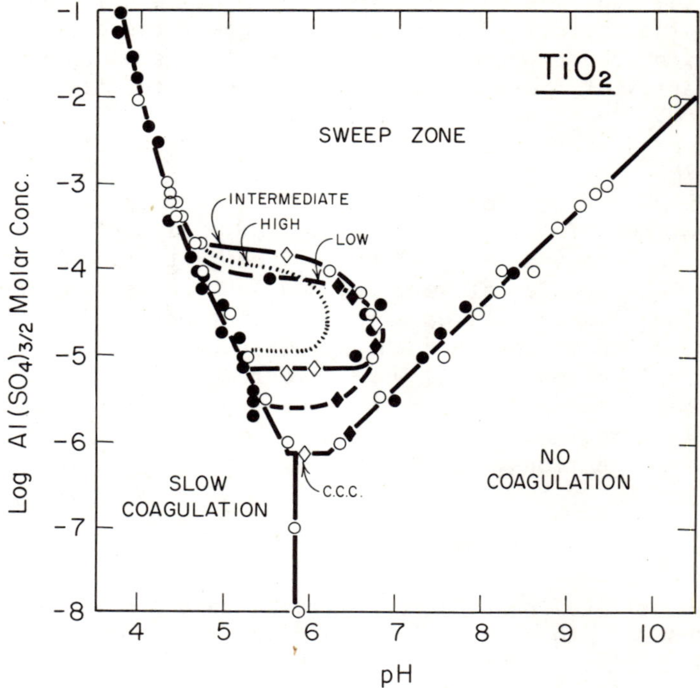

Figure 8.8. *Aluminum sulfate concentration--pH stability limit diagram for titanium dioxide showing the restabilization zone at three sol concentrations ●- - -◆ Low TiO_2 concentration; ●——◇ Intermediate TiO_2 concentration; ••••••• High TiO_2 concentration; ●○ Critical pH data; ◆◇ Critical concentration data.*

floc formation, the times required for clarification are much less than for the slow coagulation zone.

The vertical line below the c.c.c. and the left boundaries of the sweep zone form the upper pH limits of the slow coagulation zone. The lower vertical boundary was obtained by drawing the line through the critical point at 10^{-7} *M* aluminum. This gives a pH_s of about 5.8, which is in good agreement with the pH_s established in Figure 8.3. As discussed earlier, in this zone the principal coagulating species is the hydrogen ion. There was no evidence such as increased settling rates to indicate coagulation by soluble aluminum species in this zone. The extended stabilization zone indicated in Figure 8.6 is not shown because of insufficient data.

It seems fairly well established that the boundary between the sweep zone and the region of stability on the right is determined by the equilibrium between aluminum hydroxide and aluminate ion

$$Al(OH)_3(s) + H_2O \rightleftarrows Al(OH)_4^- + H^+ \quad (1)$$

where

$$K_4 = [H^+]\,[Al(OH)_4^-] \quad (2)$$

Thus, the right stability region is a zone of no coagulation due to the formation of the soluble aluminate ion. Taking negative logs and assuming that equilibrium conditions prevail, that the boundary is the limit such that aluminum hydroxide is present but only in infinitely small amounts, and that the single predominant soluble species is the aluminate ion such that its concentration is equal to the total applied (analytical) concentration of aluminum, C, then

$$\log C = pH - pK_4 \quad (3)$$

Equation 3 predicts that the slope of the boundary should be exactly +1 and its intercept is the negative of pK_4. Literature values of the latter range upwards from 12.2.[19] The experimentally determined values for the slope and intercept were +0.96 and -12.0, respectively. Corresponding values found in previous work were +1.02 and -12.3 with aluminum nitrate[9] and +0.88 and -11.0 with aluminum sulfate.[10] There is evidence for slight sulfate complexing with aluminum hydroxide along this boundary[6] that may account for the slightly decreased values of the slope and intercept.

Several polynuclear hydroxo species of aluminum have been proposed including the $Al_8(OH)_{20}^{4+}$ ion.[20] Hayden and Rubin[6] have demonstrated with aluminum nitrate solutions that $Al_8(OH)_{20}^{4+}$ and $AlOH^{2+}$ are the only two soluble hydrolyzed species in equilibrium with the hydroxide precipitate that are present in significant concentrations at low pH. For the octameric ion this equilibrium is given by

$$Al_8(OH)_{20}^{4+} + 4H_2O \rightleftarrows 8Al(OH)_3(s) + 4H^+ \quad (4)$$

and therefore, as before

$$\log C = -4pH + (pK - \log 8) \quad (5)$$

If the octamer were predominant, Equation 5 would predict a slope of -4 for the boundary on the left between the sweep zone and the slow coagulation zone. On the other hand, for the monohydroxy species, the equilibrium at the boundary is

$$AlOH^{2+} + 2H_2O \rightleftarrows Al(OH)_3(s) + 2H^+ \quad (6)$$

or

$$\log C = -2pH + pK' \quad (7)$$

Thus, if the monohydroxy species were predominant a slope of -2 for the boundary would be predicted.

Experimentally determined values for the slopes of the boundary between the sweep and slow coagulation zones were -3.53 and -1.98, respectively, above and below the applied aluminum concentration of about 10^{-4} *M*. With *E. coli* the values for the upper slope were -3.50 and -3.46 with aluminum nitrate and aluminum sulfate, respectively. The data in those studies were not adequate to obtain the lower slope, although qualitatively it was less negative.

In the presence of polynuclear species the distribution of hydrolyzed ions is dependent upon both the solution pH and the analytical concentration, with the concentration of polymeric ions decreasing with a lowering of the total metal concentration. Thus, the slope of -3.5 indicates that a highly charged, presumably four-plus species is present but is not predominant. In essence the upper boundary is a very flat curve. This slope can be calculated by taking into account the relative concentrations of the metal species.[6] It would be expected, then, that $AlOH^{2+}$ plays an increasingly important role in the formation of precipitate in dilute aluminum solutions. This is evidenced by the gradual transition in slope from -3.53 to -1.98, the latter being very nearly identical with the predicted value, although a higher value might have been expected since Al^{3+} is still an important species over the pH range.

An approximate value for the formation constant for the octamer may be obtained by the following approach

$$8\ Al^{3+} + 20\ H_2O \rightleftarrows Al_8(OH)_{20}{}^{4+} + 20\ H^+ \quad (8)$$

where

$$\beta_{8,20} = [H^+]^{20}[Al_8(OH)_{20}{}^{4+}]/[Al^{3+}]^8 \quad (9)$$

and in the presence of precipitate

$$K_{so}/K_w{}^3 = [Al^{3+}]/[H^+]^3 \quad (10)$$

where K_{so} is the solubility product for aluminum hydroxide and K_w is the ion product of water. Taking the negative logarithms of Equations 9 and 10 and substituting Equation 5 (assuming a slope of -4)

$$p\beta_{8,20} \equiv 24pK_w - 8pK_{so} - pK \qquad (11)$$

The intercept of the upper boundary between the sweep and slow coagulation zones had a value of +12.1; therefore pK = 12.1 + log 8 = 13, and $p\beta_{8,20}$ would have a value of about 67. This is in very good agreement with the value of 68.7 determined from potentiometric data at an ionic strength of 0.15.[6]

The usual assumption is that the lower boundary of the sweep zone is determined by coagulation with $Al_8(OH)_{20}^{4+}$. Further, it is assumed that the octamer is the predominant aluminum species at the concentration and pH of the c.c.c. Based on calculations with hydrolysis constants and the arguments above these assumptions would seem to be considerably in error. However, if the octamer were the predominant species at the c.c.c., its concentration would be one-eighth of 8×10^{-7} total applied aluminum or approximately 1×10^{-7} *M*.

In addition to determining the critical coagulation concentration for aluminum, corresponding values were determined for the nitrate salts of sodium and calcium. The c.c.c. values were 1.05×10^{-2} *M* and 6.8×10^{-4} *M*, respectively, for the univalent and divalent cations. The value for calcium was essentially independent of pH in the range 6 to 10. The c.c.c. for sodium was determined at pH 6.59.

As demonstrated by Težak and coworkers[17] and applied to bacteria in previous work, a plot of the logarithm of the c.c.c. against the counterion charge yields a straight line. Figure 8.9 shows such a plot of the c.c.c. data for TiO_2 along with similar results for *E. coli*[9,11] and AgBr sol[21] demonstrating the applicability of the Schulze-Hardy rule to colloidal titanium dioxide. The linearity of the plot suggests that these three ions destabilize the sol by "simple coagulation." That is, at the c.c.c. they coagulate without adsorption or other than electrostatic interaction with the sol. As stated before, comparison with the pH_s for hydrogen ion (c.c.c. about 10^{-6} *M*) suggests that H^+ strongly interacts with the sol surface leading to "adsorptive coagulation."

The apparent contradiction in results, $Al_8(OH)_{20}^{4+}$ completely negligible in one case and predominant in the other,

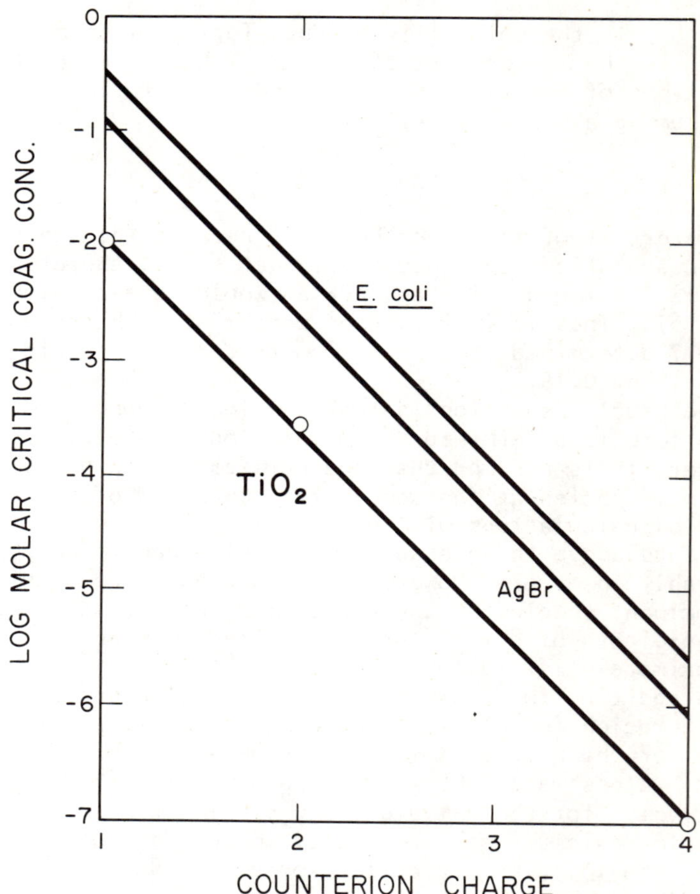

Figure 8.9. Težak plot of the Schulze-Hardy rule showing the relationship between log c.c.c. and the charge of the coagulating cation.

is indicative of the state of affairs in explaining coagulation data. Further studies in the 10^{-4}—10^{-6} *M* aluminum concentration region are needed. Unfortunately, analytical and experimental techniques remain to be developed.

Restabilization

It has been postulated that hydrolyzed species, in particular the octameric $Al_8(OH)_{20}^{4+}$ in aluminum systems, is responsible for restabilization of sols.[7,8] Supposedly the octamer exists in high enough concentrations to adsorb onto the sol surface to such an extent that charge reversal occurs, resulting in a positively charged sol. Isotherms for the adsorption of hydrolyzed iron(III) on silica sols have been obtained by O'Melia and Stumm,[22] and other workers have shown that sols are reversed in charge in the presence of hydrolyzed metal species.[23] Thus, the lower, sol concentration independent c.c.c. and the subsequent c.s.c. and c.c.c. that define the lower and upper boundaries of the central stability region at higher aluminum concentrations give rise to the phenomenon known as the "irregular series." That is, the sol is destabilized by the metal at a given concentration; increasing the concentration leads to charge reversal and restabilization if the metal is adsorbed. A further increase causes destabilization again, being coagulation of the sol by the anion added with the metal salt.

Based on the adsorption mechanism, dependence of the lower limit of the restabilization zone on sol concentration would be expected. An increase in sol concentration increases the total surface area available for adsorption, consequently increasing the amount of metal required to effect charge reversal. Figure 8.8 shows that this boundary, as delineated by c.s.c. data, is in fact dependent on the concentration of the sol. This zone is saturated with aluminum hydroxide, and streaming current measurements indicate that the precipitate is positively charged.

Since the concentration of octamer is greatest at the left boundary of the sweep zone decreasing at higher and lower pH, the right boundary of the restabilization zone would be expected to move toward lower pH with increasing sol concentration. This is evident upon comparing the right limits for the high and low sol concentration. However, aging of the solutions also appears to have an effect since the solubility of aluminum hydroxide and, consequently, the concentration of soluble hydrolyzed species decreases with time. Therefore, as extended periods of time were needed to observe settling with the low sol concentration series, some movement of its boundary might also be anticipated. Apparently it is entirely coincidental that the restabilization zone for the sol at low concentration extended to the same pH as with the intermediate sol series.

At the top boundary of the zone, coagulation of the recharged sol is by sulfate counterions. Although the higher c.c.c. would be expected to be sol concentration independent, some movement of the boundary was observed.

In part, this could be due to aging of the solution during settling resulting in less stable limits (see Figure 8.7), or possibly because of sulfate complexing. The spike in the restabilization region is indicated by the upper limit of the zone near the slow coagulation zone boundary. Since extensive $AlSO_4^+$ ion-pairing is to be expected,[24] species such as $AlOHSO_4(s)$ could account for this spike. The extent of the left boundary has been shown to be greatest at stoichiometric concentrations of aluminum and sulfate.[6,10]

The implications of the restabilization zone to practice are obvious since it is these concentration and pH ranges that are of greatest concern in water treatment. The restabilization phenomena explain the occasional need for "coagulant aids" and other adjustments in treatment. These observations are summarized in Figure 8.10 which shows a rather idealized version of the stability limit diagram.

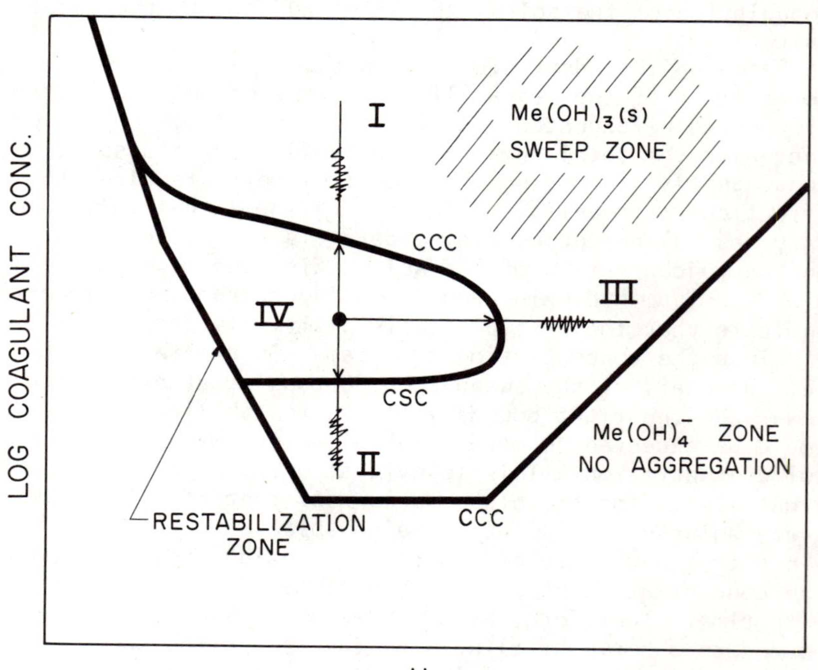

Figure 8.10. How to overcome the problem of restabilization.
I. Increase coagulant dose
II. Decrease coagulant dose
III. Raise pH - alkalinity
IV. Add coagulant aids

To overcome the problem of restabilization some of the alternatives are as follows:

1. Increase the coagulant dose, which has the effect of adding sufficient sulfate to coagulate the recharged (restabilized) sol and gives a more settleable floc. An alternative here is to add sodium sulfate. Probably, however, alum (or ferric sulfate) is as cheap a source of sulfate and the sodium salt is rarely, if ever, used in practice.
2. Decrease the coagulant dose to get below the c.s.c. A less settleable floc is, of course, formed than would be if the dose were increased. Frequently this is done when turbidity is low, and where the water temperature is low so that the increased viscosity interferes with settling, as in New England. Here settling is bypassed and the floc is removed by the filters.
3. Raise pH, add alkalinity, to get to the right of the zone. Very critical adjustment needed as with 2. Depends on the buffer capacity of the water and the presence of phosphates which interfere.
4. Add coagulant aids such as polyelectrolytes to flocculate the recharged sols; silica and bentonite to add surface area to prevent restabilization. All help to produce a more settleable and denser floc.

CONCLUSION

As with other sols the stability of colloidal titanium dioxide is strongly dependent on solution pH and aluminum sulfate concentration. The sol is slowly coagulated by very small concentrations of hydrogen ion. Above pH 6 it is stable in aluminum solutions less concentrated than 8×10^{-7} *M*. This critical coagulation concentration is lower than has been reported for other sols and is, therefore, characteristic of titanium dioxide. The concentration of aluminum required to restabilize was found to increase with increasing sol concentration and, apparently, is also related to some property of the sol.

The left and right stability limits defining the zone of rapid clarification are sol concentration independent and are essentially the same as found with other colloid systems. Indeed, these stability limits are identical with the solubility limits of aluminum hydroxide.[6] The limits of the central restabilization zone fall within the concentration-pH ranges of aluminum hydroxide formation and are adequately accounted for by the adsorption of the octameric ion. There can be no doubt that aggregation in the sweep zone is controlled by the parameters affecting the solubility of the hydroxide precipitate.

ACKNOWLEDGMENTS

T. W. K. was supported in part by a U. S. Environmental Protection Agency graduate traineeship and a research grant from the Office of Water Resources Research.

Streaming current measurements were taken by D. C. Haberkost and P. L. Hayden and the c.c.c. determinations for sodium and calcium ion were made by G. W. Dunkelberger, whose assistance is also gratefully acknowledged.

REFERENCES

1. Rubin, A. J. *J. Amer. Water Works Assoc., 60*, 832 (1968).
2. Milburn, R. M. and W. C. Vosborg. *J. Am. Chem. Soc., 77*, 1352 (1955).
3. Hedström, B. O. A. *Ark. Kemi, 6*, 1 (1953).
4. Stumm, W. and C. R. O'Melia. *J. Amer. Water Works Assoc., 60*, 515 (1968).
5. Matijević, E. and G. E. Janauer. *J. Colloid Interface Sci., 21*, 197 (1966).
6. Hayden, P. L. and A. J. Rubin. Paper submitted for publication.
7. Matijević, E., G. E. Janauer, and M. Kerker. *J. Colloid Sci., 19*, 333 (1964).
8. Matijević, E. and L. J. Stryker. *J. Colloid Interface Sci., 22*, 68 (1968).
9. Rubin, A. J. and G. P. Hanna. *Environ. Sci. Technol., 2*, 358 (1968).
10. Hanna, G. P. and A. J. Rubin. *J. Amer. Water Works Assoc., 62*, 315 (1970).
11. Rubin, A. J., P. L. Hayden, and G. P. Hanna. *Water Research, 3*, 843 (1969).
12. Füredi, H. In *The Formation and Properties of Precipitates*. A. G. Walton, Interscience, New York, 1967.
13. Težak, B. *Discussions Faraday Soc., 42*, 187 (1966).
14. Packham, R. F. *J. Colloid Sci., 20*, 81 (1965).
15. Packham, R. F. *Proc. Soc. Water Treat Exam., 12*, 15 (1963).

16. Cardwell, C. H. *J. Colloid Interface Sci., 22,* 430 (1966).
17. Težak, B., E. Matijević, and K. F. Schultz. *J. Phys. Chem., 59,* 769 (1955).
18. Black, A. P. and S. A. Hannah. *J. Amer. Water Works Assoc., 53,* 438 (1961).
19. Sillén, L. G. and A. E. Martell. *Stability Constants of Metal-Ion Complexes,* Special Publication No. 17, The Chemical Society, London, 1964.
20. Matijević, E., K. G. Mathai, R. H. Ottewill, and M. Kerker. *J. Phys. Chem., 65,* 826 (1961).
21. Matijević, E., D. Broadhurst, and M. Kerker. *J. Phys. Chem., 63,* 1552 (1959).
22. O'Melia, C. R. and W. Stumm. *J. Colloid Interface Sci., 23,* 437 (1967).
23. Matijević, E., S. Kratohvil, and L. J. Stryker. *Discussions Faraday Soc., 42,* 187 (1966).
24. Stryker, L. J. and E. Matijević. *J. Phys. Chem., 73,* 1484 (1969).

CHAPTER 9

MODEL SAND FILTERS FOR THE REMOVAL OF COLLOIDAL MANGANESE OXIDES USING SELECTED CATIONS AS FILTER AIDS

Stephen R. Jenkins, Jogeir Engeset
and Victor R. Hasfurther*

INTRODUCTION

Manganese usually enters a water treatment plant as soluble Mn^{2+}. Treatment for manganese removal entails either an oxidation of the Mn^{2+} to an insoluble manganese oxide or "stabilizing" Mn^{2+} with sodium hexametaphosphate. "Stabilization" is usually ineffective since the Mn^{2+} is not actually removed from the solution and may, therefore, re-enter it as a soluble ion later via equilibrium reactions. Thus, the soluble ion may reach the distribution system where it may be oxidized to an insoluble oxide, precipitate, and cause incrustation on the insides of water mains.

Removal of Mn^{2+} by oxidation with air, oxygen, ozone, or MnO_4^- is a more effective method. If conditions are such that the oxide particles formed are large, they are easily removed by coagulation, sedimentation, and filtration mechanisms. Unfortunately, the particles formed are often colloidal in nature and possess a negative charge. Subsequently, these particles can be difficult to remove.

Many varieties of manganese oxides exist in nature. However, it is felt by Giovanoli and co-workers[1,2] that the most predominant form precipitated on oxidation of Mn^{2+} during treatment of manganese bearing waters is a mixture of $Na_4Mn_{14}O_{27}$ and Mn_7O_{13}. This mixture has previously been identified as δ-MnO_2 and is referred to as such in this text.

---

*S. R. Jenkins, Assistant Professor; V. R. Hasfurther, Assistant Professor; J. Engeset, Graduate Assistant, Department of Civil and Architectural Engineering, Water Resources Research Institute, University of Wyoming, Laramie, Wyoming.

In the present study, this form of manganese oxide was prepared by drop-wise addition of 10^{-2} *M* Mn^{2+} solution prepared from G. Frederick Smith reagent grade $Mn(ClO_4)_2$. Standardization was by the method proposed by Lingane and Karplus[3] to a solution containing a purified laboratory grade (Fisher) $NaMnO_4$ which had been standardized by the Fowler and Bright method.[4] Stoichiometric quantities of MnO_4^- and Mn^{2+} were added to yield a MnO_2 concentration of 4×10^{-3} *M*. It was necessary to add the Mn^{2+} very slowly for if the concentration of Mn^{2+} in solution had become too high the incipiently formed MnO_2 would be aggregated by the Mn^{2+}. The oxide formed had an average oxidation grade of $MnO_{1.98}$ and pH of 6.5 or 8.5. The pH was held constant by addition of 10^{-1} *M* reagent grade NaOH with a Fisher automatic titrator.

Samples of the manganese oxides were filtered and air dried, and spindles were made for X-ray analysis. The d-spacings measured from the X-ray photographs compared favorably with those reported by Bricker[5] for δ-MnO_2.

Surface Characteristics of δ-MnO_2

Stumm, *et al.*[6] have shown that δ-MnO_2, in the same form utilized in this study, behaves like other hydrous metal oxides. That is, it may act as an acid or a base. It is convenient to visualize the surface sites as converted upon hydration to surface hydroxo groups or -Mn-OH groups. These groups can either dissociate to

$$-\text{Mn-OH} + H_2O = -\text{Mn-O}^- + H_3O^+ \tag{1}$$

or accept protons as

$$-\text{Mn-OH} + H_3O^+ = -\text{Mn-OH}_2{}^+ + H_2O \tag{2}$$

where -Mn- stands for a manganese central ion.

The acidity and zero point of charge of the -Mn-OH groups depends on the acidity of the central ion, is influenced by electrostatic field and induction effects of the solid, and is modified by the structural ordering of the water layer immediately adjacent to the solid surface. These -Mn-OH groups act like other polyvalent acids; thus, the acidity constants of these groups can be determined by comparing the titration curve of a suspension of the metal oxide with that of the medium alone. The hydroxide ion consumption by the solid phase is measured as a function of pH and, if the surface areas are known, the surface charge for each pH is calculated. That is, the number of negative or positive surface sites per unit area are determined using Equations 1 and 2 above, following which the acidity constants for these reactions can be determined.[6]

Since the manganese oxide surface is hydrated, calcium ions may also associate with the surface sites. The Ca^{2+} ions can act to displace the bound H^+ ions by

$$-Mn\text{-}OH + Ca^{2+} + H_2O = -Mn\text{-}O\text{-}Ca^+ + H_3O^+ \quad (3)$$

or

$$-Mn\text{-}OH + 1/2Ca^{2+} + H_2O = 1/2(-Mn\text{-}O)_2Ca + H_3O^+ \quad (4)$$

This displacement of bound H^+ ions will produce a shift in the alkalimetric titration curve of the δ-MnO_2 suspension as shown in Figure 9.1.

If a modified Bjerrum plot[6,7] is used stability constants can be determined as shown in Figure 9.2. By this method the ligand number ($\bar{n}$), which is the number of negative surface sites bound to a Ca^{2+}, is plotted against the negative log of the protonated surface sites divided by the hydrogen ion concentration. Figure 9.2 documents that for δ-MnO_2 under conditions of these experiments, only values of $\bar{n} \leq 0.5$ were obtained. This indicates that only monodentate associations ($-Mn\text{-}O\text{-}Ca^+$) were being formed. The stability constant for such associations was found to be

$$K_{Stability} = \frac{[-Mn\text{-}O\text{-}Ca^+]\ [H^+]}{[-Mn\text{-}OH][Ca^{2+}]} = 10^{-5.5} \quad (5)$$

It has been previously shown[8] that the above equation can be used to predict the concentration of calcium necessary to aggregate δ-MnO_2 as well as predict the concentration of Ca^{2+} necessary to cause maximum δ-MnO_2 deposition onto glass surfaces. The Ca^{2+} absorbs on the surface of the δ-MnO_2 to produce $-Mn\text{-}O\text{-}Ca^+$ surface sites. The surface charge is thus reduced until van der Waal's attractive forces overcome electrostatic repulsion and particle aggregation or particle deposition occurs. Because of the specific nature of the chemical interactions between Ca^{2+} and the surface site of the oxide surface, a stoichiometry relationship should exist between the critical coagulation concentration (ccc) of Ca^{2+} and the oxide concentration.

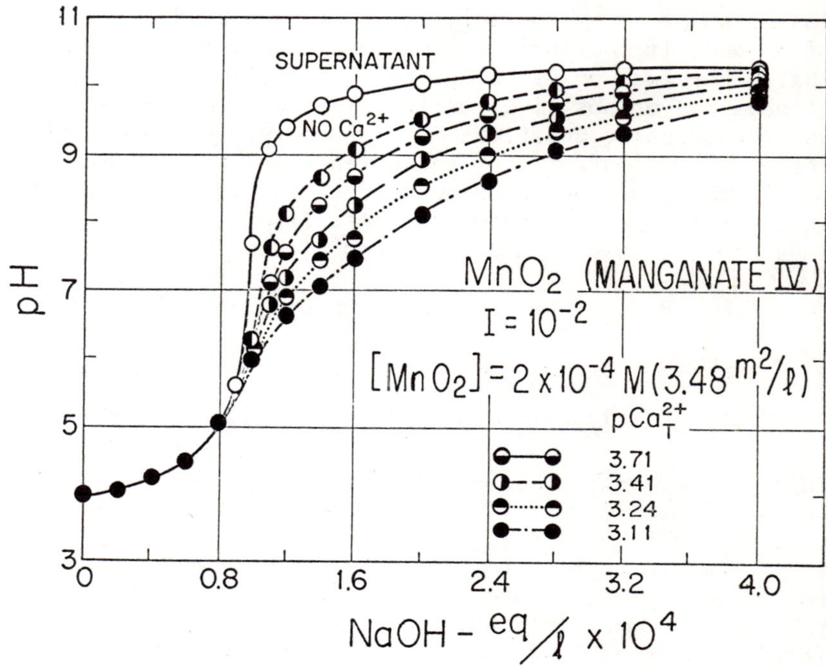

Figure 9.1. Alkalimetric titration curves of suspensions of δ-MnO_2 in presence of various concentrations of $Ca(NO_3)$. The samples originally contain some HNO_3. (Stumm, et al.[6])

In Figure 9.3, a linear relationship is shown to exist at each pH level with respect to ccc of Ca^{2+} and MnO_2 concentration. These curves show little effect of pH on the ccc of Ca^{2+} producing aggregation of a particular molar concentration of MnO_2. It is apparent from these curves, however, that at very low concentrations of MnO_2 the ccc of calcium needed for aggregation decreases with an increase in pH. This is in agreement with the studies of Morgan and Chen[9] and Posselt, *et al.*[10,11] who used MnO_2 concentrations below 0.5×10^{-3} *M*. At higher concentrations the ccc of Ca^{2+} appears to increase as the pH increases.

If it is assumed that some Na^+ also absorbs on the surface of the MnO_2 and that some Mn^{2+} possibly absorbs during oxide formation, the ccc of Ca^{2+} will be less than if Ca^{2+} was solely responsible for charge reduction. That is, at incipient aggregation the number of negative surface sites is not equal to the number of -Mn-O-Ca^+ sites.

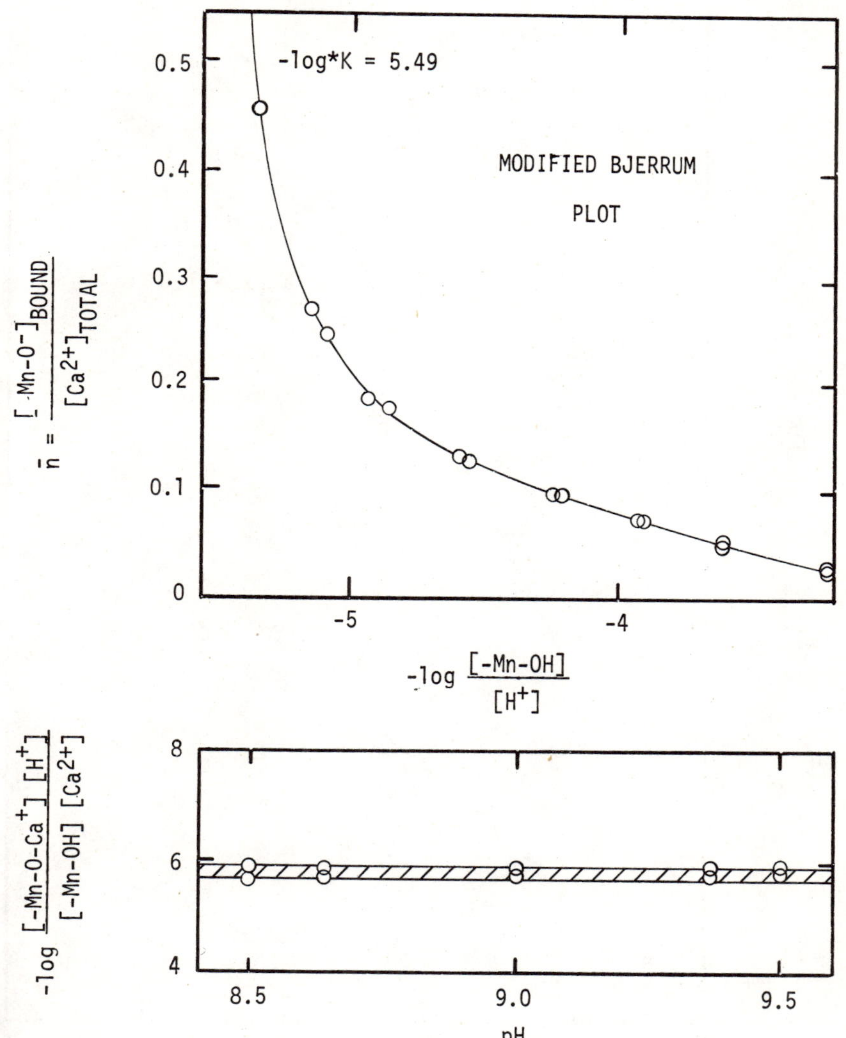

Figure 9.2. *Determination of Ca- complexes of δ-MnO_4. Stability constants are calculated. The shaded area shows limits of values for the stability constant as determined by the Modified Bjerrum Method. (Stumm, et al.*[2]*)*

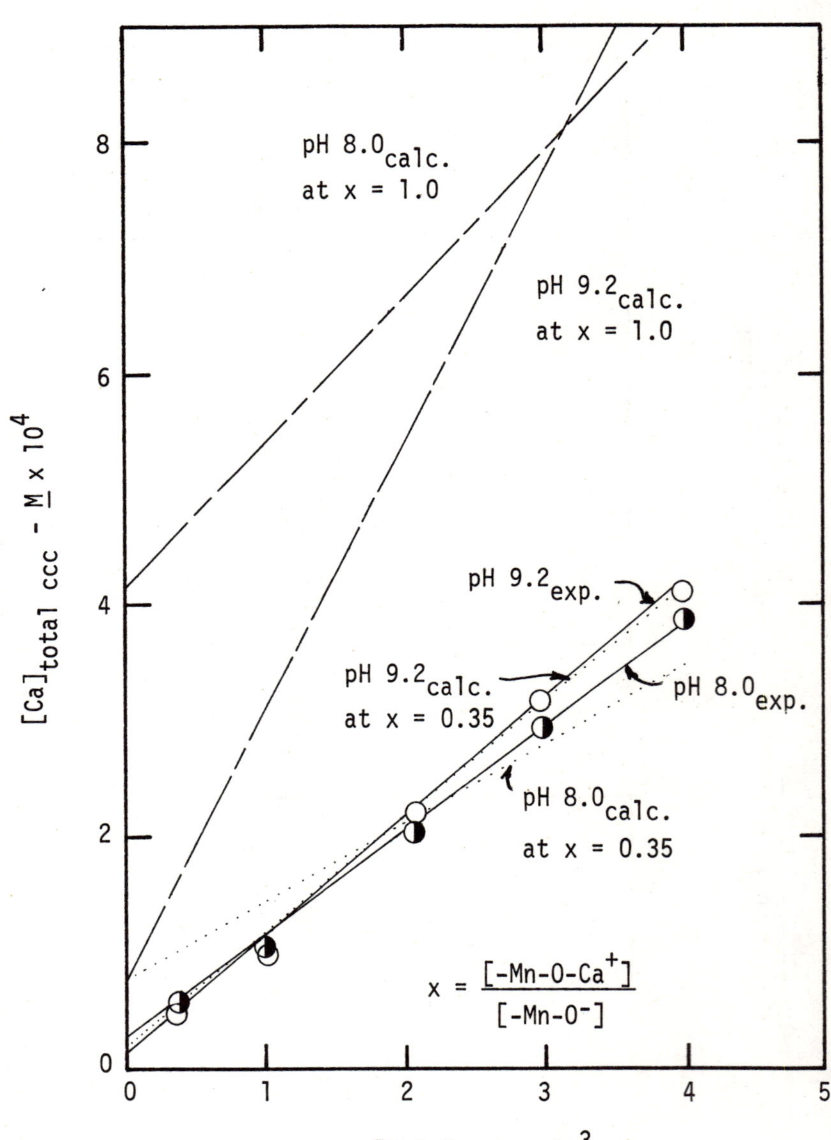

Figure 9.3. Comparison of ccc of calcium as determined experimentally with that calculated if a complex formation is assumed. (Jenkins[8])

Instead, the concentration of $-Mn-O-Ca^+$ sites will be some fraction, x, of the concentration of $-Mn-O^-$ sites

$$[-Mn-O-Ca^+] = x[-Mn-O^-] \tag{6}$$

Using the stability constant (Equation 5) to calculate the $[-Mn-O-Ca^+]$ for maximum aggregation for various amounts of $Ca^{2+}_{TOTAL\ ADDED}$ and comparing these with experimentally determined values, x is found to be about 0.35 (Figure 9.3). The experimental evidence thus confirms that calcium causes aggregation of MnO_2 colloids by reducing the surface charge. Similar evidence has been obtained using a rotating disc apparatus to show that δ-MnO_2 deposition on glass surfaces may be predicted using the stability relations given in Equations 5 and 6.[8]

In the study of Jenkins,[8] the rotating disc, similar in design to that developed by Marshall and Kitchner,[12] was utilized to simulate a uniformly accessible sand surface and to produce reproducible modes of particle transport. The rotating disc, which contains a replaceable standard surface (*e.g.*, glass), was submerged into a dispersion of MnO_2 colloids. The quantity of MnO_2 that became attached to the surface within a given time span for different solution conditions and different rotating speeds was determined analytically.

This rotating disc method provided very reproducible results even though application of the fluid dynamic flow characteristics to particle transport was not in accord with the theory proposed by Levich[13] for transport of particles from the bulk solution to the deposition surface. Although the rotating disc provides a uniformly accessible surface for deposition, in instances where the flux is independent of position the tangential velocity of the fluid increases relative to the disc with increasing distance normal to the disc surface. Hydrodynamic forces caused by this velocity gradient act to shear the larger particles from the disc surface. If, after aggregation of a colloid has occurred, one large particle is sheared from the glass surface, a sizable error (15%) in the measured amount of MnO_2 deposited on the surface can occur.

If either calcium or manganese is added to a previous colloidal δ-MnO_2, the amount of cation required to affect maximum adhesion of MnO_2 (*i.e.*, the maximum adhesion concentration, mac) is related to the concentration of MnO_2. As shown in Figure 9.4, at the pH levels indicated there is an exact stoichiometry between the concentration of cation needed for maximum deposition and number concentration of MnO_2 particles.

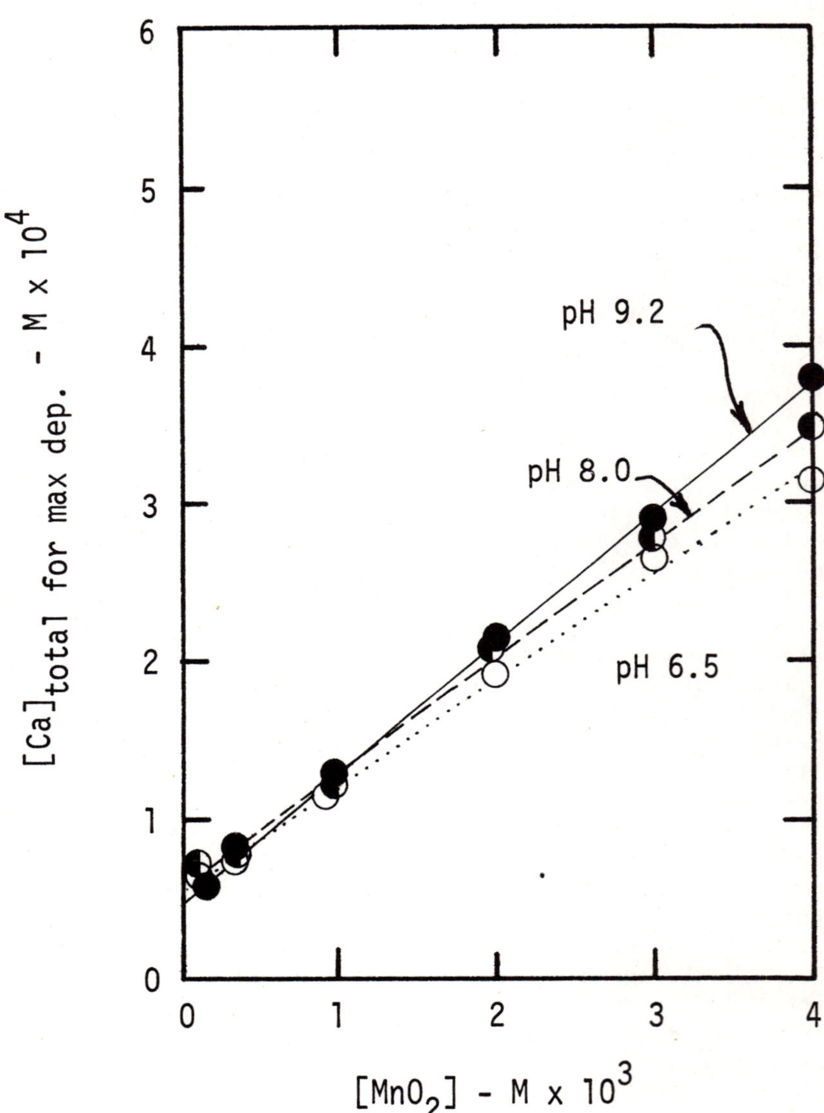

Figure 9.4. Stoichiometric relationship between concentration of cation (Ca^{2+}) necessary for maximum adhesion of MnO_2 (mac) on a rotating glass surface and concentration of MnO_2 if I = 6.5×10^{-3}.

EXPERIMENTAL APPARATUS AND METHODS

Model Sand Filters

In the present study the mathematical-chemical formulation of the association of Ca^{2+} with δ-MnO_2 (Equation 5) is also used to predict the removal of manganese oxide by model sand filters. The modes of particle transport in such packed beds are even less predictable by theory than with the rotating disc. Furthermore, these modes of particle transport are not reproducible. The laboratory simulation of common operating practice was thus undertaken to bridge the gap between the theoretical laboratory measurements and what actually occurs.

The model sand filters (Figure 9.5) used were one-inch plexiglass tubes filled to a depth of four inches with Ottawa Sand. These filters were equipped with manometers which were used to measure the total headloss across each of the filters. The sand was added to the plexiglass tubes and soaked with water overnight before each use. The rate of filtration was set at 2 gpm/ft^2 in accordance with current practice for operation of rapid sand filters. This flow rate was maintained throughout each filter run by adjustment of valve B which is shown in Figure 9.5.

Procedures

The stock 4×10^{-3} M δ-MnO_2 suspensions were prepared well in advance of use. Before each filter run, this stock manganese oxide solution was diluted to the desired concentration. While vigorous stirring conditions were maintained, the ionic strength, Ca^{2+} concentration, and pH were adjusted to the desired levels. Ionic Strength (I) was maintained at the desired level with additions of 10^{-1} M $NaNO_3$. Calcium and pH levels were adjusted by additions of 10^{-2} M $Ca(NO_3)_2$ and 10^{-1} M NaOH, respectively. These solutions were stirred slowly (100 rpm) for 20 minutes. The filter run was then begun. The manganese oxide solutions were passed through the filters at a constant rate of 2 gpm/ft^2 as stated above for 45 minutes. During the filter run the oxide solution was stirred continuously at a low rate to prevent sedimentation of any aggregate particles. The amount of δ-MnO_2 passing the filter was determined as a function of time using the o-tolidine procedure of Morgan and Stumm.[14] After each 45 minute filter run the manganese

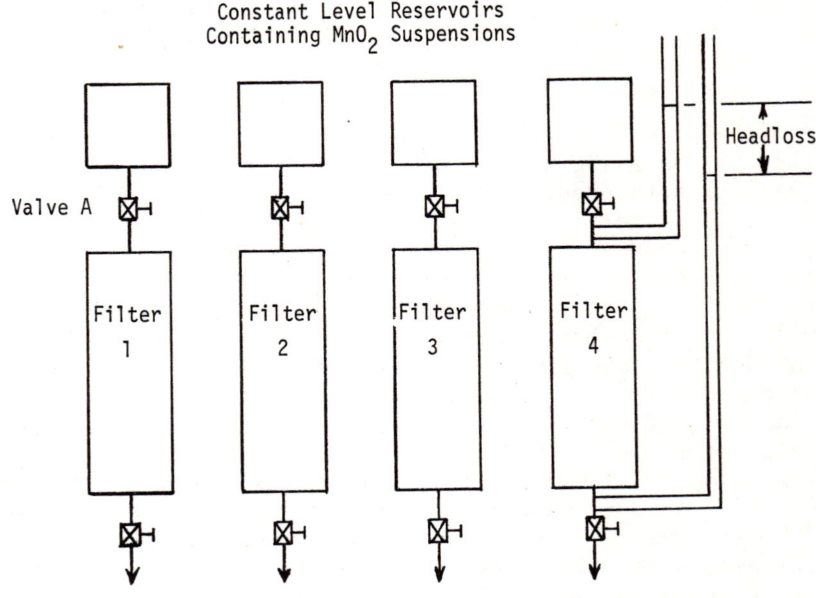

Figure 9.5. Schematic of sand filter apparatus.

oxide solution was removed and distilled water was passed through the filter for 15 minutes. During the filter runs each time a sample of the filter effluent was taken the headloss across the filter was determined. At the termination of each filter run the sand was removed from the plexiglass tube. The amount of δ-MnO_2 adhering to the sand was determined by washing the sand with distilled water and measuring the manganese oxide contained in the wash water with o-tolidine.

RESULTS

The per cent amount of δ-MnO_2 remaining in the effluent can be calculated from the o-tolidine measurement of the concentration of the oxide passing the filter and a knowledge of the influent concentration of δ-MnO_2. A plot of per cent MnO_2 remaining in the effluent versus time can then be made as shown in Figure 9.6.

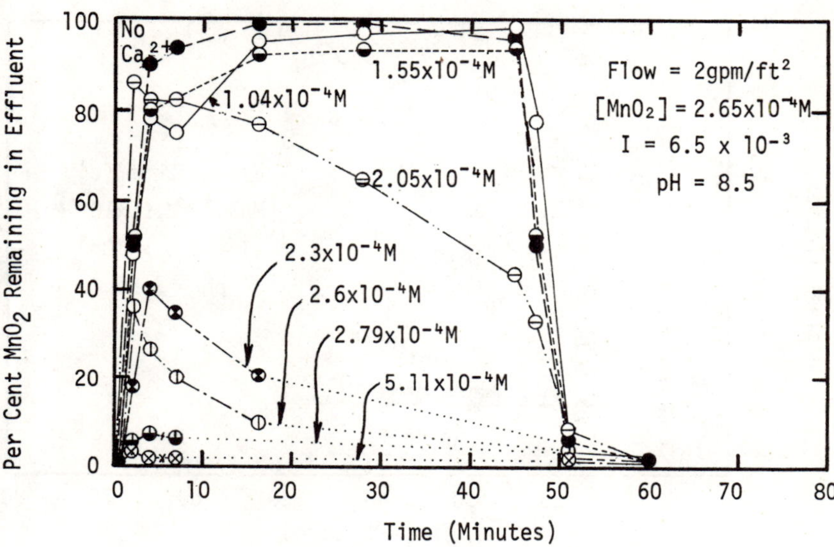

Figure 9.6. The per cent of MnO_2 which remains in the effluent versus time. The filter run was terminated after forty-five minutes and distilled water was passed through the filter. Total calcium concentrations are shown. Similar curves were obtained at pH 6.5 and other MnO_2 concentrations.

For very small doses of Ca^{2+} (1.55×10^{-4} *M*) little appreciable increase in filter effectiveness was noted. However, with subsequent increases in Ca^{2+} a dramatic increase in filter effectiveness was observed until a maximum effectiveness (a minimum per cent MnO_2 remaining in effluent) was reached. The first concentration of Ca^{2+} necessary to produce the maximum effectiveness is termed the maximum effectiveness concentration (mec) of the filter aid. Although concentrations of Ca^{2+} greater than the mec (2.3×10^{-4} *M* for a $[MnO_2] = 2.65 \times 10^{-3}$ *M*) may have increased the removal slightly, the increase was minimal. Also, at these mec of Ca^{2+}, the headloss became so large that it was impossible to maintain a constant flow rate of 2 gpm/ft^2 (Figure 9.7).

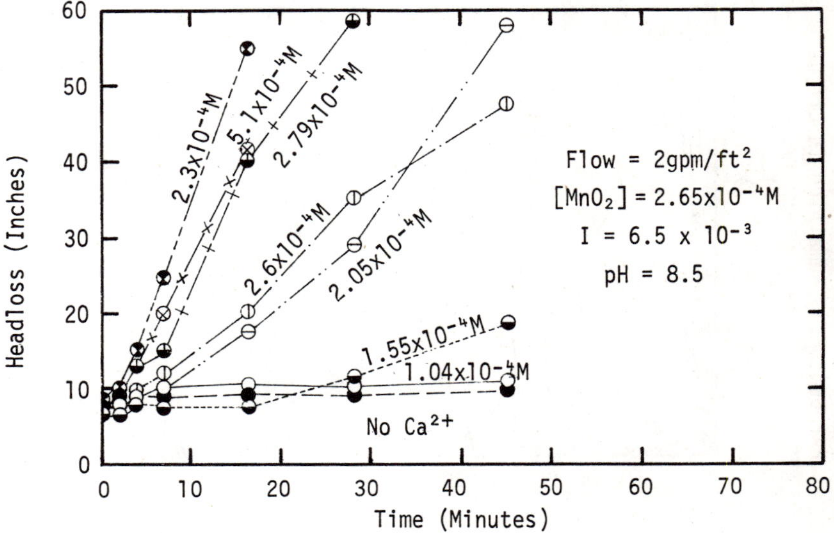

Figure 9.7. Typical data showing the headloss across the filters as a function of time.

An increase in headloss during filtration indicates an effectiveness of the filter for particle removal. Figure 9.7 indicates that with increasing concentrations of Ca^{2+} in the δ-MnO_2 solutions, the headloss increases faster than with lesser concentrations of Ca^{2+}, thus indicating a better removal effectiveness. The minimum concentration of Ca^{2+} necessary to produce the maximum filter effectiveness (mec) from Figure 9.6 is the same $[Ca^{2+}]$ necessary to produce the maximum rate of headloss buildup. For a concentration of δ-MnO_2 of 2.65×10^{-4} *M* at pH 8.5, the mec of Ca^{2+} is 2.3×10^{-4} *M*. The removal effectiveness was so great at this Ca^{2+} concentration that all 56 inches of available headloss was used in 16 minutes and the run had to be terminated prematurely. Similarly, all runs made with a δ-MnO_2 concentration of 2.65×10^{-3} *M* and a calcium concentration greater than 2.3×10^{-4} *M* had to be terminated prematurely because of clogging.

In Figure 9.8, the total amount of MnO_2 adhering to the 4-inch column of sand is plotted versus the concentration of Ca^{2+}. It can be seen from this plot that the effectiveness of the filter for removal of MnO_2 is substantially higher when the concentration of Ca^{2+} is about 2.3×10^{-4} *M*.

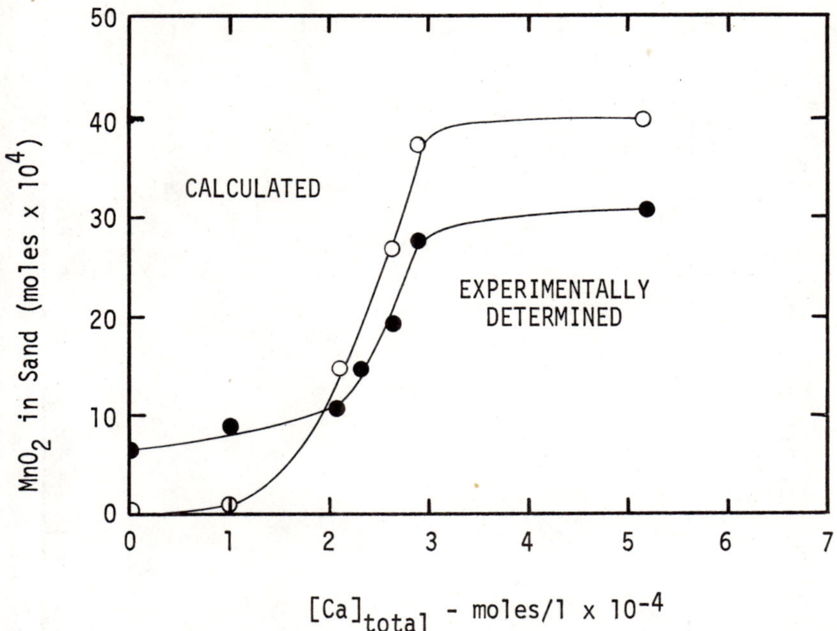

Figure 9.8. The total amount of MnO_2 adhering to the sand versus the total calcium concentration. Data shown is for $[MnO_2] = 2.65 \times 10^{-4}$ M, pH 8.5 and $I = 6.5 \times 10^{-3}$. Typical data.

The blackened circles in Figure 9.8 represent the amount of MnO_2 adhering to the sand as determined by direct measurement with the o-tolidine method. The open circles represent the amount of oxide unaccounted for by taking the difference of the influent and effluent concentrations of MnO_2 over the period of the entire filter run. Little appreciable difference in these results can be seen, indicating that the per cent MnO_2 in the effluent results are reliable. If the mec of Ca^{2+} is plotted versus each MnO_2 concentration, a stoichiometric relationship results (Figure 9.9).

The mec of Ca^{2+} for removal of δ-MnO_2 by these model sand filters are similar to the ccc and mac discussed earlier for coagulation and adherence of δ-MnO_2 particles to silica surfaces. Therefore, the mathematical-chemical model determined for predicting the concentrations of Ca^{2+}

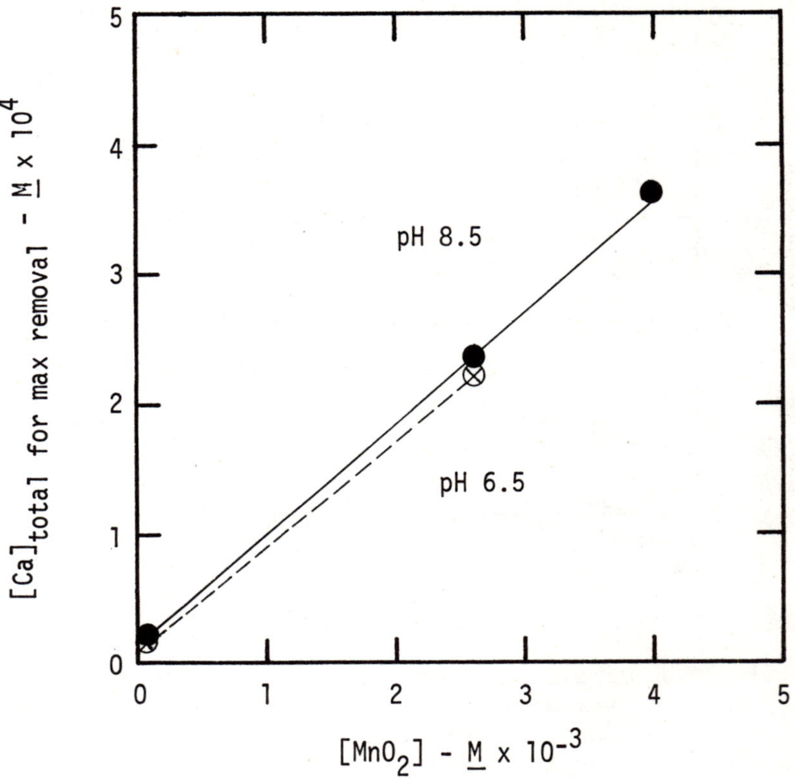

Figure 9.9. Stoichiometric relationship between the concentration of calcium necessary to produce a maximum filter effectiveness (mec) of model sand filters for removal of δ-MnO_2 and the concentration of δ-MnO_2 as a function of pH. $I = 6.5 \times 10^{-3}$ and the flowrate is 2 gpm/ft^2.

necessary to cause removal of each concentration of δ-MnO_2 by coagulation or adhesion can be used to predict the mec of Ca^{2+} by rapid sand filters. That is, using the condition

$$[-Mn-O-Ca^+] = 0.35\ [-Mn-O^-] \tag{7}$$

for each mec of Ca^{2+}, the total amount of Ca^{2+} needed can be determined as described by Stumm, *et al.*[6] and Jenkins.[8]

Effect of pH

The influence of pH on the effectiveness of the filter to remove δ-MnO_2 is minimal (Figure 9.9). This is in agreement with earlier findings.[8] At higher oxide concentrations, the mec dose of Ca^{2+} increases slightly with increasing pH.

Effect of Precoating

Precoating the sand filters with Ca^{2+} has a detrimental effect. If the filter is soaked overnight with a solution containing the same concentration of Ca^{2+} as the MnO_2 solution to be filtered, a decrease in efficiency of the filter to remove δ-MnO_2 results (Figure 9.10). It is believed that Ca^{2+} also reacts with the -Si-OH surface sites of the sand during filtration. The Ca^{2+} can thus act as a bridge between the δ-MnO_2 and the sand surface during filtration. By precoating the filter with Ca^{2+} the surface

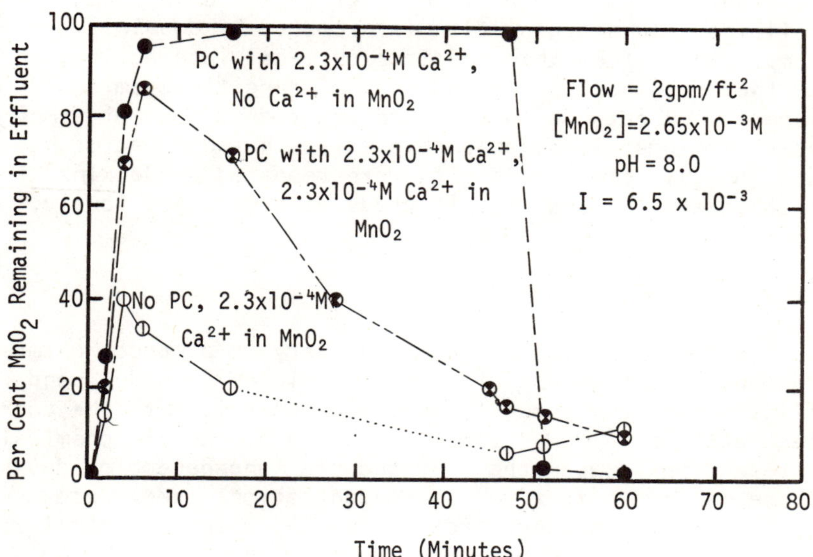

Figure 9.10. The per cent of MnO_2 which remains in the effluent versus time. PC indicates that the sand was soaked overnight with the indicated $[Ca^{2+}]$.

sites available to $-Mn-O-Ca^+$ will be occupied before filtration begins. As δ-MnO_2 is passed through the filter the efficiency of the filter is reduced because there are fewer available -Si-OH sites for the $-Mn-O-Ca^+$ complexes to bridge.

Effect of Filtration Rate

A four-fold increase in flow rate (2 gpm/ft^2 to 8 gpm/ft^2) results in the mec of Ca^{2+} increasing slightly (Figure 9.11). This is in agreement with Yao *et al.*[15] who showed that latex particles were more easily removed by low rates of filtration. That is, the mec of Ca^{2+} should go up as the rate of filtration goes up. This data also supports the theory of O'Melia and Stumm[16] who indicate that both chemical and physical phenomena must be considered when studying filtration.

Effect of Cation Selection

If Na^+ is used instead of Ca^{2+} as the filter aid, the mec (Figure 9.12) of Na^+ agrees well with values reported by others (*e.g.*, Tezak[17]) for aggregation of hydrophobic colloids. This aggregation is promoted by double-layer compaction rather than specific chemical interaction. Furthermore, because of the lack of specific chemical interaction between this cation and the surface of δ-MnO_2, no stoichiometric relationship exists between mec of Na^+ and [MnO_2]. Similar results were reported by Jenkins[8] for the aggregation and deposit study discussed previously.

CONCLUSIONS AND SUMMARY

The concentration of Ca^{2+} necessary to produce maximum removal effectiveness of δ-MnO_2 particles by model sand filters has been shown to be stoichiometric with the concentration of δ-MnO_2, thus indicating a specific chemical interaction between the Ca^{2+} and the manganese dioxide. These maximum effectiveness concentrations (mec) were similar to those concentrations of Ca^{2+} reported earlier as necessary to produce both aggregation of δ-MnO_2 and deposition of this oxide on a rotating disc surface.[8] The mathematical-chemical relationship, derived to predict the most effective concentrations of Ca^{2+} for coagulation

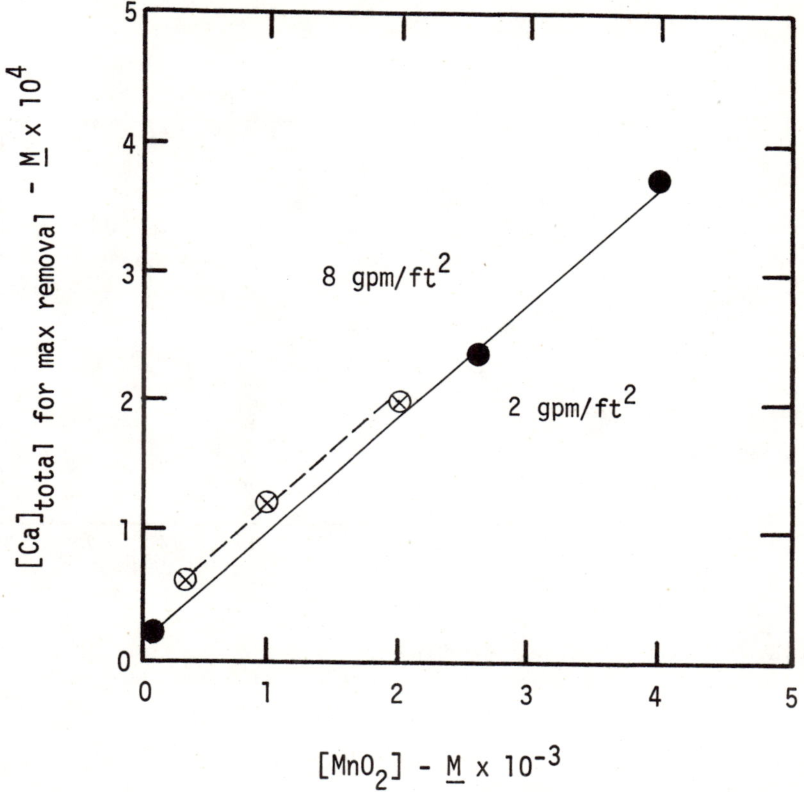

Figure 9.11. Stoichiometric relationship between the concentration of calcium necessary to produce a maximum filter effectiveness (mec) of model sand filters for removal of δ-MnO_2 and the concentration of δ-MnO_2 as a function of flowrate. $I = 6.5 \times 10^{-3}$ and pH 8.5 .

of manganese oxide particles and deposition on silica surfaces, can be used to predict effective concentrations of Ca^{2+} for removal of δ-MnO_2 by model sand filters.

Differences in pH effect the mec of Ca^{2+} only slightly. However, precoating the sand with calcium ion causes considerable change in the effectiveness of Ca^{2+} in aiding δ-MnO_2 removal. It is most likely that the Ca^{2+} reacts with the negative -Si-O^- surface sites reducing the number of sites available for a bonding with the -Mn-O-Ca^+ surface sites.

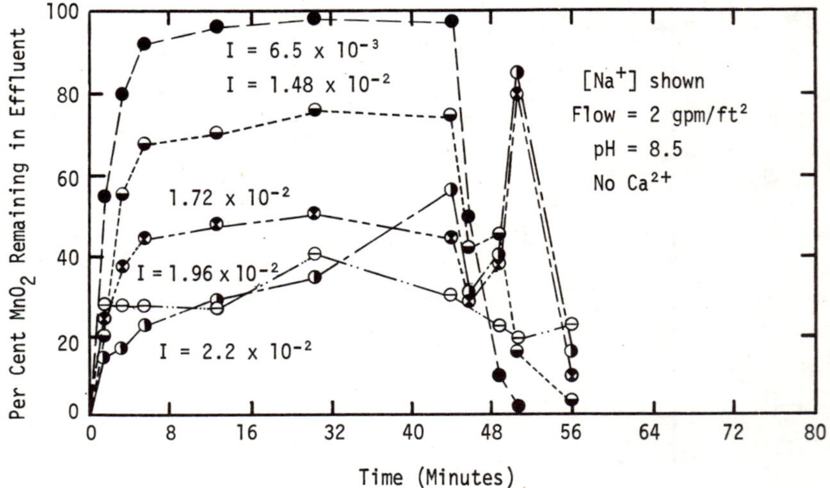

Figure 9.12. The per cent of MnO_2 *which remains in the effluent as a function of time when sodium ion is used as the filter aid.*

If Na^+ is used as a filter aid instead of Ca^{2+}, the mec required indicates that sodium ion compresses the diffuse part of the double layer. No stoichiometry exists between the mec of Na^+ and the δ-MnO_2 concentration. Thus, the destabilization of MnO_2 by Na^+ does not occur by a specific chemical interaction.

ACKNOWLEDGMENT

This research was supported in part by the Office of Water Resources Research, U. S. Department of the Interior, under the Water Resources Research Act of 1964 (PL 88-379) acting through the Wyoming Water Resources Research Institute.

REFERENCES

1. Giovanoli, R. "Strukturchemi und Um Setzungen der Oxide und Oxidhydroxide Des Drei - und Vierwertigen Mangans," Dissertation, University Bern, 1969.
2. Giovanoli, R., E. Stahli, and W. Feitknecht. *Chimia, 23,* 264 (1969).
3. Lingane, J. J., and R. Karuplu. *Eng. Chem., 18,* 191 (1946).
4. Skoog, D. A., and D. A. West. *Fundamentals of Analytical Chemistry.* Holt, Rinehart and Winston, New York, 1963.
5. Bricker, O. *The American Mineralogist, 50,* 1296 (1965).
6. Stumm, W., C. P. Huang, and S. R. Jenkins. *Croat. Chem. Acta, 42,* 223 (1970).
7. Gregor, H. P., L. B. Luttinger, and E. M. Loebl. *J. Phys. Chem., 59,* 34 (1955).
8. Jenkins, S. R., *Environ. Sci. Technol., 7,* 43 (1973).
9. Morgan, J. J., and C. Chen. "The Chemistry of Manganese in Natural Waters," Div. of Water, Air and Waste Chemistry, 155th meeting, ACS, San Francisco, Calif., March-April 1968.
10. Posselt, H. S., A. H. Reidies, and W. J. Weber. *J. Amer. Water Works Assoc., 60,* 48 (1968).
11. Posselt, H. S., F. J. Anderson, and W. J. Weber. *Environ. Sci. Technol., 2,* 1087 (1968).
12. Marshall, J. K., and J. A. Kitchner. *J. Colloid Interface Sci., 22,* 342 (1966).
13. Levich, V. G. *Physiochemical Hydrodynamics.* Prentice-Hall, 1962.
14. Morgan, J. J., and W. Stumm. *J. Amer. Water Works Assoc., 57,* 107 (1965).
15. Yao, K. M., M. T. Habibian, and C. R. O'Melia. *Environ. Sci. Technol., 5,* 1105 (1971).
16. O'Melia, C. R., and W. Stumm. *J. Amer. Water Works Assoc., 59,* 1393 (1967).
17. Težak, B. *J. Phys. Chem., 57,* 301 (1953).

CHAPTER 10

INFLUENCE OF SURFACE OXIDES ON ADSORPTION AND CATALYSIS WITH ACTIVATED CARBON

Chanel Ishizaki and John T. Cookson, Jr.*

INTRODUCTION

The production of synthetic organic compounds has a direct effect on the quality of surface and ground water courses. Many synthetic compounds are biologically resistant and include pesticides, carcinogenic compounds and toxic metal organic complexes. The number of organic compounds that have been synthesized now exceeds well over half a million, and some 10,000 new compounds are added each year.

Since surface and ground water courses are often relied on for municipal water supplies, a potential health hazard could be developing. Further, one must consider the fact that present water purification plants were not designed to remove these organics. As illustrated by several recent studies, this potential hazard is not just speculative. A survey by the United States Bureau of Water Hygiene indicated that pesticide pollution of ground waters is more widespread than commonly realized.[1] Recent identification of organic compounds in the municipal water supply for Evansville, Indiana revealed the presence of toluene, hexachlorobenzene, tetrachloroethylene, bromoform, and bis(2-chloroethyl)ether to mention a few.[2] Some of the compounds are relatively toxic. However, very little is being done to remove or monitor these chemical contaminants in public water supplies.

One of the most promising processes for removal of these organics is carbon adsorption. The ability of

---

*C. Ishizaki and J. T. Cookson, Jr., Department of Civil Engineering, University of Maryland, College Park, Maryland.

activated carbon to adsorb various organics and inorganics from solution is well recognized. However, the mechanism by which these compounds adsorb is not clearly understood. The optimum use of carbon in wastewater treatment and water purification cannot be assured without knowledge of the adsorption mechanism. As this study demonstrates, the nature of the adsorbate and carbon surface chemistry influences adsorption rates. In addition to the efficient use of carbon in treatment processes, it is important to understand adsorption mechanisms for effective regeneration of spent activated carbon. It may be possible to regenerate carbon for efficient adsorption of specific types of organics.

Activated carbon is produced by high temperature treatment of carbons of diverse forms, ranging from well defined crystalline materials to amorphous or microcrystalline materials. The temperature and type of gaseous atmosphere used during the activation process has a significant effect on the surface chemistry of activated carbon. In addition to the activation process, gases will chemisorb on carbon at room temperature, influencing adsorption characteristics.

A great deal of attention has been directed at the carbon-oxygen surface complexes on activated carbon. It was probably in 1863 that the first investigator suggested that oxygen undergoes a chemical change with adsorption on carbon.[3] Since that time, the nature of these surface complexes has been studied, but their influence on adsorption has only been touched on by a few investigators. This study evaluates the influence of surface oxides on activated carbon with respect to the adsorption of several types of organic compounds.

The adsorbates studied include: n-butyl mercaptan, butyl disulfide, decane and p-hydroxybenzaldehyde. These organics have particular significance to water supplies. Mercaptans and disulfides have been found responsible for taste and odor problems in water. The hydrocarbon is representative of nonpolar paraffinic molecules and its adsorption characteristics are important in understanding the influence of sulfur in the adsorption of disulfides. The p-hydroxybenzaldehyde is an aromatic compound which is typical of degradation products of natural color-producing compounds in water.

Oxygen Surface Complexes

When oxygen is adsorbed on a carbon surface, it forms a physicochemical complex of variable composition. The complex decomposes upon heating, giving a carbon monoxide and carbon dioxide mixture.

It is a common practice to classify the surface oxides as basic or acidic. Basic surface oxides are formed when a surface is first freed from all surface oxides by heating in a vacuum or in an inert atmosphere and then allowed to adsorb oxygen at low temperatures, usually room temperature. Acidic surface oxides are formed when carbon is exposed to oxygen at temperatures in the neighborhood of 400° C, or by the action of aqueous oxidizing solutions.[4]

Acidic surface oxides are described as carboxyl, lactone, and phenolic functional groups. Quinones and hydroquinones are also reported to exist on the surface of activated carbon. The nature of the basic surface oxides has not yet been elucidated to an entirely satisfactory degree. Chromene-like, pyrone-like, and isoviolanthrene-like structures have been proposed to account for the basic nature. Detailed discussions on most of these groups have been presented in the literature and will not be repeated here.[5-10]

Upon vacuum outgassing, the surface oxides come off as carbon dioxide, carbon monoxide and water vapor. When outgassed at temperatures below 500° C, carbon dioxide is the main product. Near 500° C carbon monoxide starts to come off, and the ratio of CO_2/CO decreases with increasing temperature.[7,11] It has been suggested that carboxylic groups and their derivatives such as lactones decompose to give carbon dioxide; quinone groups give carbon monoxide; and hydroquinones and phenols give carbon monoxide and water.[4,7]

MATERIALS AND METHODS

Carbons

The activated carbon used in this study was Filtrasorb 200 (Calgon Corporation), a granular carbon produced from coal by high temperature steam activation. Granular carbon particles of size 20/30 (particles that passed a standard U.S. No. 20 sieve with 0.84-mm mesh opening but were retained on a standard U.S. 30 sieve with 0.59-mm mesh opening) were obtained by sieving. The mean particle diameter was 616 microns. Dust was removed by rinsing several times with distilled water. The carbon particles were then oven-dried in thin layers at 105° C for 24 hours and stored in airtight glass jars. The specific surface of these uniform size particles was measured by the BET technique using nitrogen as the adsorbate at liquid nitrogen temperature. A specific surface area of 714 m^2/g was obtained for the commercial activated carbon. Electrophoretic mobility measurements at pH 7 indicated that this carbon has a positive zeta potential.

Standard neutralization techniques were used to determine the distribution of the surface oxides. The neutralization capacities are presented in Table 10.I, and the acidic surface group distribution according to Boehm[6] is given in Table 10.II.

Table 10.I

Acid and Base Neutralization Capacities for Filtrasorb 200

Neutralization by	*Normality*	*Amount Neutralized in meq/100 g*
NaHCO	0.025	2.18
NaHCO	0.25	6.52
Na CO	0.025	3.71
Na CO	0.25	13.00
NaOH	0.025	10.27
NaOH	0.25	37.20
HCl	0.025	29.40
HCl	0.25	45.00

Table 10.II

Distribution of Acidic Surface Groups Obtained From the Neutralization Data

	Amount in % of Total Acidic Groups	
Group	*From 0.025 N Solutions*	*From 0.25 N Solutions*
Carboxylic	21	18
f-Lactones	15	17
Phenolic	64	65

The surface of the commercial activated carbon was modified by chemical and physical treatments, including methylation, oxidation with aqueous solution of ammonium persulfate, vacuum outgassing, and dry oxidation with oxygen at several temperatures.

The carbon was reacted with diazomethane to methylate carboxyl, f-lactone and phenolic groups. The methylation technique has been presented elsewhere.[12]

In order to oxidize its surface, the carbon was allowed to react with a 0.1 *N* aqueous solution of ammonium persulfate for one week under constant agitation at room temperature. Carbon treated in this manner will be labeled CHOX. It is believed that the oxidation with oxidizing solutions leads to the formation of surface complexes which, upon outgassing, come off as carbon dioxide,[7,13] that is, carboxylic groups and their derivatives.

To remove the surface oxides originally present, carbon was outgassed at various temperatures from 200 to 900° C under a 10^{-6} Torr vacuum for 24 hours and then cooled to room temperature in a nitrogen atmosphere. The outgassed carbon will be labeled OG, followed by the outgassing temperature.

The surface of the carbon was also modified by oxidizing with oxygen gas at several temperatures. For this treatment, the carbon was first outgassed at 900° C for 24 hours and then cooled in a nitrogen atmosphere until the temperature desired for oxidation was reached. At that point nitrogen was displaced with oxygen and the oxidation accomplished under constant temperature at an oxygen flow rate of 40 ml/min. After the oxidation process, the carbon was cooled to room temperature under a nitrogen atmosphere. These carbons will be labeled OG 900 OX, followed by the oxidation temperature in degrees centigrade.

Specific surface area measurements by the BET technique were made on carbon outgassed at 900° C and oxidized with oxygen at 300° C. The specific surface area of this carbon was found to be 970 m^2/g, an increase of 36 per cent over the original commercial carbon.

Adsorbates

The adsorbates studied include n-butyl mercaptan, butyl disulfide, decane, and p-hydroxybenzaldehyde. These compounds were chosen for the following reasons: (1) Mercaptans and disulfide have been found responsible for taste and odor problems in water.[14-16] (2) The hydrocarbon was studied to understand the influence of sulfur in the adsorption of the former compounds and as representative of non-polar paraffinic molecules; and, (3) The p-hydroxybenzaldehyde is an aromatic compound which represents degradation products of larger molecules identified as natural color-producing compounds in water.[17]

All of the adsorbates were reagent grade chemicals supplied by Eastman Kodak Organic Chemicals, used without further purification.

Analytical Procedures

Analytical determinations of mercaptan-disulfide mixtures in water were performed with a Perkin Elmer, Model 900, gas chromatograph equipped with a flame ionization detector. The detection conditions were as follows:

Column:	material	-	stainless steel
	length	-	6 feet
	diameter	-	1/8 inch
Packing:	support	-	chromosorb W-AW-DMCS
	mesh size	-	80/100
	stationary phase	-	carbowax 1540
	loading	-	5 per cent
Gases:	carrier	-	helium, 30 ml/min
	flame	-	hydrogen, 50 ml/min, 22 psi
		-	air, 30 psi
Temperatures:	manifold	-	140° C
	injector	-	175° C
	program	-	initial temperature 60° C for 4 min
		-	final temperature 95° C for 16 min
	heating rate	-	24° C/min

The injection volume was 10 µl. Retention times were one minute and twelve seconds for the n-butyl mercaptan and thirteen minutes and sixteen seconds for the butyl disulfide. Concentrations below the µmol/l range could be measured.

The column just described was used for the analysis of butyl disulfide. The only modification was that temperature was maintained at 95° C for isothermal conditions. Under these conditions the retention time for the disulfide was six minutes and fifteen seconds.

Decane was measured with the same column held at 80° C for isothermal conditions. The retention time for the decane was two minutes and eleven seconds.

The analysis of aqueous solutions of p-hydroxybenzaldehyde was performed by ultraviolet spectrophotometry at 285.0 mμ using a Beckman DU-2 spectrophotometer. Detection below the μmol/l range was possible.

Copper and iron identification and determinations were made using a Varian-Techtron atomic absorption spectrophotometer. The instrumental conditions are given in Table 10.III.

Table 10.III

Instrumental Conditions for the Determination of Copper and Iron

Instrument Parameters	*Metals*	
	Cu	*Fe*
Lamp current, mA	3	5
Wavelength, nm	327.4	372.0
Fuel	Acetylene	Acetylene
Support	air	air
Spectral Band Pass, nm	0.1	0.1

Equilibrium and Kinetic Studies

All the adsorption studies were conducted in batch systems. Solutions of carbon and adsorbate under controlled conditions of pH, ionic strength, temperature, and reactant concentration were mixed in glass vials by a wrist-action shaker for a period of 24 hours. For the mercaptan, disulfide and decane experiments "Mininert" teflon sealing valve caps were used. Blanks without carbon were run for reference. Buffer capacity and ionic strength were controlled with a potassium phosphate solution.

Kinetics studies were also performed in batch systems. The samples were shaken at a speed sufficient to eliminate the influence of mixing on the adsorption rate. Adsorbate concentration was monitored as a function of time. Controls without carbon were run in all cases.

EXPERIMENTAL RESULTS

Catalytic Effect of Carbon on Oxidation

During the preliminary adsorption tests for n-butyl mercaptan, a secondary peak was observed in the chromatograms. This compound was identified as butyl disulfide, which resulted from the oxidation of mercaptan to disulfide. The oxidation of the mercaptan by air in the absence of carbon at pH 7 is insignificant.

The adsorption of mercaptan on carbon is complex because its adsorption as well as disulfide on the carbon parallels the conversion of the former, according to the following equation:

$$2\ BuSH + \frac{1}{2}\ O_2 \rightarrow (BuS)_2 + H_2O \tag{1}$$

where BuSH represents n-butyl mercaptan and $(BuS)_2$ is butyl disulfide.

In order to study the influence of the surface oxides on the catalytic effect, carbon was outgassed at 900° C and mixed with mercaptan, maintaining all other experimental conditions constant (Table 10.IV). This table illustrates the influence of activated carbon on the formation of disulfide. The magnitude of the catalytic effect cannot be evaluated without knowledge of the equilibrium relationships between free mercaptan and disulfide and their adsorbed components. To obtain this information, a mass balance is used with equilibrium relationships. A mass balance on BuSH gives

$$[BuSH]_{is} = [BuSH]_{fs} + [BuSH]_{ads} + 2[(BuS)_2]_{prod} \tag{2}$$

The subindices indicate: is = initial in solution, fs = final in solution, ads = adsorbed on the carbon, and prod = produced.

Equation 2 is valid at any time, but, for simplicity, the relative amounts of disulfide produced are studied at equilibrium. The modified Langmuir equations for competitive

Table 10.IV

Catalytic Effect of Activated Carbon on the Oxidation of Mercaptan (pH = 7, Ionic Strength = 10^{-3}, Temperature = 21 ± 1° C)

Test Conditions		*Carbon Concentration*	*Equil. BuSH Concentration*	*Equil. $(BuS)_2$ Concentration*
Carbon	*Oxygen*	*g/ℓ*	*mmol/ℓ*	*mmol/ℓ*
Untreated	Absent	1.03	5.6	0.0
Untreated	Present	1.00	2.1	1.25
OG 900	Present	1.16	4.4	0.42

adsorption presented by Jain and Snoeyink[18] were used in order to calculate the amounts of mercaptan and disulfide adsorbed. The Jain and Snoeyink[18] equations are:

$$q_1 = \frac{(Z_1 - Z_2)K_1 \ Ce_1}{1 + K_1Ce_1} + \frac{Z_2K_1Ce_1}{1 + K_1Ce_1 + K_2Ce_2} \qquad (3)$$

$$q_2 = \frac{Z_2K_2Ce_2}{1 + K_1Ce_1 + K_2Ce_2} \qquad (4)$$

for $Z_1 > Z_2$ where Ce is the equilibrium adsorbate concentration, and Z_1 and Z_2, and K_1 and K_2 are the monolayer coverages and equilibrium constants, respectively, for components 1 and 2 obtained from the Langmuir equation for single solute systems. In order to obtain the single solute parameters for the mercaptan, oxygen was purged from the system with nitrogen. The values of K and Z for the single solute systems are presented in the next section.

Using Equations 2 and 3, the amount of butyl disulfide produced was calculated and compared with the amount obtained from a mass balance on butyl disulfide given by Equation 5.

$$[(BuS)_2]_{prod} = [(BuS_2)]_{fs} + [(BuS)_2]_{ads} \qquad (5)$$

Good agreement was found between the amount of butyl disulfide produced as calculated from Equations 2 and 3 with that calculated from Equations 4 and 5, as shown in Table 10.V. The oxidation of mercaptan to disulfide is

Table 10.V

Butyl Disulfide Production:
Comparison of Values Obtained by Equations 2 and 5

Treatment	*Carbon Conc.* [*g/ℓ*]	*Butyl Disulfide Produced in mmol/ℓ Calculated Value from Equation 2*	*Equation 5*
Untreated	1.0	2.74	2.99
OG 900	1.16	1.16	1.14

catalyzed by activated carbon, and this catalytic effect is significantly reduced by outgassing at elevated temperatures.

Two possible mechanisms can explain the catalytic nature of activated carbon. One mechanism involves the participation of quinone groups. The presence of quinone groups on the surface of activated carbon has been demonstrated by polarographic and infrared techniques,[9,19,20] and such groups are known to catalyze the oxidation of mercaptans to disulfide.[21] In the proposed mechanism, shown in Figure 10.1, the disulfide formation arises via a redox reaction involving the formation of thiyl radicals (RS·) and semiquinone radical anions.

The second mechanism results from metals on the surface of the carbon. Outgassing the carbon at elevated temperatures (900° C) produced deposits at the ends of the heating chamber. This deposit was collected by glass filters, dissolved in a known amount of 0.1 *N* hydrochloric acid, and analyzed by an atomic absorption spectrophotometer. The metals present were identified as copper and iron. The amount of copper released was calculated to be 0.041 mg Cu per gram of carbon. The amount of iron was considerably lower.

It is well known that very small amounts of iron, copper and other metals catalyze the oxidation of thiol compounds to disulfides[22-27] but no definite mechanism has been proposed. In the case of cysteine, it has been shown that the rate of oxidation at pH 7.3 is proportional to the amount of iron present, but for copper there is a limiting concentration after which the rate no longer increases.[24]

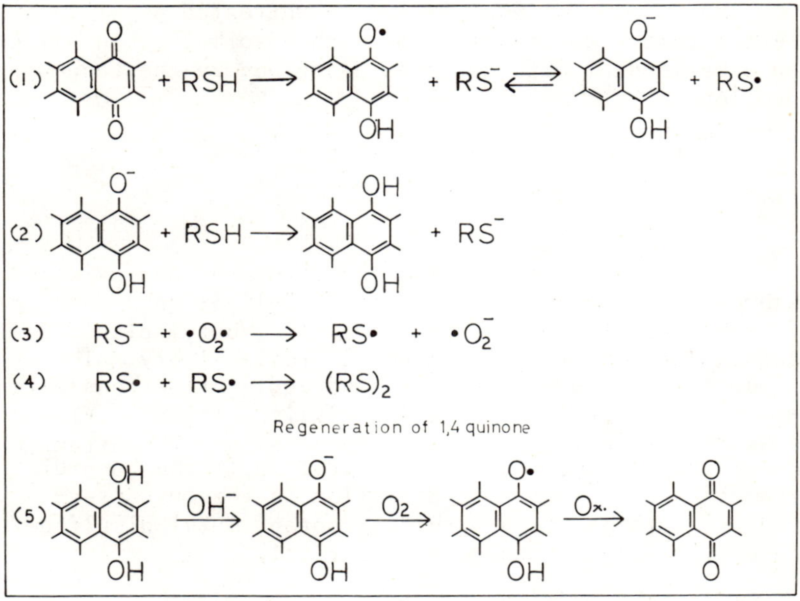

(3) $RS^- + \cdot O_2\cdot \longrightarrow RS\cdot + \cdot O_2^-$

(4) $RS\cdot + RS\cdot \longrightarrow (RS)_2$

Regeneration of 1,4 quinone

Figure 10.1. Proposed catalytic action of activated carbon in the oxidation of mercaptan.

Many authors have concluded that a free radical reaction is involved, as well as oxidation-reduction of the metal and complex formation with the thiol and disulfide. Schubert[25] proposed the following steps: (1) Formation of a complex of the thiol with the metal in the low oxidation state; (2) Oxidation of this complex by means of oxygen, to a complex in which the metal is in a higher oxidation state; and (3) Autoreduction of this complex in which the metal is reduced while the thiol sulfur is oxidized to disulfide with simultaneous breaking up of the complex.

It is not the matter of this paper to go further into the discussion of the actual steps involved in catalytic oxidation of thiols by iron and copper, but rather to point out that the possibility of these metals being responsible for the catalytic activity of activated carbon is likely. Outgassing at 900° C removes both the metals and quinone groups from carbon and correspondingly reduces the catalytic potential.

Adsorption Equilibrium

The adsorption equilibrium on untreated as well as treated carbon was studied for each adsorbate. In all cases the experimental data could be described by the Langmuir equation

$$q = \frac{K\,Z\,C_e}{1 + KC_e} \tag{6}$$

Adsorption isotherms for the outgassed and oxidized carbons are presented in Figures 10.2-10.5, on a "per gram of carbon" basis for n-butyl mercaptan, butyl disulfide, decane, and p-hydroxybenzaldehyde, respectively. The values for monolayer coverages and equilibrium constants were obtained by least square analysis. Table 10.VI shows the values for all the adsorbates on the untreated commercial carbon. In Tables 10.VII-10.IX the Langmuir parameters are given for adsorption on the outgassed-oxidized carbons for n-butyl mercaptan, butyl disulfide and p-hydroxybenzaldehyde, respectively.

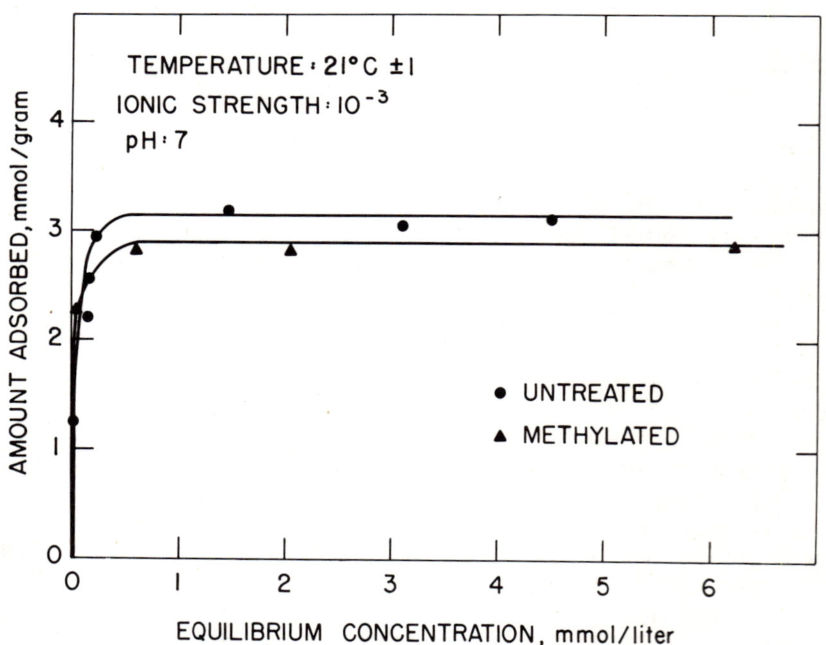

Figure 10.2. Adsorption isotherms for n-butyl mercaptan.

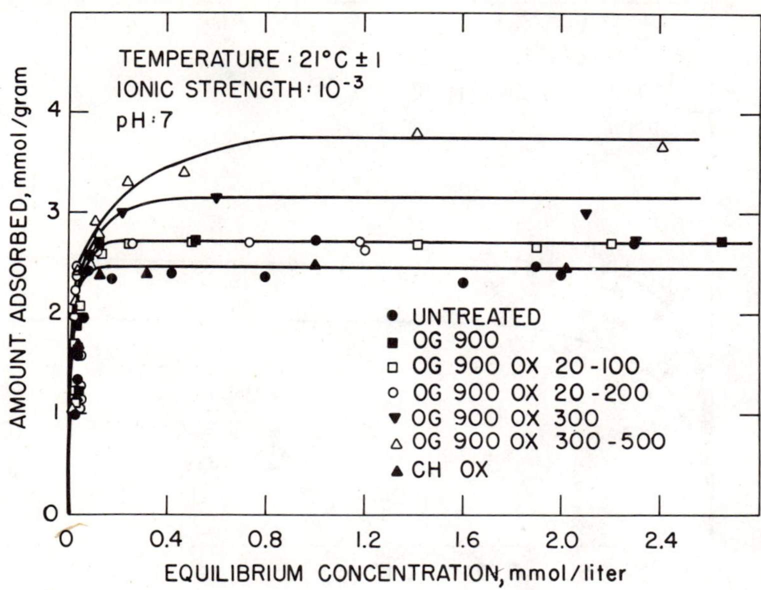

Figure 10.3. Adsorption isotherms for butyl disulfide.

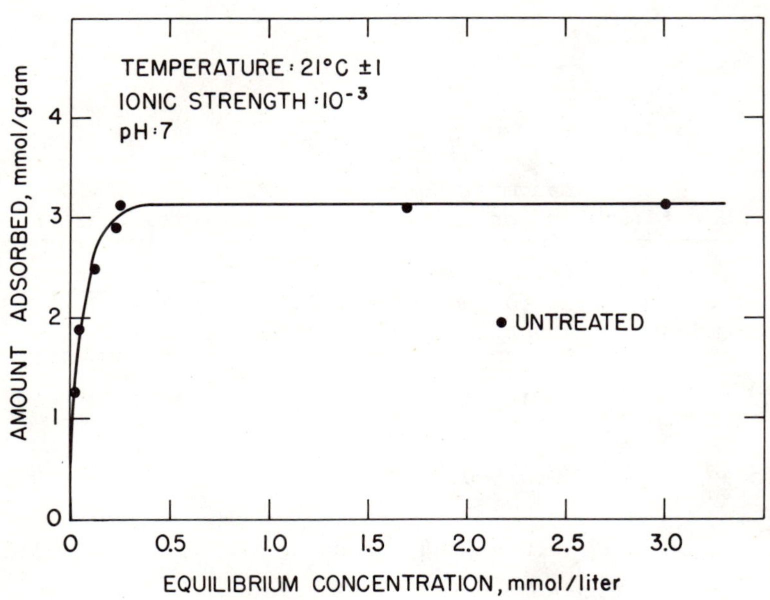

Figure 10.4. Adsorption isotherm for decane.

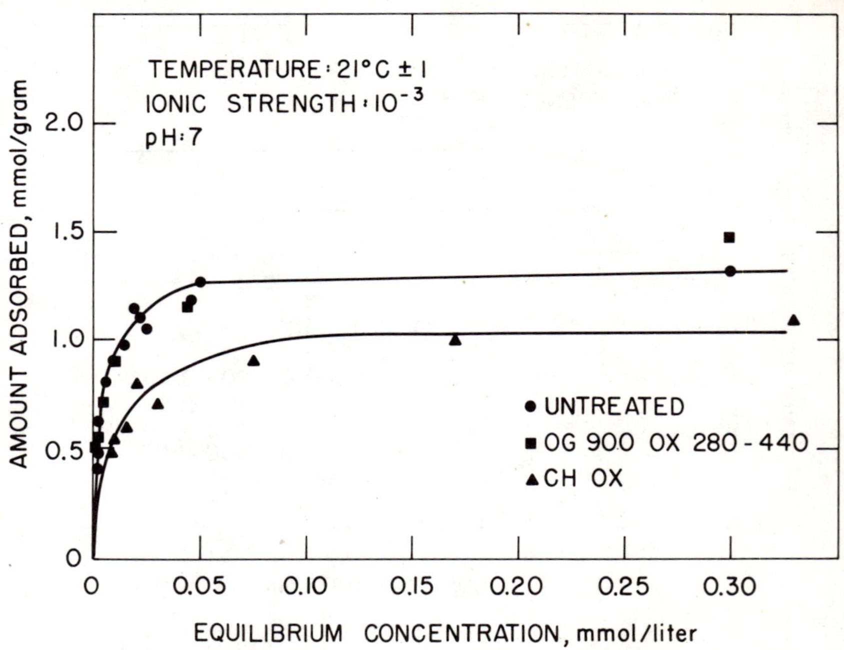

Figure 10.5. Adsorption isotherms for p-hydroxybenzaldehyde.

Table 10.VI

Langmuir Equation Constants for Adsorption on Filtrasorb 200 Activated Carbon

		Langmuir Parameters		
Adsorbate	*Equilibrium Conc. Range, mmol/ℓ*	*K, ℓ/mmol*	*Z, mmol/g*	*Z', μmol/m²*
n-butyl mercaptan	0.156-1.47	18.56	3.31	4.64
Decane	0.025-0.25	25.92	3.53	4.94
Butyl Disulfide	0.008-0.45	86.23	2.40	3.37
p-hydroxy-benzaldehyde	0.0007-0.3	198.38	1.34	1.87

Table 10.VII

Langmuir Parameters for the Adsorption of n-Butyl Mercaptan on Treated and Untreated Carbons

Carbon	*Equilibrium Conc. Range, mmol/ℓ*	*Langmuir Parameters*	
		K, ℓ/mmol	*Z, mmol/g*
Untreated	0.156-1.47	18.56	3.31
Methylated	0.015-6.24	68.215	2.895

Table 10.VIII

Langmuir Parameters for the Adsorption of Butyl Disulfide on Untreated and Treated Carbons

Carbon	*Equilibrium Conc. Range, mmol/ℓ*	*Langmuir Parameters*		
		K, ℓ/mmol	*Z, mmol/g*	*Z', μmol/m²*
Untreated	0.008-0.45	86.23	2.40	3.37
OG 900	0.014-0.52	34.44	2.85	
OG 900 OX 20-100	0.025-0.51	28.35	2.93	
OG 900 OX 20-200	0.0195-0.26	31.08	2.78	
OG 900 OX 300	0.044-2.10	26.77	3.07	3.17
OG 900 OX 300-500	0.026-0.47	13.46	3.64	
CH OX	0.022-0.32	60.57	2.54	

Table 10.IX

Langmuir Parameters for the Adsorption of p-hydroxybenzaldehyd on Untreated and Treated Carbons

Carbon	*Equilibrium Conc. Range, mmol/ℓ*	*Langmuir Parameters* *K, ℓ/mmol*	*Z, mmol/g*	*Z μmol/m²*
Untreated	0.0007-0.3	198.38	1.34	1.87
OG 900 OX 280-440	0.0013-0.3	120.25	1.52	1.64
CH OX	0.008-0.33	74.11	1.11	1.55

The disulfide and the p-hydroxybenzaldehyde were extensively studied to understand the influence of the surface oxides on the adsorptive capacities and adsorption energies. As can be seen in Table 10.VIII and Figure 10.3, upon outgassing of the carbon the capacity for disulfide increased on a mmol per gram basis. However, the outgassing and dry oxidation of carbon causes an increase in surface area, as previously mentioned. Outgassing at 900° C results in an increase of the available surface area over that available in the commercial carbon. Oxidation of the outgassed carbon does not appear to increase surface area at temperatures below 200° C. At higher temperatures, 300-500° C, there is a measurable increase in surface area as determined by N_2 adsorption and application of the BET isotherm.

Although the adsorption capacity for disulfide increased on a gram basis, when corrected for the increase due to greater specific surface area, the difference in capacities becomes insignificant (Table 10.VIII).

Carbon oxidized by chemical treatment had no significant change in adsorption capacity for disulfide on a gram or surface area basis. Chemical oxidation does not increase the BET-(N_2) specific surface area of carbon, but it does increase surface functional groups.[28] Several studies have indicated that chemical oxidation can cause a decrease in BET-(N_2) specific surface area.[28-30] This appears to be especially significant for carbons with an initially high specific surface area. The specific surface area is increased to its original value after outgassing at high temperature.[30] This point is important when considering the influence of chemical oxidized carbons and dry oxidized carbons on their respective adsorption capacities, as will be discussed below.

These findings indicate that the acidic surface oxides are not specific sites for the adsorption of disulfide, and the capacity appears to be a function of the available surface area.

Methylation of surface oxides showed a slight decrease in adsorption capacity for mercaptan (Figure 10.2 and Table 10.VII). This decrease, however, is believed to reflect a reduction in specific surface area as a result of solution-chemical treatment and not methylation of surface oxides. Solution-chemical treatments appear to reduce the BET-(N_2) specific surface area, possibly by blockage of pores. A second point is that this particular commercial carbon has an insignificant number of oxides of the carboxylic and lactone types, and methylation would not be expected to have a significant effect.

Surface oxides do influence the energy of adsorption. The variations in the adsorption energies will be discussed in connection with the rates of adsorption.

The results for p-hydroxybenzaldehyde show a reduction in capacity for carbon oxidized with a solution of ammonium persulfate (Figure 10.5 and Table 10.IX). Carbon outgassed at 900° C and oxidized at 280-440° C increased capacity on a gram basis and decreased capacity on an area basis. Again, it should be noted that chemical oxidation appears to block some surface area, since this treatment gave the lowest coverage reported as $\mu mol/m^2$ (Table 10.IX).

On an area basis, the presence of acidic surface oxides reduced the capacity of carbon for p-hydroxybenzaldehyde. This conclusion agrees with the findings of Coughlin, *et al.*[29,30] for phenol, nitrobenzene, and sodium benzene sulfonate, and Mattson, *et al.*[20,31] for nitrophenol. Using an ammonium persulfate solution to oxidize carbon, Coughlin found a reduction of capacity for phenol, nitrobenzene and sodium benzene sulfonate, concluding that the acidic surface oxides hinder the adsorption of these molecules. On the other hand, Mattson, *et al.* studied the adsorption of nitrophenol on sugar carbons activated at different temperatures, finding a maximum in the adsorption capacity (on a gram basis) at 500° C activation temperature. If the data of Mattson, *et al.*[31] is translated to an "area basis," the maximum is much less pronounced. Puri[7] has also reported the increased preference for benzene adsorption on carbons when outgassed at 750° C. When the carbonyl oxygen was removed by outgassing at 1000° C, the preference for benzene decreased.

Epstein, *et al.*[32] studied the adsorption of p-nitrophenol on isotropic pyrolytic carbon, for which oxidized and reduced forms of the surface functionality may correspond to quinone and hydroquinone groups. Their results indicate that the adsorption of p-nitrophenol is favored on the oxidized surface.

These findings are all consistent since the surface oxides developed by chemical treatment and by dry oxidation at temperatures below 400° C are mainly the carboxylic and lactone types. These groups hinder adsorption of many aromatics.[29,30,32] At temperatures above 400° C, the dry oxidation treatment yields carbonyl groups in the form of quinone and hydroquinone structures. These groups may aid the adsorption of aromatics by involving the formation of an electron donor-acceptor complex of the aromatic ring with the surface carbonyl groups.

Unfortunately, neither Coughlin nor Mattson reported the values of the equilibrium constant, but the data presented in Table 10.IX shows that the lowering in the affinity is more pronounced than the reduction of monolayer coverage, although both follow the same trend.

The reduction in affinity seems to indicate that there is an interaction between the carbonyl-like groups and the aromatic adsorbates. However, it should be noted that those sites are not the only means of adsorption; otherwise, the reduction in the monolayer coverages should be more pronounced.

Rates of Adsorption

The kinetics of adsorption was studied for all the adsorbates on untreated as well as treated carbons. The experimental data followed second-order reversible kinetics, being first-order with respect to both adsorbent and adsorbate concentrations.

The second-order reversible reaction can be represented as

$$a + b \underset{k_2}{\overset{k_1}{\rightleftarrows}} X \qquad (7)$$

with the solution

$$t\, k_1\, m = \ln \left\{ \left[\frac{0.5\ (p+m) - X}{0.5\ (p-m) - X}\right] \left[\frac{p-m}{p+m}\right] \right\} \qquad (8)$$

where $p = a_o + b_o + K^{-1}$ and

$$m = [(a_o - b_o)^2 - K^{-1}(2a_o + 2b_o + K^{-1})]^{1/2},$$

where a_o and b_o are the initial concentrations of adsorbate and sites, respectively.

The carbon concentration is expressed in terms of available sites (mmol/g) as determined from the Langmuir isotherm for each outgassing-oxidation temperature studied. The Langmuir equilibrium constant, K, was likewise determined for the appropriate treated carbon.

The values obtained for the forward, k_1, and reverse, k_2, rate constants for all the adsorbates on the untreated commercial carbon are shown in Table 10.X. The results show that the forward rate constants do not follow the same sequence as the equilibrium constants, nor is the rate of adsorption solely influenced by molecular size or solubility. For example, the diffusion coefficient of butyldisulfide is less than n-butyl mercaptan, but the disulfide adsorbs at ten times the rate of the mercaptan. The same comparison can be made between the disulfide and decane. They have comparable molecular size, but the disulfide adsorbs approximately 20 times faster. It should also be noted that the decane is less soluble than the disulfide which should promote its adsorption. Para-hydroxybenzaldehyde, a more soluble compound than either decane or n-butyl mercaptan, adsorbs at a faster rate than either. As the following illustrates, the rate of adsorption is greatly influenced by the carbon's surface chemistry.

The influence of surface oxides on the rate of adsorption of butyl disulfide was investigated by outgassing and oxidizing the surface of the carbon. The experimental results are shown in Figures 10.6-10.8 and in Table 10.XI. The rates of adsorption on untreated carbon and on carbons outgassed at different temperatures up to 700° C did not decrease upon outgassing, but did decrease when the outgassing temperature was raised to 900° C (Table 10.XI). This reduction in adsorption rate resulted from the evolution of metals from the commercial carbon. The deposition of evolved copper and iron was not observed until temperatures of 900° C. Figure 10.7 shows the combined effect of outgassing and oxidation. The rate of adsorption of butyl disulfide on a carbon that was outgassed at 300° C and oxidized at the same temperature did not decrease. On the other hand, the rate of adsorption was strongly reduced on a carbon that was outgassed at 900° C and then oxidized at

Table 10.X

Rate Constants for Adsorption on Filtrasorb 200 Activated Carbon from Aqueous Solution at 21° C

Adsorbate	Equilibrium Conc. Range mmol/ℓ	Equilibrium Const., K ℓ/mmol	Forward Rate Const., k_1 ℓ/mmol-min $x\ 10^2$	Reverse Rate Const., k_2 l/min $x\ 10^4$
Butyl Disulfide	0.008-0.45	86.23	7.84	9.10
p-hydroxy benzaldehyde	0.0007-0.30	198.38	1.27	0.643
n-butyl mercaptan	0.156-1.47	18.56	0.76	3.78
Decane	0.025-0.25	25.92	0.38	1.47

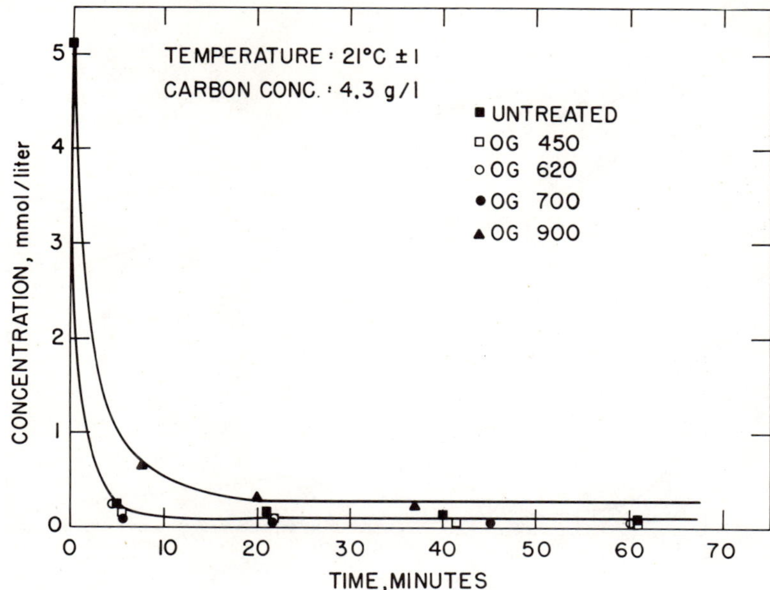

Figure 10.6. Rates of adsorption of butyl disulfide on untreated and outgassed carbons.

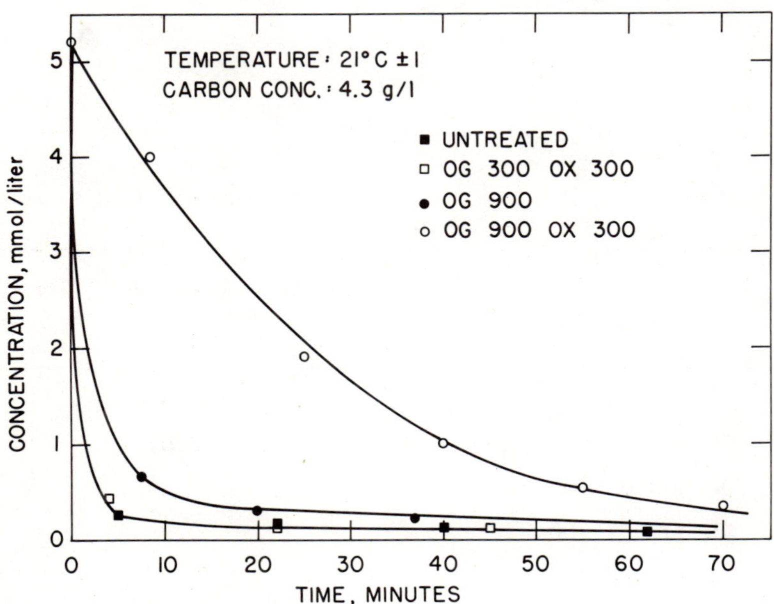

Figure 10.7. Influence of metals on the rate of adsorption of butyl disulfide.

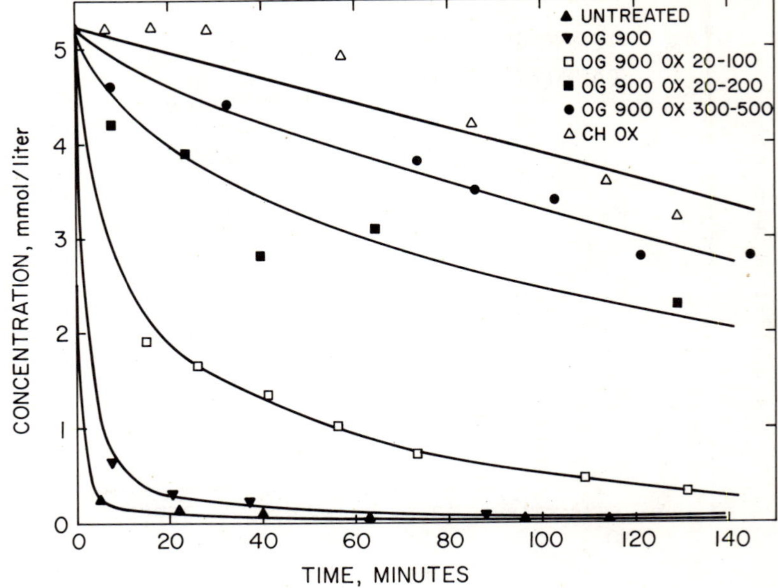

Figure 10.8. Influence of surface oxides on the rate of adsorption of butyl disulfide, temperature 21° C.

Table 10.XI

Influence of Surface Oxides on the Rate of Adsorption of Butyl Disulfide

Carbon	*Forward Rate Constant, k_1 ℓ/mmol-min x 10^3*
Untreated	78.4
OG 900	6.67
OG 900 OX 100-20	2.24
OG 900 OX 200-20	0.94
OG 900 OX 300-500	0.28
CH OX	0.38

300° C. These observations, together with those presented in Figure 10.6, indicate that the presence of metals on the carbon surface have a strong enhancing effect on the adsorption rate, which can even overcome the hindering effect of the acidic surface oxides, at least for this oxidation temperature. The hindering effect of the acidic oxides is revealed by the results presented in Figure 10.8 and Table 10.XI. After outgassing at 900° C the rate of adsorption is further decreased by increasing the oxidation temperature up to 500° C. The forward rate of adsorption is decreased from 7.84×10^{-2} ℓ/mmol-min for untreated carbon to 2.8×10^{-4} ℓ/mmol-min for a carbon outgassed at 900° C and oxidized by pure oxygen in the range 300-500° C. The same effect is observed with a chemically oxidized carbon, which is reported to have the same type of surface oxides as those obtained by oxygen in the same temperature range. The forward adsorption rates for butyl disulfide on these two carbons are 3.8×10^{-4} and 2.8×10^{-4} ℓ/mmol-min, respectively.

The influence of the metals and surface oxides was also investigated for the rate of adsorption of decane. The experimental results are given in Figure 10.9 and Table 10.XII. For decane, the presence of metals on the

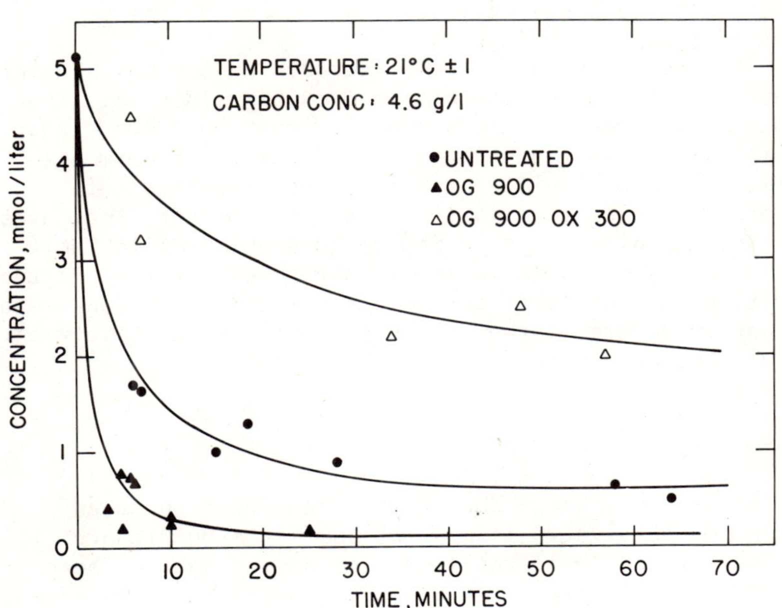

Figure 10.9. Influence of surface oxides on the rate of adsorption of decane.

Table 10.XII

Influence of Surface Oxides on the Rate of Adsorption of Decane

Carbon	*Forward Rate Constant, k_1 ℓ/mmol-min x 10^3*
Untreated	3.81
OG 900	16.5
OG 900 OX 300	1.18

surface does not seem to have a great influence on the rate of adsorption, but the surface oxides appear to have the same hindering effect shown for the disulfide. The rate of adsorption increased when the carbon surface was freed from the surface oxides and decreased when the carbon was oxidized. This behavior would be expected since, in a competition for sites with water molecules, the nonpolar decane will be preferentially adsorbed on a surface free of acidic surface oxides.

Finally, the influence of surface oxides was studied regarding the adsorption of p-hydroxybenzaldehyde. The experimental results are shown in Figure 10.10 and Table 10.XIII. For this adsorbate neither of the treatments seem to have a significant effect upon the rates of adsorption. The equilibrium constant, however, indicates a lower affinity for oxidized carbons as shown in Table 10.IX. This tends to indicate that pore diffusion is the limiting step and that the acidic surface oxides and metals do not interfere as they do in the adsorption of butyl disulfide and decane.

SUMMARY AND CONCLUSIONS

It is well known that carbon can function as an oxidation catalyst. In most cases, the significance of surface oxygen complexes in the oxidation has not been demonstrated. However, the maximum activity is observed in the case of $Fe(CN)_6^{4-}$, NO_2^-, AsO_3^-, S_n^{2+}, and quinol, when carbons contain the maximum amounts of surface oxides.[34]

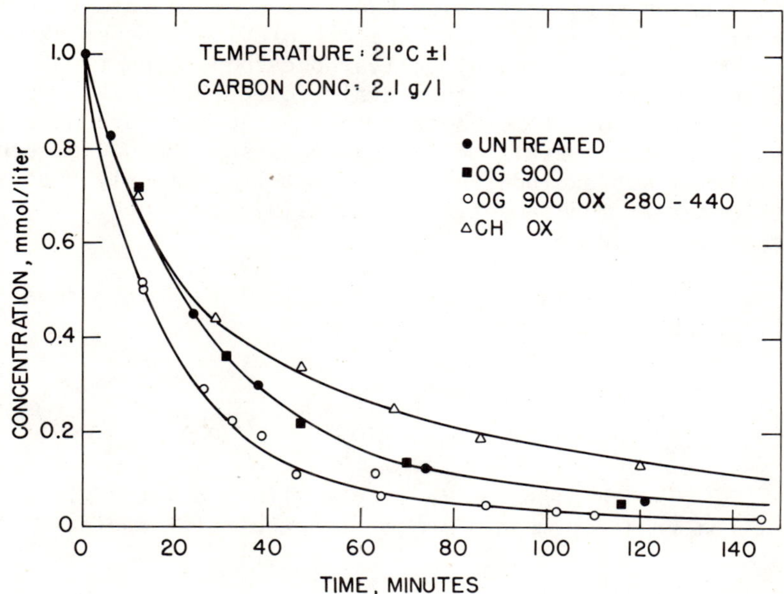

Figure 10.10. Influence of surface oxides on the rate of adsorption of p-hydroxybenzaldehyde.

Table 10.XIII

Influence of Surface Oxides on the Rate of Adsorption of p-hydroxybenzaldehyde

Carbon	*Forward Rate Constant, k_1 ℓ/mmol-min x 10^3*
Untreated	12.7
OG 900	11.6
OG 900 OX 280-440	16.1
CH OX	11.0

Ash, such as metals, is also known to promote catalysis.[34] The studies presented in this paper provide another example of oxidation catalyzed by activated carbon. The two mechanisms proposed for the catalytic activity of carbon in the oxidation of mercaptans involve quinone groups and the presence of metals on the carbon's surface. One mechanism does not exclude the other, but it is likely that both occur at the same time. A possible combined mechanism could be as follows:

1. The thiolate anion, (SH^-), is formed in a reversible step with the quinone groups,

O O O⁻

+ RSH ⟶ + RS^- ⇌ + RS

O OH OH

2. The thiolate anion reacts with the metal via an electron transfer mechanism to give the thiyl radical,

$$RS^- + M^{+n} \rightarrow RS\cdot + M^{+(n-1)}$$

3. Dimerization of the thiyl radicals leads to disulfide,

$$2\ RS\cdot \rightarrow RSSR$$

4. The metal in the low oxidation state reacts with oxygen,

$$2M^{+(n-1)} + O_2 \rightarrow 2M^{+n} + O_2^{-2}$$

5. The peroxide ion reacts with water producing hydroxyl ions,

$$O_2^{-2} + H_2O \rightarrow \frac{1}{2}\ O_2 + 2OH^-$$

The mechanism presented in Figure 10.1 can perfectly parallel this combined process.

The presence of metals on the surface of carbon has a significant effect on the adsorption of butyl disulfide. The sulfur-sulfur bond of the disulfide molecule is a point of localized excess electrons; therefore, the positively charged metal ions on the surface of the carbon will provoke an induced polarization in the disulfide molecule, increasing the attraction energy. This explanation is corroborated by the finding that the adsorption energy for butyl disulfide is considerably higher than that for mercaptan and decane. After removal of the surface metals by outgassing, the adsorption energy for butyl disulfide on carbon compares with those for mercaptan and decane.

The attraction forces exerted by the surface metals also increase the rate of adsorption for the disulfide. As expected in light of the preceding postulation, the adsorption of decane, a molecule of the same characteristics as the disulfide except for the two sulfur atoms, is not influenced by the presence of metal ions on the surface of the carbon.

The experimental results show that, for the adsorbates under consideration, the adsorption capacities on an area basis are almost the same for the commercial untreated carbon and any of the treated carbons. It can be concluded that the amount adsorbed is to a great extent a function of the available surface area. The minor differences encountered, on the order of ten per cent, can be accounted for as follows: The acidic surface oxides will adsorb water molecules preferentially to any of the adsorbates. Upon outgassing, the acidic groups are removed and those sites become available to the adsorbate. When the carbon is oxidized, those groups build up again, reducing the sites available to the adsorbate as a result of competitive adsorption with water.

The most significant effect of the surface oxides on adsorption is found in the adsorption energies and the rates of adsorption. The values for the equilibrium constants, shown in Table 10.VI, indicate that the affinity of p-hydroxybenzaldehyde for carbon is greater than that of the other adsorbates. As mentioned previously, several studies have indicated a strong interaction of aromatic adsorbates involving the formation of an electron donor-acceptor complex of the ring with carbonyl groups on the carbon's surface. This interaction is enhanced by substitutions in the aromatic ring, such as NO_2 groups, or in the case of the p-hydroxybenzaldehyde by the aldehyde group. The presence of acidic surface oxides on the surface of carbon reduces the affinity of aromatic adsorbates. As Table 10.IX shows, the values of the equilibrium constant for the adsorption of p-hydroxybenzaldehyde on a chemically oxidized carbon is reduced almost by a factor of 3 with respect to the commercial-untreated carbon. This effect is not observed on the rate of adsorption. As Table 10.XIII indicates, the rates of adsorption of p-hydroxybenzaldehyde on the commercial-untreated carbon, as well as treated carbons, are of the same magnitude. This would tend to indicate that pore diffusion is the limiting step.

The acidic surface oxides also reduce the adsorption energy for butyl disulfide as seen in Table 10.VIII. It should be pointed out that the chemically oxidized carbon (CHOX) was not previously outgassed, and the presence of some metals may explain the higher K value. Of particular significance is the influence of acidic surface oxides on the adsorption rate of butyl disulfide. The results (Table 10.XI) show a decrease of over three hundred times as compared with the commercial-untreated carbon, or a decrease of 25 times with respect to the outgassed carbon. Although the rates of adsorption changed, the equilibrium capacities are comparable. A possible cause for the reduction could be the high affinity of water molecules for the acidic surface oxides, blocking the pores and therefore hindering the diffusion process of nonpolar molecules. Another possibility is that adsorbed water reduces the attractive forces by hindering the development of induced dipole moments, as discussed above.

As a general conclusion it should be stated that an examination of the adsorption equilibrium alone does not provide enough information on the influence of carbon-surface chemistry on adsorption. In several cases, the equilibrium conditions are only slightly affected by surface oxides, but the rate of adsorption is reduced by more than 2 logs. A magnitude of this size can have appreciable influence on the contact time required for efficient treatment. Thus, the activation and regeneration procedure can have a significant influence on the efficiency of carbon for adsorption.

This study indicates that adsorption of these adsorbates on carbon is non-specific in that the surface coverage on an area basis varied little with surface chemistry. Acidic oxides hinder the equilibrium capacity to a small degree as a result of greater hydrogen bonding of water with the carbon's surface. Of a greater significance is the effect of acidic oxides and surface metals on the rate of adsorption.

NOMENCLATURE

a_o = initial concentration of adsorbate, mmol/l

b_o = initial concentration of sites, mmol/l

C_e = adsorbate concentration in solution at equilibrium, mmol/l

K = equilibrium constant, $(K = \frac{k_1}{k_2})$, $\frac{1}{mmol}$

k_1 = forward rate constant, l/mmol-min

k_2 = reverse rate constant, l/min

m = $[(a_o - b_o)^2 - K^{-1}(2a_o + 2b_o + K^{-1})]^{1/2}$, mmol/l

p = $a_o + b_o + K^{-1}$, mmol/l

q = amount of adsorbate adsorbed per unit weight of carbon at equilibrium, mmol/g

t = time, min

x = concentration of adsorbed adsorbate at time t, mmol/l

Z = monolayer coverage, mmol/g

Z' = monolayer coverage, $\mu mol/m^2$

ACKNOWLEDGMENTS

The authors thank the Water Resources Research Center, University of Maryland, for support under Project No. A-014-Md.

REFERENCES

1. "Report of the Secretary's Commission on Pesticides and Their Relationship to Environmental Health." U.S. Department of Health, Education, and Welfare, December, 1969, p. 117.
2. Kleopfer, R. D. and B. J. Fairless. *Environ. Sci. Technol.*, *6*, 1036 (1972).
3. Smith, R. A. *Proc. Roy. Soc.* (London), *12*, 424 (1863).
4. Boehm, H. P., E. Diehl, W. Heck, and R. Sappok. *Angew. Chem. Intern. Edit.*, *3*, 669 (1964).
5. Snoeyink, V. L. and W. J. Weber, Jr. *Environ. Sci. Technol.*, *1*, 228 (1967).

6. Boehm, H. P. In *Advances in Catalysis, 16*, D. D. Eley, H. Pines, and P. B. Weisz, Eds., Academic Press, New York, 1966, p. 179.
7. Puri, B. R. In *Chemistry and Physics of Carbon, 6*, P. L. Walker, Jr., Ed. Marcel Dekker, New York, 1970, p. 191.
8. Boehm, H. P. and M. Voll. *Carbon, 8,* 227 (1970).
9. Studebaker, M. L. and R. W. Rinehart, Jr., presented at the Tenth Biennial Conference on Carbon, Bethlehem, Pa., June, 1971.
10. Mattson, J. S. and H. B. Mark, Jr. *Activated Carbon,* Marcel Dekker, New York, 1971.
11. Coltharp, M. T. and N. Hackermann. *J. Phys. Chem., 72,* 1171 (1968).
12. Cookson, J. T., Jr., C. Ishizaki, and C. R. Jones. *Water-1971 AIChE Symp. Series No. 124, 68,* 157 (1972).
13. Puri, B. R., S. Singh, and O. P. Mahajan. *J. Indian Chem. Soc., 42,* 427 (1965).
14. Jenkis, D., L. L. Madsker, and J. F. Thomas. *Environ. Sci. Technol., 1,* 731 (1967).
15. Taft, C. E. *Water and Algae—World Problems.* Educational Publishers, New York, 1965.
16. Kawahara, F. K. *Environ. Sci. Technol., 5,* 235 (1971).
17. Christman, R. F. and M. Ghassemi. *J. Amer. Water Works Assoc., 58,* 723 (1966).
18. Jain, S. J. and V. L. Snoeyink. Presented at the 45th Annual Conference of the Water Pollution Control Federation, Atlanta, Georgia, October, 1972.
19. Hallum, J. V. and H. V. Drushel. *J. Phys. Chem., 62,* 110 (1958).
20. Mattson, J. S., L. Lee, H. B. Mark, Jr., and W. J. Weber, Jr. *J. Colloid Interface Sci., 33,* 284 (1970).
21. Oswald, A. A. and T. J. Wallance. In *The Chemistry of Organic Sulfur Compounds, 2,* N. Kharasch and C. Y. Meyers, Eds., Pergamon Press, New York, 1966, p. 205.
22. Michaelis, L. *J. Biol. Chem., 84,* 777 (1929).
23. Cannan, R. K. and G. M. Rieharson. *Biochem. J., 23,* 1242 (1929).
24. Elliott, K. A. C. *Biochem. J., 24,* 310 (1930).
25. Schubert, M. *J. Am. Chem. Soc., 54,* 4077 (1932).
26. Leussing, D. L. and T. N. Tischer. In *Advances in Chemistry Series 37.* American Chemical Society, Washington, D.C., 1963.

27. Stern, E. W. In *Transition Metals in Homogeneous Catalysis*. G. N. Schrauzer, Ed., Marcel Dekker, New York, 1971.
28. Avgul, N. N. and A. V. Kiselev. In *Chemistry and Physics of Carbon, 6,* P. L. Walker, Jr., Ed. Marcel Dekker, NewYork, 1970, p. 1.
29. Coughlin, R. W. and F. S. Ezra. *Environ. Sci. Technol., 2,* 291 (1968).
30. Coughlin, R. W. and T. R. Tan. *Chem. Eng. Progr. Symp. Ser. No. 90, 64,* 207 (1968).
31. Mattson, J. S., H. B. Mark, Jr., M. D. Malbin, W. J. Weber, Jr., and J. C. Crittenden. *J. Colloid Interface Sci., 31,* 116 (1969).
32. Epstein, B. D., E. Dalle-Molle, and J. S. Mattson. *Carbon, 9,* 609 (1971).
33. Graham, D. *J. Phys. Chem., 54,* 896 (1955).
34. Smith, R. N. *Quart. Rev. (London), 13,* 287 (1959).

CHAPTER 11

ACTIVE CARBON: DECHLORINATION AND THE ADSORPTION OF ORGANIC COMPOUNDS

Vernon L. Snoeyink, Hon Tong Lai,
James H. Johnson and J. Francis Young*

INTRODUCTION

A process to remove excess chlorine may be necessary when superchlorination is used to destroy odor- and taste-causing compounds[1] and/or to disinfect, or following prechlorination when it is necessary to prevent chlorine compounds from attacking ion exchange resins.[2] In disinfection of treated wastewaters, dechlorination is also desirable prior to discharge of the effluent to a receiving water, or to land if the effluent is used for irrigation, because of undesirable effects of the chlorine-containing compounds.

One of the available dechlorination processes[3] consists of chemical reduction of the chlorine-containing compounds via reaction with active carbon. While the reaction of carbon with HOCl and OCl^- was the concern of earlier studies,[3,4] some recent studies have also shown the ability of carbon to destroy NH_2Cl and $NHCl_2$.[5-8]

There is also increasing concern over the presence of organic compounds in water supplies.[9] Odor, color, and taste have been the principal water quality parameters for organic matter used in the past by the water treatment industry. However, this listing of parameters is being rapidly expanded. The use of carbon in wastewater treatment

---

*V. L. Snoeyink, H. T. Lai, and J. F. Young, Department of Civil Engineering, University of Illinois at Urbana-Champaign, Urbana, Illinois. J. H. Johnson, Department of Civil Engineering, Howard University, Washington, D.C.

is also becoming more common. The demonstrated ability of carbon to remove organic compounds from water is well known and it is reasonable to question whether active carbon can be used efficiently for both dechlorination and adsorption of organic compounds. This paper is concerned with that question and deals with the effect of the dechlorination reaction on the characteristics of the carbon and on its ability to adsorb organic compounds. The effect of the presence of organic compounds on the ability of carbon to function as a dechlorinating agent is an equally important area of study though it is not considered here.

Chlorine-Active Carbon Reactions

According to Magee[4] and others[3,10] hypochlorous acid will react with active carbon as indicated by the following equation

$$HOCl + C^{*} \longrightarrow H^+ + Cl^- + C^{*}O \qquad (1)$$

active carbon — surface oxide

Active carbon serves as the reducing agent with the net result being that H^+, Cl^-, and surface oxides are produced. In the event that HOCl is produced by adding gaseous chlorine to water

$$Cl_{2(g)} + H_2O \rightleftharpoons HOCl + Cl^- + H^+ \qquad (2)$$

additional H^+ and Cl^- are produced by the hydrolysis reaction. In situations where buffer capacity of the water is low, a pH adjustment step may be necessary to compensate for the H^+ which has been added to the water by the chlorination-dechlorination process. If hypochlorite, OCl^-, reacts instead of HOCl in Equation 1, no additional H^+ is produced. However, a pH decrease may be noted if the oxide formed is a carboxyl group which ionizes.

In a study of the rate at which the reaction given by Equation 1 took place in a packed bed reactor, Magee[4] found two distinct stages. The first stage involved a very rapid rate of HOCl removal which he attributed to a diffusion-controlled step. During this stage a net buildup of Cl-containing species took place on the carbon surface as indicated by the fact that fewer of these species were in the column effluent than in the influent. Continued application of HOCl resulted in development of a second stage slower reaction. During this second stage the HOCl removed was equal to the Cl^- released in accordance with Equation 1. Magee hypothesized that this stage was initiated when the surface of the carbon was nearing saturation with HOCl and the rate limiting step became one of the reduction of HOCl at the surface. The type of active carbon was also found to have a significant effect on the rate of reaction in the second stage.

Magee[4] also observed a significant decrease in the rate of reaction in the second stage as the predominant aqueous chlorine species changed from HOCl to OCl^-. Interpreting the rate of reaction to be first order with respect to HOCl + OCl^- concentration, which he found to be consistent with his data, he observed that the rate constant at pH 8.5-10 was one-fourth the value at pH 5-6. Because the ionization constant for HOCl is 3.7×10^{-8}, [11] OCl^- predominates about pH 7.4 while HOCl is the predominant species below this pH. The mechanism of reaction which would explain this decrease in the rate constant is not apparent.

In a study of the reactions of chloramines with carbon it was found that monochloramine, NH_2Cl, was destroyed by carbon.[6,7] Ammonia was either reformed or oxidized in the reaction depending on the state of the carbon surface. It was hypothesized[6] that dichloramine, NH_2Cl, would react by either Equation 3 or 4

$$NH_2Cl + H_2O + C \longrightarrow NH_3 + Cl^- + H^+ + CO \qquad (3)$$

$$2\ NH_2Cl + CO \longrightarrow N_2 + H_2O + 2\ H^+ + 2\ Cl^- + C \qquad (4)$$

However, the production of surface oxides according to Equation 3, and the reduction of surface oxides according to Equation 4, have yet to be verified by direct analysis. It was also found that carbon readily destroyed $NHCl_2$,[6,7] and that the experimental data which were collected were consistent with Equation 5.[6]

$$2\ NHCl_2 + H_2O + C \longrightarrow N_2 + 4\ H^+ + 4\ Cl^- + CO \qquad (5)$$

Acidic Surface Oxides

The characteristics of the oxides on carbon surfaces following reaction with oxidizing agents, such as HOCl, OCl^-, MnO_4^-, $CrO_4^=$, and $S_2O_8^=$ in aqueous solution, have been the subject of much study (see references 10, 12 and 13 for reviews). A primary characteristic of at least a large fraction of these oxides is that they are acidic in nature. Boehm and co-workers[12] developed a titration procedure which permitted categorization of the oxides into the following four groups:

1. Strongly acidic carboxyl groups which will react with bicarbonate or stronger bases,
2. Weakly acidic carboxyl groups which will react with carbonate or stronger bases but not with bicarbonate,
3. Phenolic-OH groups which will react with hydroxide or stronger bases but not with carbonate, and
4. Carbonyl groups which will react with ethoxide but not with hydroxide.

The surface oxides produced in accordance with Equation 1 are either unstable to a certain extent, or some free CO and CO_2 are produced directly as endproducts. Magee,[4] for example, observed the production of CO and CO_2 as the reaction between carbon and free chlorine proceeded. The production of such endproducts prevents the available carbon surface from becoming saturated with oxides, thus likely resulting in extended life of the dechlorination bed. The fact that some of those groups are unstable, however, is contrary to the work of Puri, *et al.*[14] who found that essentially all of the oxygen from Equation 1 was fixed on the surface in the form of carboxyl groups. The reason for the differences in findings of these investigators is not apparent.

Even though some of the oxides are evolved during the reaction, significant amounts still accumulate on the carbon surface; this accumulation is probably the factor which poisons the carbon surface and results in the eventual passage of aqueous chlorine by the bed. According to Boehm[12] and Puri,[10] heating of the carbon to temperatures in excess of 1000° C in an inert atmosphere is required to completely evolve all fixed oxides from the surface. However, original dechlorination capacity is restored by heating to 500-700° C.[4] The gas given off at these lower temperatures is primarily CO_2, which is considered to be derived from surface carboxyl groups.[10]

The presence of the surface oxide has effects other than those on the dechlorination reaction. The increased oxygen content makes carbon less hydrophobic, and this increased polarity results in selective adsorption of the more polar component of a binary liquid system.[15] Coughlin and co-workers[16-18] demonstrated that extensive oxidation in aqueous systems with excess ammonium persulfate significantly decreased the amount of phenol, nitrobenzene, and benzene-sulfonate which could be adsorbed. The capacity for phenol was decreased by a factor of eight on a weight basis and a factor of four on the basis of BET-nitrogen surface area. The decrease in adsorption capacity could be attributed to both increased surface acidity and the 50 per cent decrease in BET-nitrogen surface area. Possible explanations for the decrease in surface area derived from a study of ozonated carbons are that the micropores are enlarged and are converted into a fewer number of larger diameter pores, and that blockage of micropores occurs as a result of formation of oxides at the pore entrance.[19] Pore blockage would be the most undesirable because of the much greater effect which a limited amount of oxidation could have on adsorption of organic molecules. A more detailed study of surface area and pore size distribution of oxidized carbons is needed before a more definitive statement can be made as to the validity of these hypotheses for carbons which have been oxidized with chlorine.

If carbon is to be used for the dual purpose of dechlorination and removal of organic compounds, there is more which must be learned about the reaction. Specifically, the study reported herein was designed to yield information on the number of surface oxides which result as a function of the extent of treatment with free chlorine, and to relate the extent of buildup of surface oxides to the extent of decrease in ability of the carbon to adsorb model organic compounds.

EXPERIMENTAL

Materials

A commerical 12 x 40 U. S. Standard Mesh size active carbon, Filtrasorb 400 (Calgon Corporation), was used in all studies. The carbon was prepared in the laboratory by grinding to produce smaller particles, sieving to produce a uniform particle size, washing with deionized water to remove fines, and drying at 105-110° C. Carbons with the sieve sizes of 40x50 and 60x80 were used in this study. The manufacturer reports an ash content of 8.5 per cent and a surface area of 1000 to 1200 m^2/g for this carbon. The chlorine solutions were prepared by appropriate dilutions of a household bleach (Clorox) with deionized water. p-Nitrophenol (Matheson, Coleman and Bell White Label Grade) and analytical reagent grade phenol were used as model compounds in the study of the removal of organic compounds from water. These are small aromatic compounds and are representative of a classification of molecules that are troublesome in water treatment.

Procedure

The active carbon-chlorine reactions were conducted by placing 0.2 g carbon, 100 ml deionized water, and a known amount of standard NaOCl solution in a small bottle. These bottles were then placed on a mechanical shaker, covered to avoid exposure to light, and agitated for 18 to 24 hours. At the end of the agitation period, an immeasurable amount of chlorine remained. The initial pH of 9.5-10.5 decreased to approximately 5-7 in the course of the reaction unless otherwise noted.

Chlorine was quantitatively determined by using the DPD Ferrous Titrimetric Method and Cl^- was analyzed using the Argentometric prodedure, both as outlined in Standard Methods.[20]

Following reaction with free chlorine, the carbon was analyzed for its content of surface oxides and its adsorption capacity after (1) no further treatment (non-dried), (2) washing with several volumes of deionized water (washed), (3) drying at 105-110° C (dried), and (4) drying and outgassing for 12 days at 105-110° C (dried and outgassed). It should be noted that the non-dried carbon and, to a lesser extent, washed carbon potentially have oxidation products adsorbed on their surfaces which may affect the outcome of subsequent tests.

Acidic surface oxides were determined by titration with aqueous solutions of HCO_3^-, $CO_3^=$, and OH^-. Ethoxide was not used because of the water molecules which were potentially on the surface after the various carbon treatment procedures. Known quantities of carbon were contacted with these bases by adding a known volume of standard base to a bottle containing the carbon such that the concentration of the base was approximately 5×10^{-3} *M*. Following 24 hours of agitation a known volume of supernatant solution was back-titrated with 0.05 or 0.1 *M* HCl to determine the concentration of remaining base.[3] A different sample bottle was used for each base and for each carbon sample.

Adsorption isotherms for phenol and p-nitrophenol were determined at room temperature (20-22° C) using non-buffered solutions of adsorbate at pH 3.5-3.8. The adsorbates were contacted with the carbon using continuous agitition for approximately three days, following the procedure of Coughlin and Tan.[18] In some instances, standard adsorbate solution was added directly to the bottle containing the carbon and the associated water from which the free chlorine had been removed. In other instances, the carbon was dried at 105-110° C prior to addition of the adsorbate solution.

Phenol and p-nitrophenol were quantitatively determined at wavelengths of 270 and 316 nm, respectively, using the Beckman DU and ACTA III UV-visible spectrophotometers.

RESULTS

Surface Oxide Production

The first series of experiments was designed to show the extent of buildup of acidic surface oxides as a function of the degree of treatment with free chlorine. In these experiments, carbon was reacted with a known volume and concentration of free chlorine; subsequent to this a known volume of standard base was added directly to bottles containing the carbon and dechlorinated solution. On occasion a known volume of dechlorinated solution was decanted from the bottle prior to addition of the base. No significant difference was noted in the amount of base reacted using the two different procedures, however, possibly due to the extensive scatter in the data. The data shown in Figure 11.1 are the results of these experiments. As can be observed, the quantity of titratable substances reached a plateau as the degree of treatment increased. The large amount of scatter in the data, particularly for the NaOH titratable oxides, could possibly be caused by the production of CO_2, some of which escaped from the reaction vessels.

Several samples of solution which remained after the free chlorine had reacted with carbon were analyzed for Cl^-. It was found that within the limits of the Argentometric procedure the free chlorine was converted to an equivalent amount of Cl^- which was then released to the solution.

When more than 15 mmoles/g of free chlorine had reacted, a very distinct brown color developed in the reaction vessel. This substance was not identified or characterized in the course of this study but it was very likely the same brown colloidal matter which, according to Boehm,[12] is an indication that the carbon is completely oxidized.

Because of the way in which the titration was performed for the data shown in Figure 11.1, some of the titratable substances may well have been adsorbed reaction products or acidic reaction products in the dechlorinated solution. In order to evaluate this possibility, the carbon was treated by various procedures prior to titration with the bases. In Figure 11.2, data are presented which show the NaOH and Na_2CO_3 titratable oxides on dried carbon and NaOH titratable oxides on carbon which had been dried and outgassed. The curve for titration of non-dried carbon with NaOH is reproduced from Figure 11.1 for comparison. As can be seen there is very little difference between the NaOH titratable oxides on the dried, and the dried and outgassed carbons, though a significant difference is noted between the curves for non-dried carbon and these two carbons. Washing of non-dried carbon with two to five 200-ml volumes of deionized water followed by titration with NaOH resulted in a curve midway between those for dried and non-dried carbon, also as shown in Figure 11.2. This indicated that many of the acids which were present during the titration of non-dried carbon were either in the dechlorinated solution or were weakly adsorbed on the carbon.

Figure 11.3 shows the NaOH titratable oxides that result as a function of the number of cycles of treatment with 15 mmoles free chlorine per gram of carbon; each cycle included drying at 105-110° C for seven days after treatment.

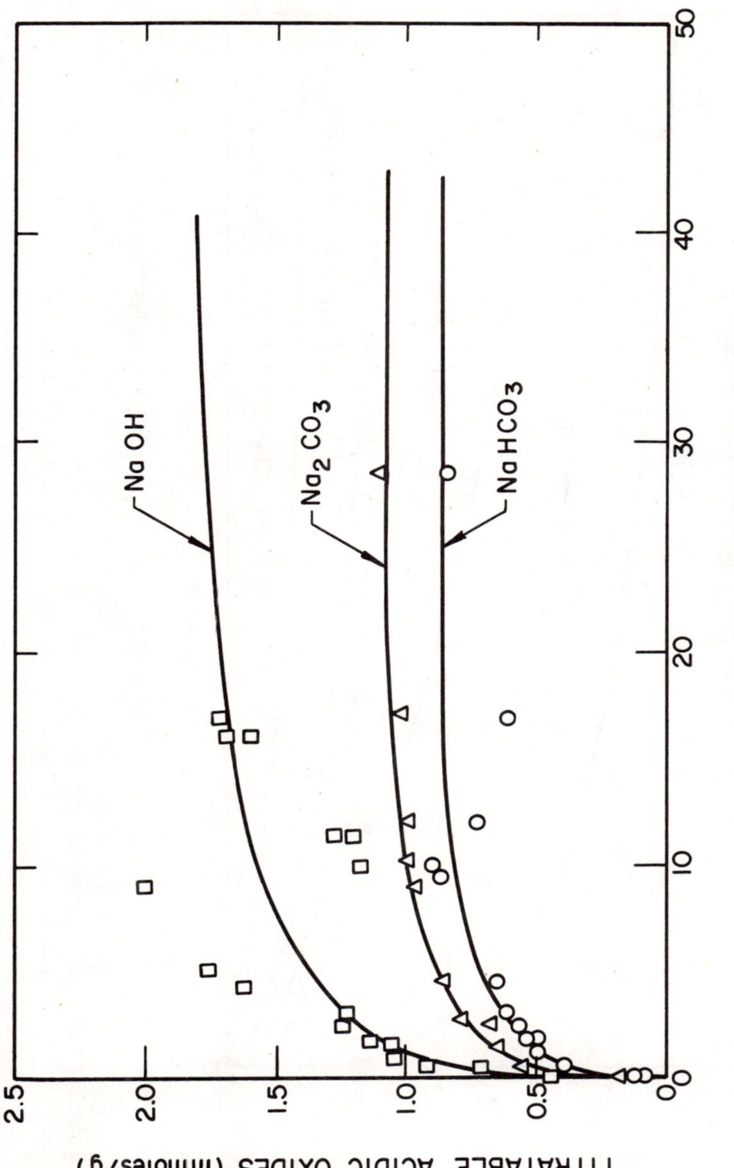

Figure 11.1. Titratable oxides on non-dried carbon after reaction with free chlorine.

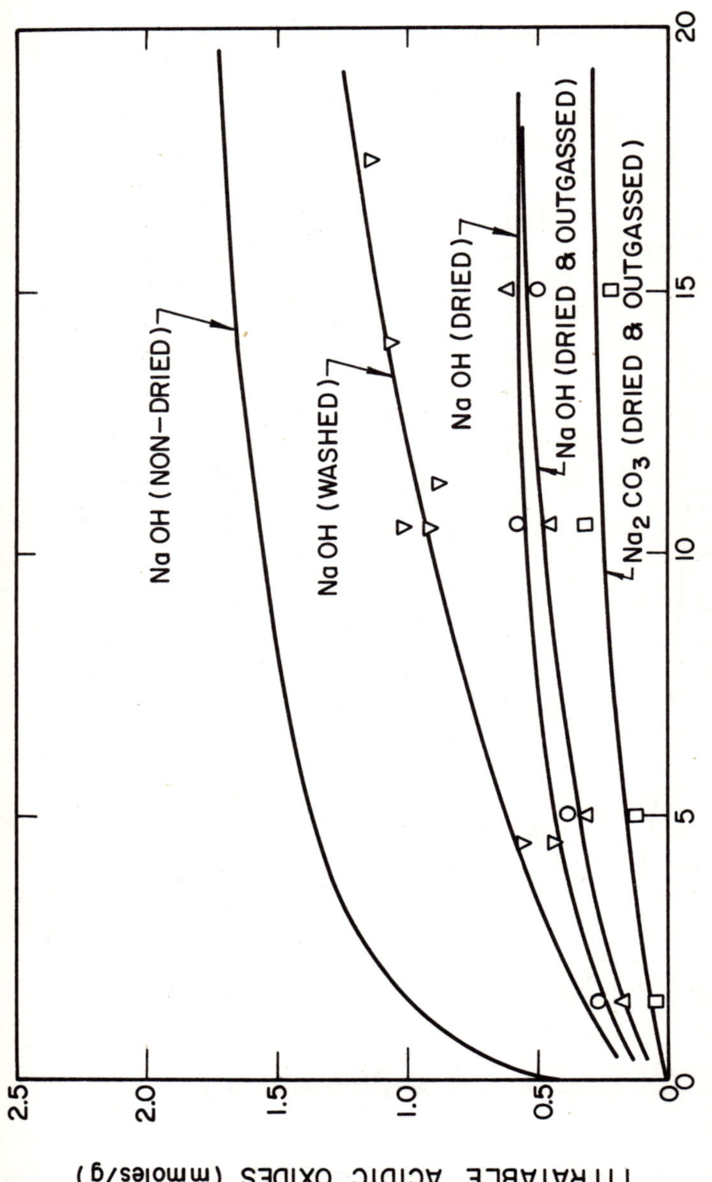

Figure 11.2. Titratable oxides on non-dried, washed, dried, and dried and outgassed carbons after reaction with free chlorine.

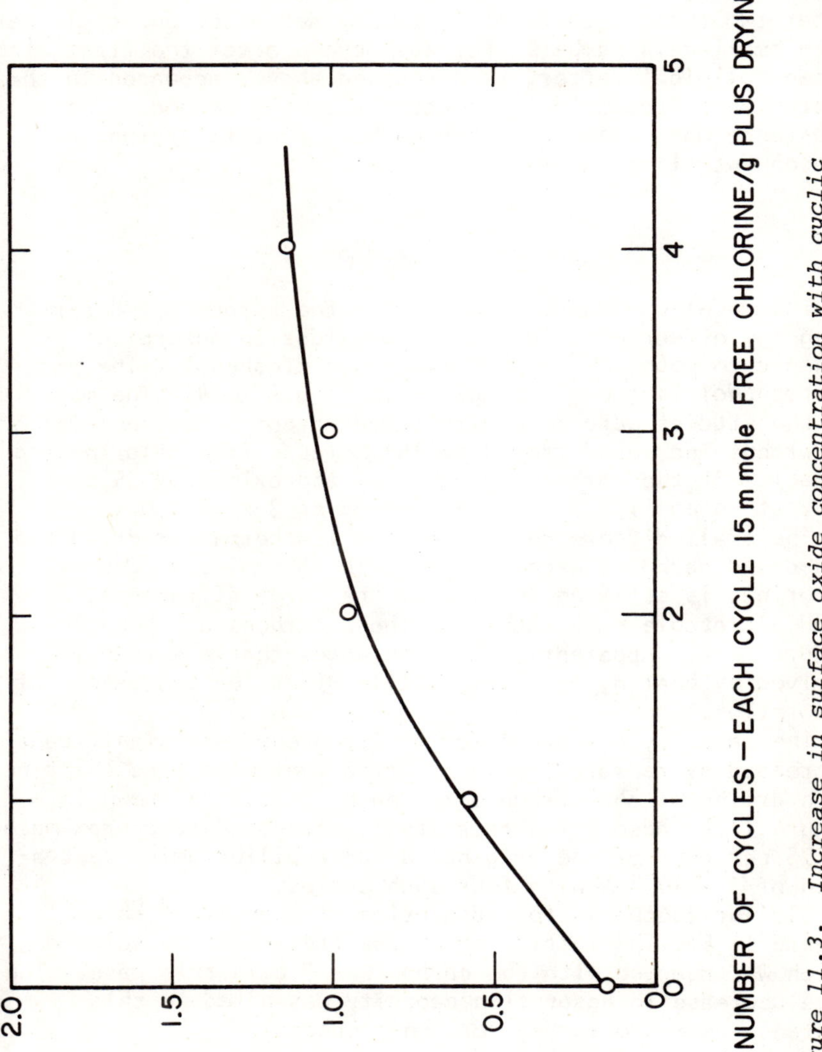

Figure 11.3. Increase in surface oxide concentration with cyclic chlorine treatment.

After each cycle a portion of the carbon was titrated. A sharp increase in oxides was noted after one cycle and again after two cycles; thereafter, the oxides added per cycle were much smaller in number. For each cycle after the first, the brown colloidal matter, as discussed above, appeared in the system when free chlorine reacted with the carbon. This substance was washed from the carbon prior to drying and subsequent titration with NaOH.

Adsorption Studies

The model compounds selected for the purpose of determining the effect of acidic surface oxides on adsorption of organic compounds were phenol and p-nitrophenol. The p-nitrophenol isotherms are given in Figure 11.4. The majority of the studies were done with dried carbon. As the level of treatment increased from 0 to 15 mmoles/g free chlorine the capacity of the carbon decreased by approximately 35 per cent at an equilibrium concentration of 3×10^{-4} *M*.

The small difference between the isotherms for dried and non-dried carbons, each treated with 15 mmoles/g free chlorine, is striking in view of the large difference in NaOH titratable substances for these carbons as shown in Figure 11.2. Apparently the substances that are readily evolved by heating have very little effect on the extent of adsorption.

The capacity of carbon for p-nitrophenol was significantly decreased by repeated cycles of treatment with free chlorine with drying of the carbon after each cycle. As shown in Figure 11.4, adsorption capacity is decreased to approximately 75 per cent of the original at an equilibrium concentration of 3×10^{-4} *M* with four such cycles.

Similar isotherms for adsorption of phenol are shown in Figure 11.5. The initial pH of the free chlorine solution which was reacted with the carbon was 7.8 in this case. The same decrease in adsorption capacity was noted in this instance as was observed for p-nitrophenol.

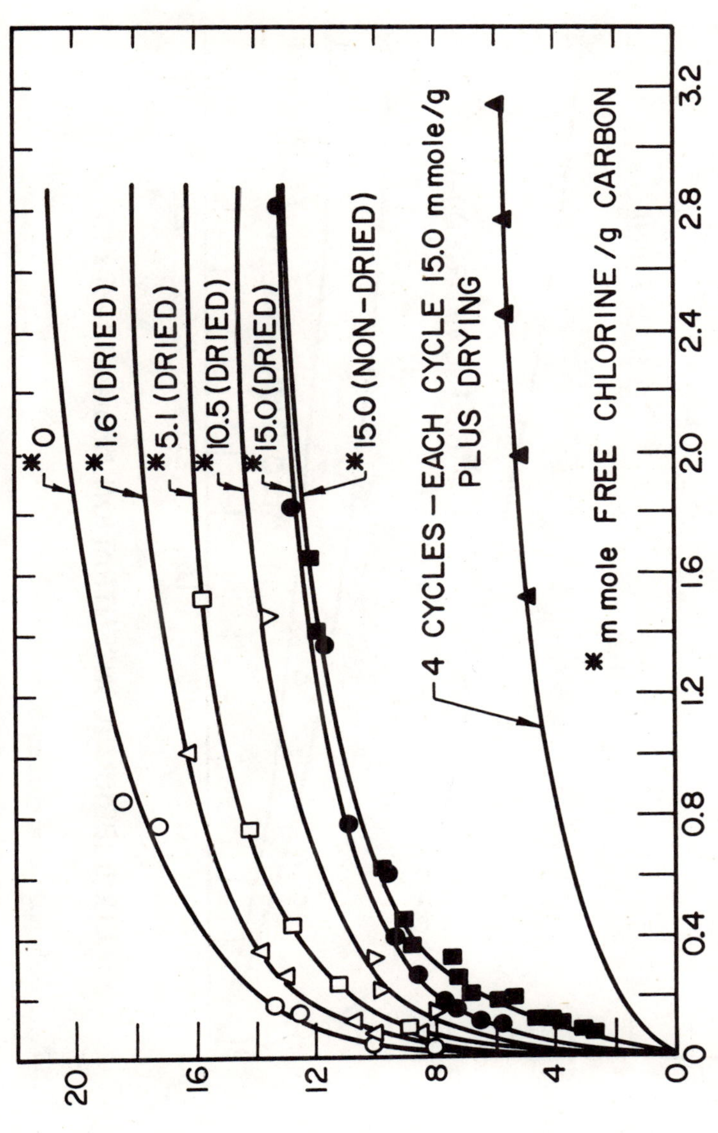

Figure 11.4. p-nitrophenol adsorption isotherms.

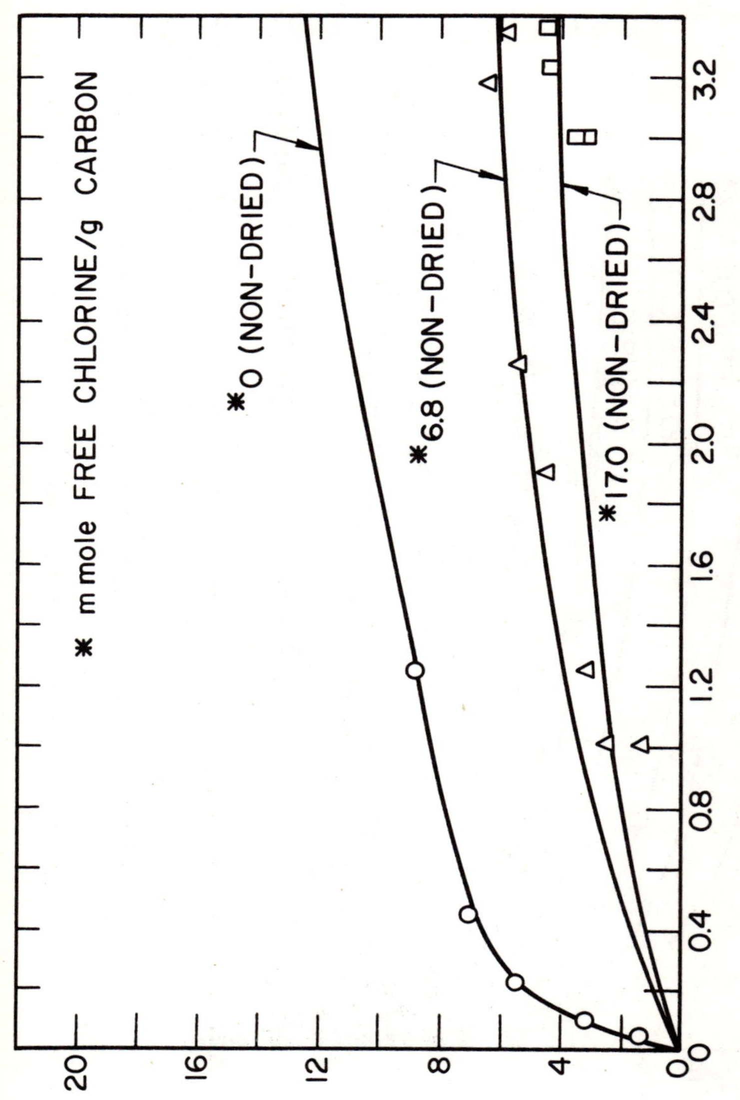

Figure 11.5. Phenol adsorption isotherms.

DISCUSSION

Titratable Surface Oxides

As shown in Figure 11.1 for non-dried carbon, the concentration of NaOH titratable substances on the carbon surface reaches a plateau at approximately 1.7 mmoles/g as degree of treatment with free chlorine is increased. It is reasonable that such a plateau be reached because the carbon surface has only a limited capacity for oxygen. Assuming the adsorbed area of a chemisorbed oxygen atom to be 10 A^2, and given the experimentally determined surface area for the non-treated carbon of approximately 1000 m^2/g, 17 mmoles of oxygen per gram of carbon represents monolayer coverage. If each of these oxygen atoms was associated with an acidic hydrogen atom, the maximum base comsumption would be 17 mmoles/g. In actuality, the maximum coverage would be somewhat less because the portion of the surface area associated with basal planes of the graphitic microcrystallites would not be expected to chemisorb oxygen. In considering this factor, it is not possible to establish a definite surface concentration as the plateau value but it is expected that this value should be significantly less than 17 mmoles/g.

As shown in Figure 11.2, many of the NaOH-consuming substances which were present on the non-dried carbon can readily be removed. Simple decanting of the dechlorinated water prior to addition of the standard base did not result in a significant reduction in NaOH consumed, while washing with several volumes of deionized water caused a decrease of 35 to 40 per cent. This indicated that most of the substances thus removed were present either on the carbon surface or within the carbon pores. The decrease of approximately 70 per cent in NaOH consumed, which occurred when the free chlorine treated carbon was dried, or dried and outgassed, is a further indication that many of the acidic substances were weakly adsorbed and volatile. By comparing Figures 11.1 and 11.2 it can be seen that the percentage decrease in the Na_2CO_3 titratable substances brought about by drying and outgassing is also approximately 70 per cent.

The NaOH titratable oxides on dried carbon of approximately 0.5 mmoles/g after reaction of 15 mmoles/g free chlorine and more than 1 mmole/g after four cycles of treatment with 15 mmoles/g free chlorine plus drying can be compared with a value reported by Coughlin *et al.*[17] of 1.1 mmoles/g for a slightly different carbon after treatment with a 0.1 *M* $(NH_4)_2S_2O_8$ solution for two weeks. These results are contrary to those of Puri *et al.*[14,21] who found a near 1:1 ratio between equivalents of chlorine reacted and $Ba(OH)_2$ titratable oxides formed. The reason for the difference in results is not known. The 15:0.5-1 mmoles chlorine reacted to NaOH titratable oxides formed per gram of carbon is consistent with the evolution of CO and CO_2 noticed by Magee.

An aspect which was not examined in this study and which requires further research concerns the effect of oxide buildup on the rate of dechlorination by carbon. Hagar and Flentje[8] give the life of dechlorination beds as 2-3 years for levels of free chlorine applied in the range of 2-4 mg/l at an application rate of 1 gpm/ft.[3] This information can be used to calculate that approximately 10-15 mmoles free chlorine react per gram carbon before breakthrough. Because this level of treatment corresponds to that at which the surface nears saturation with the NaOH titratable oxides, the possibility that the oxides are poisoning the surface sites where the dechlorination reaction takes place should be examined.

Another aspect of this study that requires further research concerns the effect of pH on the buildup of surface oxides. For the data reported in this paper, free chlorine in a solution with an initial pH of approximately 10 was reacted with carbon. During the reaction the solution pH decreased to approximately 6 and thus some HOCl reacted with the carbon as well as OCl^-. Preliminary tests performed using a chlorine solution with an initial pH of 4 and a resulting final pH of near 2 gave indication that the quantity of fixed surface oxides was higher than for the same system at pH 10. Further studies are needed to verify this, however.

The data on buildup of fixed oxides as the amount of chlorine reacted is increased are important, but care must be taken in relating these data to field dechlorination units. In normal field operations, carbon is reacted with low levels of chlorine over periods of time which are very long relative to the times used in these studies. The possibility that additional oxides are released from the surface as the length of reaction time for a given quantity of chlorine is increased should not be overlooked and warrants further study.

Effect of Surface Oxides on Adsorption

Figures 11.4 and 11.5 show a significant decrease in capacity of carbon for adsorption of the aromatic compounds, phenol and p-nitrophenol, on dried and non-dried carbon, each treated with 15 mmoles/g free chlorine. It should be noted that this level of treatment is typical of the level at which chlorine breakthrough of carbon beds is observed.[8] The sizable quantity of acidic matter present on non-dried carbon, which is evolved on heating, has essentially no effect on adsorption capacity. Increase in numbers of NaOH titratable oxides to more than 1 mmoles/g by cyclic treatment with free chlorine resulted in a 75 per cent decrease in amount of adsorption from that on non-treated carbon. This compares with a more than 90 per cent decrease for phenol and 75 per cent decrease for nitrobenzene on a different carbon as found by Coughlin *et al.*[17]. One-half of the decrease found by these workers could be attributed to a decrease in surface area, but the remaining decrease was attributed to increased surface acidity. They attributed the effect of acidity to the surface oxide withdrawing electrons from the delocalized π electron system of the basal planes of the graphitic microcrystallites which make up active carbon. This decrease in π electron density decreases the strength of the donor-acceptor bond which forms between the π electron system of the aromatic molecule and the graphitic plane. Mattson *et al.*[22] felt that a more preferable explanation of the effect of surface acidity on adsorption of aromatics, which they also observed, was that the aromatic is adsorbed to a great extent by forming a donor-acceptor bond between carbonyl surface oxides and the π electron system of the aromatic. Oxidation of the surface destroys this carbonyl oxygen[17,22] and thus reduces the amount of adsorption.[22] Puri,[10] however, agrees with Bhaaca's explanation that the aromatic compound reacts with the positive charge on the carbonyl carbon atom. Further studies are needed in order to determine which of these explanation is correct. The possibility that other factors, in addition to the effect of the carbonyl group, affect adsorption should also be examined.

More extensive work by Coughlin *et al.*[17] on the phenol isotherm showed that the effect of acidity was negligible at very high equilibrium concentrations of adsorbate, *i.e.* approximately 0.2 moles/l.

It is also interesting to note that treatment of carbon with approximately 15 mmoles/g free chlorine resulted in a reduction in capacity for both phenol and p-nitrophenol of

7×10^{-4} moles/g at an equilibrium concentration of 1.8×10^{-4} *M*, thus providing evidence for similar mechanisms of adsorption of these two compounds.

CONCLUSIONS

A definite relationship exists between the buildup of acidic surface oxides and the degree of treatment with free chlorine. The surface has a maximum level of oxides that it can accomodate, but many of the oxides or oxygen-containing reaction products formed via reaction with free chlorine are volatile. This volatility undoubtedly results in a much longer life for the dechlorination capability of active carbon beds. The brown colloidal matter that appears as a product after extensive reaction with chlorine is not of major concern because the useful life of field dechlorination units is exceeded before this product appears.[8]

The reaction products which were readily washed from the carbon surface, or volatilized during drying of the carbon, apparently had no major affect on the adsorption capacity of the carbon for the aromatics studied. The presence of fixed oxides, however, reduced adsorption capacity. This fact should be taken into account if carbon beds are to be used for both dechlorination and removal of dissolved organics.

Several topics related to this study require further investigation. The effect of acidic surface oxides on adsorption of more polar compounds should be investigated because it is possible that adsorption of these types of compounds will be increased with oxide buildup owing to dipole-dipole interactions between the molecule and the surface. The rate of oxide buildup under low but prolonged levels of free chlorine application should be studied. Also, the effect of organic molecules on the carbon surface on the rate of dechlorination and the life of the dechlorination bed needs to be evaluated if conjunctive use for dechlorination and adsorption is contemplated. The character of the reaction products, such as the acidic reaction products that are removed from the carbon when it is heated, should also be studied. It would also be desirable that more in-depth studies be undertaken relating changes in surface area and pore size distribution to the effect of acidic surface oxides on adsorption of organic compounds.

ACKNOWLEDGMENTS

The support of J. H. Johnson by an Environmental Protection Agency Traineeship and H. T. Lai by a National Science Foundation Undergraduate Research Participation Award is gratefully acknowledged.

REFERENCES

1. Laubusch, E. J. In *Water Quality and Treatment,* 3rd ed., American Water Works Assoc., McGraw-Hill, New York, 1971.
2. Seels, F. H. *Chem. Eng. Deskbook Issue, 80,* 27 (1973).
3. Kovach, J. L. *Ind. Water Eng., 8,* 30 (1971).
4. Magee, V. *Proc. Soc. Water Treat. Exam., 5,* 17 (1956).
5. Atkins, P. F., Jr., D. Scherger, and R. A. Barnes. Presented at the 27th Purdue Industrial Waste Conference, May 1972.
6. Bauer, R. C., and V. L. Snoeyink. "Reactions of Chloramines with Active Carbon," submitted for publication in *J. Water Poll. Cont. Fed.*
7. D'Agostaro, R. A. M.S. Thesis, Cornell University, Ithaca, New York, 1972.
8. Hagar, D. G., and M. E. Flentje. *J. Amer. Water Works Assoc., 57,* 1440 (1965).
9. Snoeyink, V. L., Ed., Proc. 15th Water Quality Conf., Engineering Publications, University of Illinois at Urbana-Champaign, Urbana, Illinois, 1973.
10. Puri, B. R. In *Chemistry and Physics of Carbon,* P. L. Walker, Ed., Vol. VI, M. Dekker, New York, 1970.
11. *Handbook of Chemistry and Physics*, 40th ed. Chemical Rubber Publishing Co., Cleveland, Ohio, 1958.
12. Boehm, H. P. In *Advances in Catalysis,* D. D. Eley, *et al.,* Eds. Academic Press, New York, 1966.
13. Snoeyink, V. L., and W. J. Weber, Jr. In *Progress in Surface and Membrane Science,* Vol. 5, Academic Press, New York, 1972.
14. Puri, B. R., D. D. Singh, J. Chander, and L. R. Sharma. *J. Indian Chem. Soc., 35,* 181 (1958).
15. Gasser, C. G., and J. J. Kipling. *Proc. Conf. Carbon, 4,* 51 (1960).
16. Coughlin, R. W., and F. Ezra. *Environ. Sci. Technol., 2,* 291 (1968).
17. Coughlin, R. W., F. Ezra, and R. N. Tan. *J. Colloid Interface Sci., 28,* 386 (1968).

18. Coughlin, R. W., and R. Tan. *Chem. Eng. Prog. Symp. Ser., 64,* 207 (1968).
19. Dietz, V. R., and J. L. Bitner. *Carbon, 10,* 145 (1972).
20. *Standard Methods for the Examination of Water and Wastewater,* 13th ed., American Public Health Association, New York, 1970.
21. Puri, B. R., O. P. Mahajan, and D. D. Singh. *J. Indian Chem. Soc., 37,* 171 (1960).
22. Mattson, J. S., L. Lee, H. B. Mark, Jr., and W. J. Weber, Jr. *J. Colloid Interface Sci., 33,* 284 (1970).

CHAPTER 12

INTERACTIONS OF AQUEOUS CHLORINE WITH ACTIVATED CARBON

Larry L. Olson and C. David Binning*

INTRODUCTION

The removal of aqueous chlorine from solution can be accomplished in a number of ways. One of the most practical techniques is through the use of activated carbon. This procedure was introduced to this country in 1925 from Czechoslovakia under a patent and gained widespread use after about 1928.[1]

Amount Removed

In 1929, Baylis[2] published his observations on the production of chlorides and carbon dioxide when aqueous chlorine was adsorbed on activated carbon. He was able to account for most of the chlorine removed by measuring chlorides in the effluent of the filter. After the filter was allowed to stand overnight, the remainder of the chlorides appeared in the solution in which the activated carbon was submerged. Likewise, the appearance of chlorides in the effluent lagged the application of chlorinated water to the filter by a considerable length of time. He also observed carbon dioxide in the effluent of the filter and postulated the following equation to account for the removal.

$$2Cl_2 + C + 2H_2O \rightleftharpoons 4HCl + CO_2 \qquad (1)$$

*L. L. Olson, Department of Civil Engineering, University of Delaware, Newark, Delaware. C. D. Binning, U. S. Navy, Naval Station Adak, Adak, Alaska.

This concept of adsorption followed by an oxidation reaction has been reported as fact for many years,[3,4] and in some cases, the oxidation reaction has even been ignored.[5,6]

Another view held by several researchers and reported by Hassler[7] is that the activated carbon acts merely as a catalyst for the production of chlorides and oxygen

$$2Cl_2 + 2H_2O \rightleftharpoons 4HCl + O_2 \qquad (2)$$

Hassler[7] also states that "Instances are reported in which the dechlorinating action will temporarily diminish and then recover if the carbon is given a brief rest period."

The concept of activated carbon catalyzing the reaction expressed above (Equation 2) was also presented by Puri, *et al.*[8] In their study, it was shown that the concentration of surface oxygen complexes increases in direct proportion to the amount of aqueous chlorine converted to chlorides. None of the chlorine is "fixed" on the charcoal and the capacity of the charcoal for converting chlorine to chlorides is a function of the surface area of the charcoal available for oxidation. In another study by Puri, *et al.*[9] it was found that the activated carbon reduces the chlorine, and, after that capability has been exhausted (approximately three to four days), the activated carbon acts as a catalyst for the formation of chloric acids. Both studies were for aqueous reactant concentrations far in excess of those normally encountered in water treatment applications.

A study by Puri and Sharma,[10] in which various oxidizing agents were used to treat the carbon, showed that the complex formed initially on the surface of the carbon is largely non-acidic. The complex is formed by chemsorption of the oxygen at the unsaturated sites as represented by

$$\text{surface}\left\langle \begin{matrix} C \\ \| \\ C \end{matrix} \right. + O_2 \rightleftharpoons \text{surface}\left\langle \begin{matrix} C - O \\ | \quad\ \ | \\ C - O \end{matrix} \right. \qquad (3)$$

At later stages, the complex is acidic, and it is formed by the association of two atoms of oxygen per active surface carbon atoms as shown below:

$$\text{surface}\left\langle \begin{matrix} C - O \\ | \quad\ \ | \\ C - O \end{matrix} \right. + O_2 \rightleftharpoons \text{surface} - C\left\langle \begin{matrix} O \\ | \\ O \end{matrix} \right. + CO_2 \qquad (4)$$

Once again, no chlorine was "bound" in the activated carbon structure.

Although "fixation" of chlorine was not observed in those systems, a later study by Puri, *et al.*[11] using a mixture of H_2O_2 and HCl showed that both oxygen and chlorine were fixed on the activated carbon. They found that in most cases the oxygen is fixed more readily at the unsaturated sites, and that chlorine is fixed primarily by substitution for hydrogen. Donnet,[12] using similar oxidizing agents, found that not only could surface changes be brought about but that it was also possible to solubilize a portion of the activated carbon. Johnson, *et al.*[13] showed that HOCl was a more effective regenerant for activated carbon than HOBr, but less effective than $Na_2Cr_2O_7$.

The removal of aqueous chlorine in a packed bed reactor was investigated by Magee,[14] who observed a two-stage removal process. The first stage was very rapid, and a net build up of chlorine containing species on the surface was observed. The second stage was much slower and chloride build up in solution was equal to chlorine removal when the aqueous species was OCl^- rather than HOCl.

Kovach[15] briefly discussed the removal process for aqueous chlorine on activated carbon. He showed that the two equations commonly used for predicting aqueous chlorine removal on activated carbon are slightly divergent, yet either can be used under appropriate conditions. Both equations represent first order removal with respect to aqueous chlorine.

These studies constitute the bulk of the reported work on the aqueous chlorine-activated carbon system, and illustrate the importance of the various surface groups with respect to the adsorption capacity of the activated carbon. They also illustrate the differences in observed products of the reactions, depending on the reaction conditions, from chloride production with no fixation when starting with chlorides, to dissolution of the activated carbon itself.

Because the aqueous chlorine-activated carbon system has not been extensively studied, parallel types of behavior were sought in the chlorine gas-activated carbon system. The formation of hydrochloric acid and the fixation of chlorine were reported by several authors.[16-20] It was even found possible to "fix" chlorine to diamond.[21] For these studies with gaseous chlorine and activated carbon, the amount of chlorine fixed depended on the initial hydrogen content of the carbon.[19] There is some evidence that the chlorine is fixed or substituted because the hydrogen causes formation of atomic chlorine.[16] The differential heat of adsorption for one particular charcoal was reported to be 8.86 Kcal/mole.[20] These studies illustrate some similarities between the aqueous and the gaseous systems, but because an oxygen source was not present in these gaseous systems, formation of surface oxides was not observed. However, for a system in which both oxygen gas and chlorine gas are present, chlorine is chemisorbed exclusively at the unsaturated site while oxygen is chemisorbed only to a small extent at these sites. The rest of the oxygen is chemisorbed at sites where it is possible to form structures that may evolve CO and CO_2 on high temperature evacuation. The location of "fixation" for chlorine and oxygen in the gaseous system is contrary to the aqueous work of Puri, *et al.*[10]

The effect of the surface groups on the adsorption of chlorine on activated carbon has been shown to be important; these groups are also important in controlling the adsorptive capacity of activated carbons for other adsorbates as well. For example, the adsorption of most organics is diminished by the presence of surface oxygen groups[22-24] while the adsorption of urea is increased.[22] The surface oxygen groups and their importance were discussed in several reviews and articles[25-31] and will not be discussed here.

Kinetics of Removal

The kinetics of the removal of aqueous chlorine from solution using activated carbon have been studied in an indirect fashion by Baumann.[32] Although it was not stated in the report, it was possible to calculate from Baumann's data the characteristic time for the breakthrough of aqueous chlorine from an activated carbon filter. This characteristic time was found to be inversely proportional to the influent concentration. Admittedly, this is worthless data for characterizing the exact kinetics of the removal process, but no other study could be found in the literature. However, several studies of the chemisorption of oxygen gas on activated carbon are reported[33-36] and all show multiple rate phenomena. For example, Puri, *et al.*[36] show that the oxidation occurs in two steps. First, the oxygen combines with the activated carbon and forms CO_2 evolving sites. Later, and at a slower rate, CO evolving sites are formed. The data generally fit the Elovich equation[37] and indicate from two[8,34] to five distinct rates.[33] Reported energies of activation for the chemisorption are from 10 to 12 Kcal/mole.[36]

From this brief literature review, it can be seen that little is known about the kinetics of the interaction of aqueous chlorine and activated carbon, and that the stoichiometry of the interaction depends markedly on the reaction conditions. The goal of the work reported herein is to more fully characterize both the kinetics and the stoichiometry of this interaction.

EXPERIMENTAL TECHNIQUES

The apparatus used for this study was a 5-l Nalgene reaction chamber immersed in a water bath for temperature control within ± 0.5° C of the desired value. The reaction chamber, containing the carbon-chlorine mixture, was constantly stirred with a stainless steel, low pitch, 2" diameter mixer at a speed of approximately 240 rpm. pH control during the reaction was accomplished by hand titrating a 1 *N* solution of NaOH into the reaction chamber which was being continuously monitored. The pH was maintained at a constant 10.0 ± 0.1 or 5.5 ± 0.1 units throughout the run (note - at pH 5.5 the aqueous chlorine is about 99% HOCl, and at pH 10.0 the aqueous chlorine is about 99% OCl^-).

The aqueous chlorine stock solution was prepared by slowly bubbling chlorine gas through distilled water to which enough NaOH had been added to drive the initial pH above 13. The solution was prepared to a sodium hypochlorite concentration in excess of 59,000 mg/l and then filtered through a 0.45 μ membrane filter.

Distilled water was carefully measured, placed in the reaction vessel, and allowed to equilibrate to the desired run temperature. Powdered activated carbon was weighed and placed into the distilled water in the reaction chamber and allowed to mix for a minimum of 30 minutes. The sodium hypochlorite solution was pipetted into a clean flask and placed over a Bunsen burner long enough to raise its temperature to the run temperature immediately prior to introducing it into the system. The initial chlorine concentration was determined by withdrawing a sample of the stock solution just before its addition to the reactor. The stock solution was then emptied into the reaction chamber and the pH was adjusted to 10.0 ± 0.1 units with 1 *N* NaOH solution. During the duration of the run, the amount of NaOH added to the reactor to maintain a constant pH was measured and recorded.

At predetermined intervals, 50 ml of the carbon chlorine solution-suspension was withdrawn from the reaction chamber and immediately filtered through a 0.45 μ membrane filter. All tests were run in accordance with Standard Methods.[38] Chlorine analysis on the filtrate was made using the potassium iodide thiosulfate method for chlorine concentrations of 10 mg/l and above. For chlorine concentrations less than 10 mg/l, the acid orthotolidine analysis using a 1-cm cuvette at a wavelength of 490 nm was used. Chloride analysis was performed using silver nitrate titration with a potentimetric (silver and glass electrodes) endpoint indicator.

The carbon-oxide concentrations were investigated using a spectrophotometer with a 1-cm cuvette. The wavelengths chosen for the carbon-oxide studies were based on the results of a spectral absorption analysis. This analysis was made to determine those spectrophotometer wavelengths which would produce the least absorbance due to the soluble inorganics in the system while exhibiting a maximum absorbance to the carbon-oxide products. The carbon-oxide sample used for the spectral scan was the product of a 100-minute run of 3000 mg/l activated carbon (Darco G-60), and a 4920 mg/l initial sodium hypochlorite solution. The run was maintained at a pH of 10.0 for the 100-minute period and was then acidified with concentrated H_2SO_4 to a pH of 4.5 to affix the oxides to the activated carbon surface. The resulting suspension was filtered through a 0.45 μ membrane filter and the filtrate discarded. The filtered activated carbon was washed with 200 ml of distilled water, then rinsed with 1 *N* NaOH solution. Thus, the carbon-oxides were desorbed off the surface, collected, and spectrophotometrically examined from wavelengths of 200 nm up to 800 nm.

The powdered activated carbons used for this study (Table 12.I) were pharmaceutical grade products; one, a product of Atlas Chemical Company distributed under the trade name Darco G-60, and the other, a coconut shell product of West Virginia Pulp and Paper Company distributed under the trade name Nuchar C-1000 N. Distilled water for the preparation of the hypochlorite stock was obtained from a Corning all-glass still, while the distilled water used for the carbon-chlorine runs was obtained from a Barnstead Model EL 12 stainless steel still. No buffers were used during this investigation.

The infrared scans were performed on a Perkin-Elmer 221 Spectrophotometer. The products of the reaction were dialyzed using a polythene bag, dried and pressed

Table 12.I

Properties of Powdered Activated Carbon

	Darco G-60	*Nuchar C-1000N*
Moisture content	max of 12%	max of 5%
Water-soluble surface products	max of 0.3%	3-7%
Acid-soluble surface products	max of 1%	--
Ash (approximate range)	3-4%	max of 9%
Bulk density (lb/ft^3)	22-28	9-11
Mesh Size (approximate)		
% through 100-mesh screen	95	90-99
% through 325-mesh screen	70	55-75
Surface Chloride as NaCl	less than 0.05%	--
Surface Iron as Fe_2O_3	less than 0.05%	--
Copper, Calcium and Sulfides	virtually none	--
Surface area (m^2/g)	563	1000-1100*

* Measured this investigation, 790 m^2/g.

into potassium bromide pellets. The concentration of material in the KBr pellet was about 1%. Infrared scans were performed on the products from the Nuchar C-1000N activated carbon only.

Base adsorption capacities of the activated carbon were performed after the fashion of Puri and Singh[39] using 1.0 gram of activated carbon in 100 ml of 0.25 *N* $Ba(OH)_2$. The samples were agitated for 48 hours and then allowed to settle. Next, portions of the supernatant were titrated with 0.1 *N* HCl, and the base adsorption capacity was calculated.

RESULTS

Kinetics of Chlorine Removal

Preliminary work without temperature control was performed using the Nuchar carbon. The surface area of this carbon was determined to be 790 m^2/g using nitrogen adsorption and the modified B.E.T. procedure of Nelsen and Eggersten.[40] The concentration of aqueous chlorine as a function of time for several runs at pH 10.0 is shown in Figure 12.1. The semilogarithmic plot illustrates that removal of aqueous chlorine from solution appears to be first order after about 10 minutes. The removal of aqueous chlorine is very rapid initially and then becomes first order. The slope of the linear portion of Figure 12.1 is the first order rate constant. Close examination of Figure 12.1 shows that the greater the amount of aqueous chlorine removed in the first portion of the removal process (the non-linear portion), the smaller the first order rate constant.

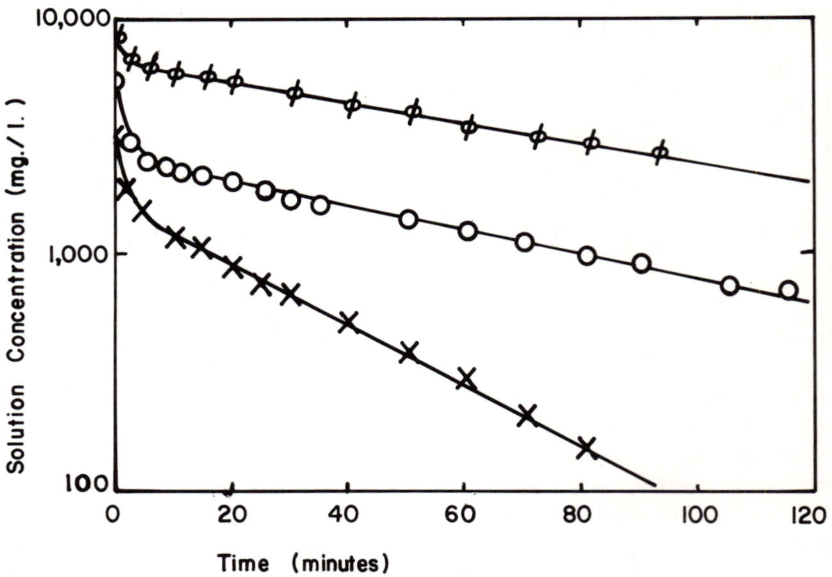

Figure 12.1. Aqueous chlorine concentration as a function of time. 3.0 g/l Nuchar C-1000N at pH 10.0.

If one considers the removal process to be made up of two different removal steps occurring sequentially, it is possible to separate the two processes by extrapolating the linear portion of Figure 12.1 to time zero. It can then be assumed that that portion of the removal represented by the difference between the initial concentration, C_o, and the concentration at the extrapolated values of the linear portion of the curve, C_o', may be used to represent the first portion of the removal process (Part I). This difference in concentration, C_o-C_o', is herein referred to as ΔC. The value of ΔC may then be divided by the concentration of activated carbon, AC, and an (x/m) apparent is obtained, where x/m is commonly used nomenclature for grams adsorbed per gram of adsorbant. This (x/m) apparent is equal to the x/m obtained from adsorption experiments, but one must assume that the phenomenon represented by Part I is at equilibrium for the equality to hold.

The value of (x/m) apparent may be plotted as a function of the first order rate constant as shown in Figure 12.2.

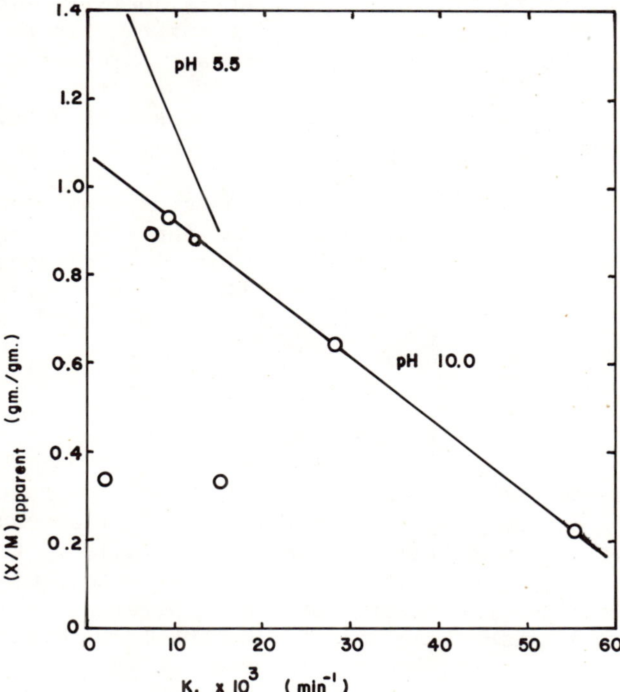

Figure 12.2. Initial removal as a function of apparent first order rate constant. 3.0 g/l Nuchar C-1000N at 25° C.

A linear relationship results. The data for the two runs at pH 5.5 are also shown in this figure. The points fall above those for pH 10.0, indicating that the amount of aqueous chlorine removed due to Part I at pH 5.5 is somewhat greater. Referring to Runs 2 and 3 of Table 12.II, it can be seen that some, but not all, of the difference can be accounted for by chloride production. In addition, it can be observed from the table that the quantity of chlorides produced per gram of activated carbon, $\frac{\Delta Cl^-}{AC}$, is approximately constant for this series of runs.

As can be seen from Figure 12.2 and Table 12.II, the data are not consistent or extensive, and it was concluded that they were inadequate to fully describe the removal kinetics. Consequently, another study was initiated in which temperature control was practiced. A different activated carbon (Darco G-60) was used. The precision of the iodometric determination of aqueous chlorine was tested in this portion of the study, and was found to be within plus or minus 2% for the concentration normally encountered in the test series. The per cent error in reproducibility of ΔC and the first order rate constant, k_1, was found to be less than 4% and 10%, respectively, for temperatures from 10° C to 60° C. Because k_1 is dependent on ΔC, it is reasonable to observe poorer reproducibility for k_1 than for ΔC.

The data from this portion of the study, which are summarized in Table 12.III, illustrated the same two-part removal process. Similarly, the plot of (x/m) apparent, or $\Delta C/AC$ versus k_1, illustrates a linear inverse relationship (Figure 12.3). The slope of the plots is dependent on the temperature.

Temperature Effects

The dependence of the rate constant on temperature may be an aid in determining whether the removal of aqueous chlorine is diffusion-controlled or reaction-controlled. Accordingly, the natural logarithm of the first order rate constant was plotted against the inverse of the temperature in Figure 12.4 for the same $\Delta C/AC$ values, and the first order rate constant was plotted against temperature in Figure 12.5 for the same $\Delta C/AC$ values. In both figures, it can be seen that an apparent linearity exists for certain temperature ranges and that the slope of the data for 30° C or less is

Table 12.II

Summary of Chlorine Uptake and Chloride Production Data for Studies with Nuchar C-1000N

Run No.	pH	C_o (mg/l)	C_o^1 (mg/l)	ΔC (mg/l)	AC (mg/l)	$K_1 \times 10^3$ (min^{-1})	$(x/m)_{app}$ (gm/gm)	ΔCl^- (mg/l)	$\Delta Cl^-/AC$ (gm/gm)
2	5.5	5740	2500	3240	3000	10.9	1.080	1370	0.456
3	10.0	5630	3000	2630	3000	12.3	0.878	1190	0.397
4	10.0	3590	1670	1920	3000	28.2	0.640	1230	0.410
6	10.0	926	595	331	1000	15.4	0.331	475	0.475
7	10.0	128	100	28	83	2.1	0.337	25	0.301
9	10.0	2420	1530	890	1000	7.3	0.890	-	-
10	10.0	1260	1260	0	3000	478	0	-	-
12	10.0	9330	6550	2780	3000	9.7	0.928	-	-
13	5.5	9050	5180	3870	3000	7.0	1.280	-	-
14	10.0	564	256	308	1000	55.7	0.229	-	-
14A	10.0	310	240	70	1000	113	0.070	-	-
15	10.0	10.2	8.8	1.4	10	0.9	0.240	-	-
16	10.0	5.4	4.0	1.4	10	4.3	0.140	-	-

Table 12.III

Data Summary of Chlorine Uptake on Darco G-60

Run	Temp (°C)	AC (mg/l)	C_o (mg/l)	C_o' (mg/l)	ΔC (mg/l)	$k_1 \times 10^3$ (min^{-1})	$\Delta C/AC$
A	20	3000	5420	4340	1080	3.48	0.346
B	60	3000	5420	4050	1370	37.51	0.457
C	3	3000	5420	4720	700	2.02	0.233
D	42	3000	5420	4100	1320	6.48	0.440
E	30	3000	4624	3146	1180	3.45	0.393
F	53	3000	4624	3146	1478	16.23	0.492
G	24	3000	4624	3554	1070	3.23	0.357
H	10	3000	5064	4247	817	2.55	0.273
I	10	3000	5064	4278	786	2.73	0.262
J	30	3000	5110	4058	1052	5.48	0.351
K	30	3000	5110	4058	1052	6.01	0.351
L	50	3000	5281	3944	1337	15.28	0.446
M	50	3000	5281	3893	1388	14.64	0.463
N	10	3000	7625	6700	925	1.53	0.308
O	10	3000	6074	5174	900	2.08	0.300
P	30	3000	7082	5902	1180	4.10	0.393
Q	50	3000	4899	3619	1280	17.05	0.427
R	50	3000	4430	3190	1240	18.08	0.413
S	30	3000	4700	3858	842	9.95	0.281
T	30	3000	4784	3744	1040	7.07	0.347
U	30	3000	5039	4145	894	5.04	0.298
V	30	3000	4806	3669	1137	5.77	0.379
W	50	3000	6155	4596	1559	9.84	0.520

X	50	3000	4110	2870	1240	18.35	0.413
Y	40	3000	6329	5019	1310	6.55	0.437
Z	40	3000	5155	3945	1210	8.55	0.403
AA	40	3000	4095	2935	1160	9.71	0.387
BB	10	3000	2500	1742	758	4.47	0.253
CC	20	3000	7710	6587	1123	1.24	0.374
DD	20	3000	2482	1747	735	8.24	0.245
EE	30	3000	1054	474	580	18.57	0.193
FF	30	3000	107	--	107	--	0.036
GG	30	499	107	86	21	22.52	0.042
HH	30	37.91	9.98	6.68	3.30	21.93	0.087
II	30	17.41	9.98	7.68	2.30	14.02	0.132
JJ	30	3000	1667	905	762	41.14	0.254
KK	50	3000	136	--	136	--	0.045
LL	50	499	136	79	57	64.52	0.114
MM	50	30	9.98	6.18	3.80	53.80	0.127
NN	50	14.98	9.98	6.99	2.99	31.21	0.200
OO	60	3000	2606	1713	893	58.90	0.298
PP	60	3000	4976	3546	1430	34.65	0.477
QQ	60	3000	7139	5622	1517	29.65	0.506
RR	30	3000	1490		incomplete	run	
SS	30	3000	4744	3569	1175	4.29	0.392
TT	30	3000	8740	7230	1510	3.12	0.503
UU	30	3000	1976	1006	970	11.08	0.323
VV	40	3000	1764	1162	602	31.80	0.201

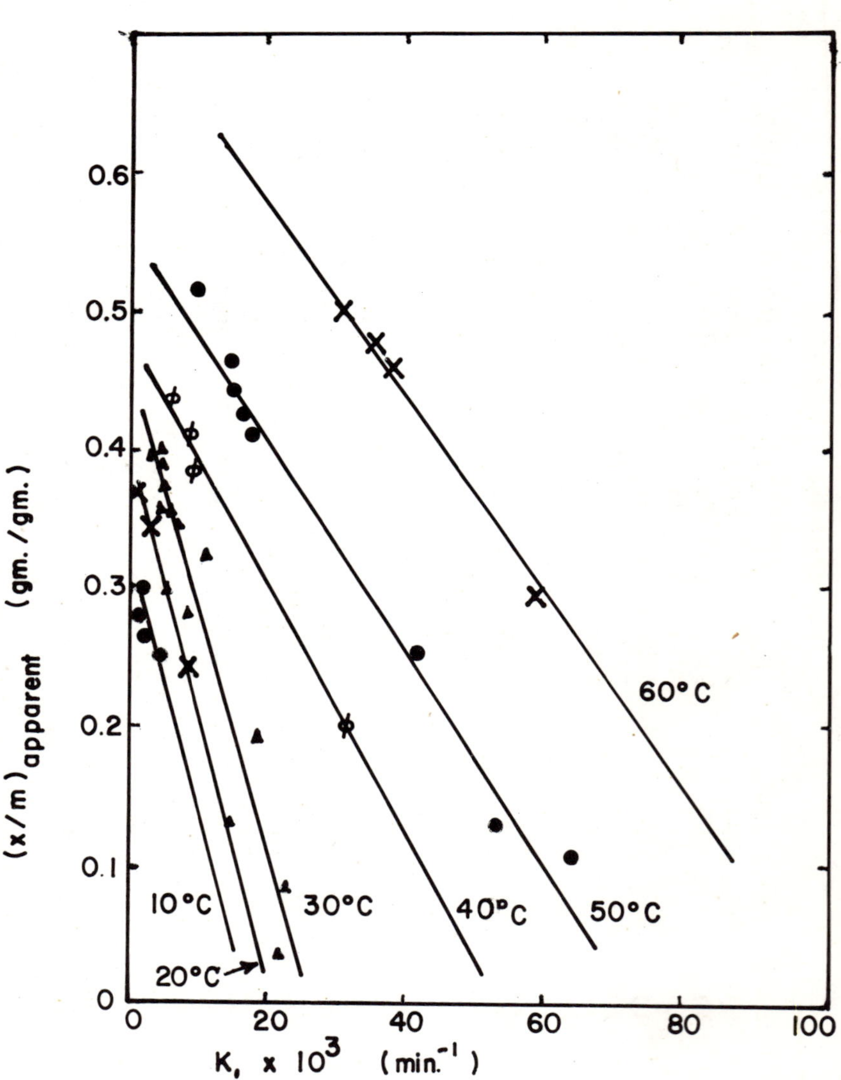

Figure 12.3. Initial removal as a function of apparent first order rate constant. Darco G-60 at pH 10.0, variable carbon concentration.

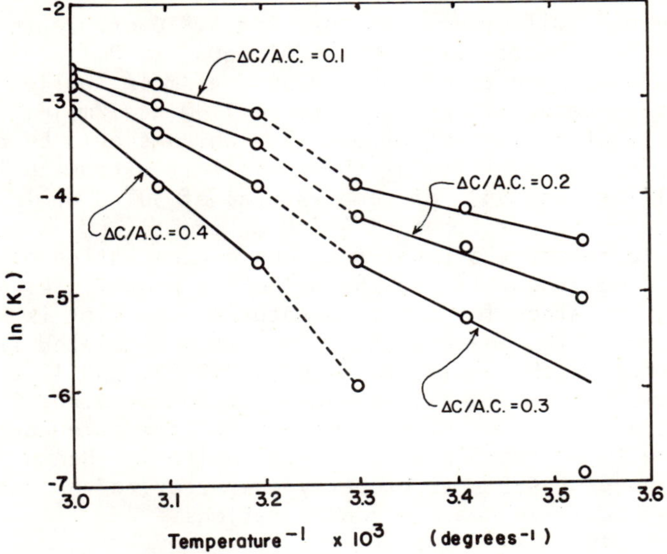

Figure 12.4. Logarithm of the apparent first order rate constant against inverse temperature for various ΔC/AC values. Darco G-60.

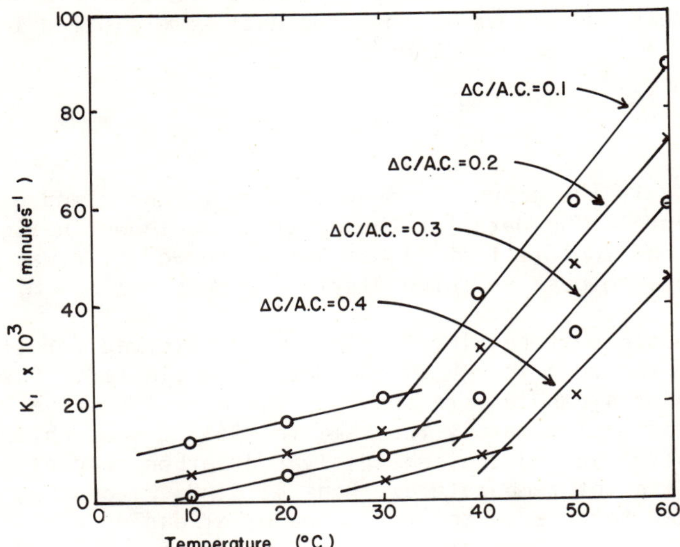

Figure 12.5. Apparent first order rate constant as a function of temperature for various ΔC/AC values. Darco G-60.

considerably different than that for 40° C or greater. The energy of activation for $\Delta C/AC$ equal to 0.1 at 30° C or less was found to be about 4.56 Kcal/mole, while for 40° C or greater it was found to be 7.40 Kcal/mole. Similarly, the slope of Figure 12.5, which should be a measure of the mobility of the species if diffusion is controlling, is 0.43 for temperatures of 30° C or less and 2.3 for temperatures of 40° C or greater.

For a constant activated carbon concentration of 3.0 g/l, a good linear fit to the Arrhenius plot ($\ln k_1$ versus $1/T$) was obtained for all temperatures (the plot is not presented). The energies of activation calculated from the slopes of these plots were 9.35, 10.52, and 12.24 Kcal/mole, respectively, for $\Delta C/AC$ values of 0.20, 0.267, and 0.333. The plot of k_1 versus T for the same data was observed to be non-linear over its entirety. Hence, the Arrhenius plot seems to properly characterize the data for high activated carbon concentrations.

Elovich Equation

Plots of solution concentration against time appeared to be linear from two minutes until about 10 to 20 minutes into the run. In an attempt to determine if there were more than two removal rates, the data were plotted in accord with the Elovich equation[37]

$$\frac{dq}{dt} = a \cdot \exp(-\alpha q) \qquad (5)$$

where q is the amount removed at time t, and a and α are constants. The data fit the equation as shown in Figure 12.6, and two distinct slopes were observed. It was not possible to observe three distinct slopes for any of the runs.

Multiple-stage Elovich plots were observed for almost all of the runs for which the data were plotted. Observed values of α_1 and α_2 are shown in Table 12.IV. Five runs that had approximately the same initial aqueous chlorine concentration and the same activated carbon concentration, but different temperatures (Runs B, C, D, E, and F), were selected to observe the variation of α_1 and α_2 with temperature. Plots of α_1 and α_2 versus the inverse of the temperature were found to be linear (Figure 12.7). The variation of α_1 appeared to be independent of initial concentration of aqueous chlorine at the same temperature

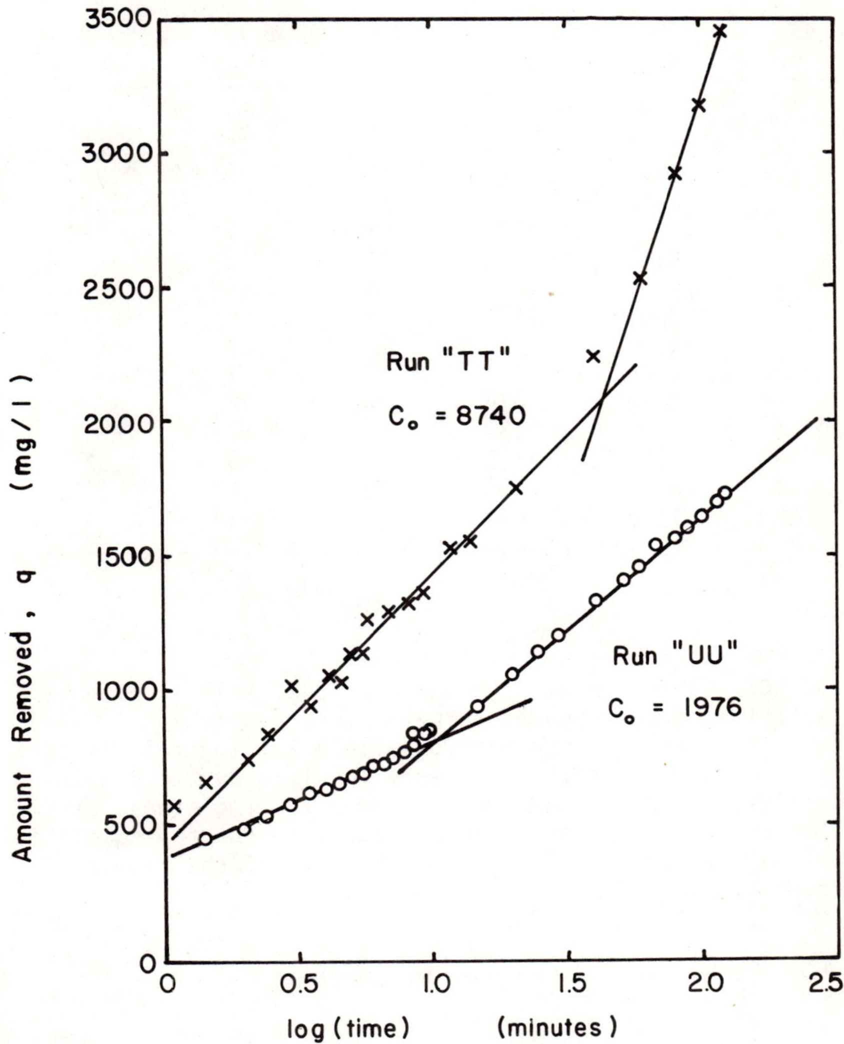

Figure 12.6. Elovich plots for runs "TT" and "UU." 3.0 g/l Darco G-60 at 30° C and pH 10.0.

Table 12.IV

Elovich Data for Removal of Aqueous Chlorine on 3.0 g/l Darco G-60

Run No.	C_o *(mg/l)*	ΔC *(mg/l)*	*Temp (°C)*	α_1 $x\ 10^3$ *(l/mg)*	α_2 $x\ 10^3$ *(l/mg)*
B	5420	1370	60	0.85	--
C	5420	700	3	5.75	2.28
D	5420	1320	42	2.20	0.76
E	4620	1180	30	2.94	1.38
F	4624	1478	53	1.36	0.75
J	5110	1052	30	3.13	--
K	5110	1052	30	3.56	--
P	7082	1180	30	3.52	--
S	4700	842	30	2.47	--
T	4784	1040	30	3.51	--
U	5039	894	30	3.74	--
V	4806	1137	30	2.88	--
SS	4744	1175	30	2.68	1.32
TT	8740	1510	30	2.34	0.77
UU	1976	970	30	5.22	2.54

(30° C), having an average value of 3.27×10^{-3}. The variation of α_2 appeared to be dependent on the initial aqueous chlorine concentration with a log-log relationship (Figure 12.8), but the data were insufficient to provide a conclusive trend. The break between the two stages of the Elovich plots was observed to occur at about the same time as the break between Part I and Part II of the removal process.

Chloride Production

Chloride production as measured in Runs A-F was very rapid initially and was complete in 20 to 40 minutes. These data are plotted according to the Elovich equation as shown in Figure 12.9. No difference in the slopes for the various runs could be detected, and α values were

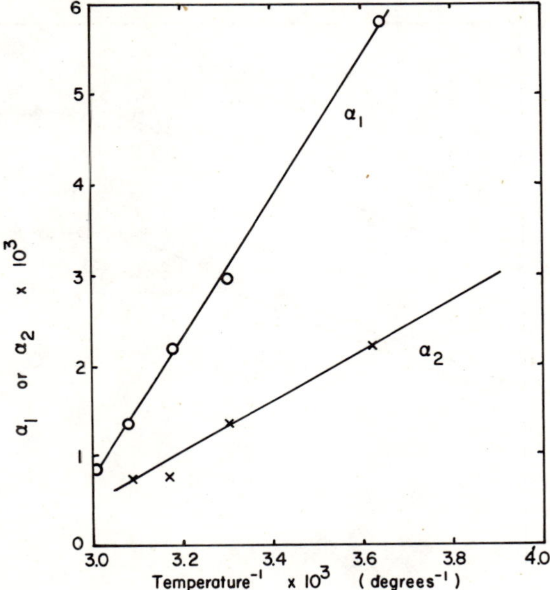

Figure 12.7. *Variation of α_1 and α_2 with temperature, Elovich data for removal of chlorine.*

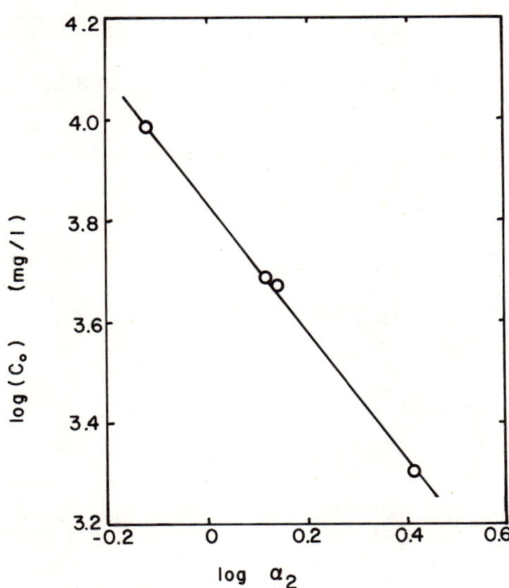

Figure 12.8. *Variation of α_2 with C_o. 3.0 g/l Darco G-60 at 30° C and pH 10.0.*

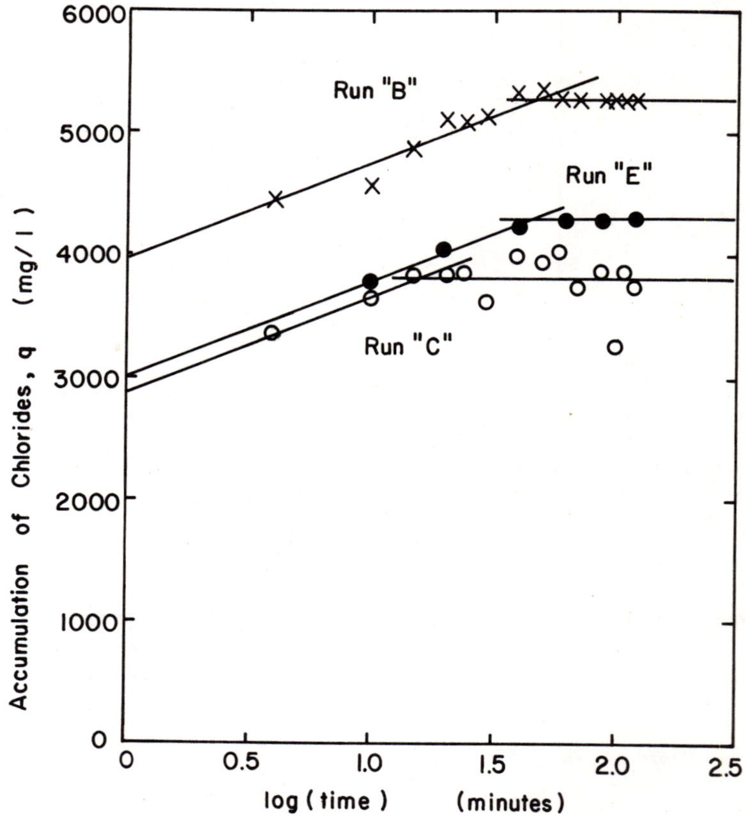

Figure 12.9. Elovich plots for chloride production. 3.0 g/l Darco G-60.

Table 12.V

Summary of Chloride Production Data, 3.0 g/l Darco G-60

Run	Temp (°C)	C_o (mg/l)	ΔCl^- (mg/l)	ΔC (mg/l)	$k_1 \times 10^3$ (min^{-1})
A	20	5420	1080	1000	3.48
B	60	5420	1370	1730	37.51
C	3	5420	700	350	2.02
D	42	5420	1320	1420	6.48
E	30	4624	1180	1210	3.45
F	53	4624	1478	1720	16.23

calculated to be about 4×10^{-3}. Each of the runs exhibited a maximum quantity of chlorides produced, and the difference between initial and final chloride concentrations is termed ΔCl^-. The values of ΔCl^- for the various runs, as well as other measured quantities, are listed in Table 12.V.

The relationship between the chlorides produced at the various temperatures and the amount of aqueous chlorine removed via Part I (ΔC) of the previously mentioned two-part removal process is shown in Figure 12.10. All of the points except Run B (60° C) fall on the line described by the equation $\Delta C = 0.566\ Cl^- + 500$. The 1200 mg/l chlorides produced using the Nuchar C-1000N at a concentration of 3 g/l (see Table 12.II) corresponds quite well to the 1210 mg/l of chlorides produced for Run 3 at 30° C.

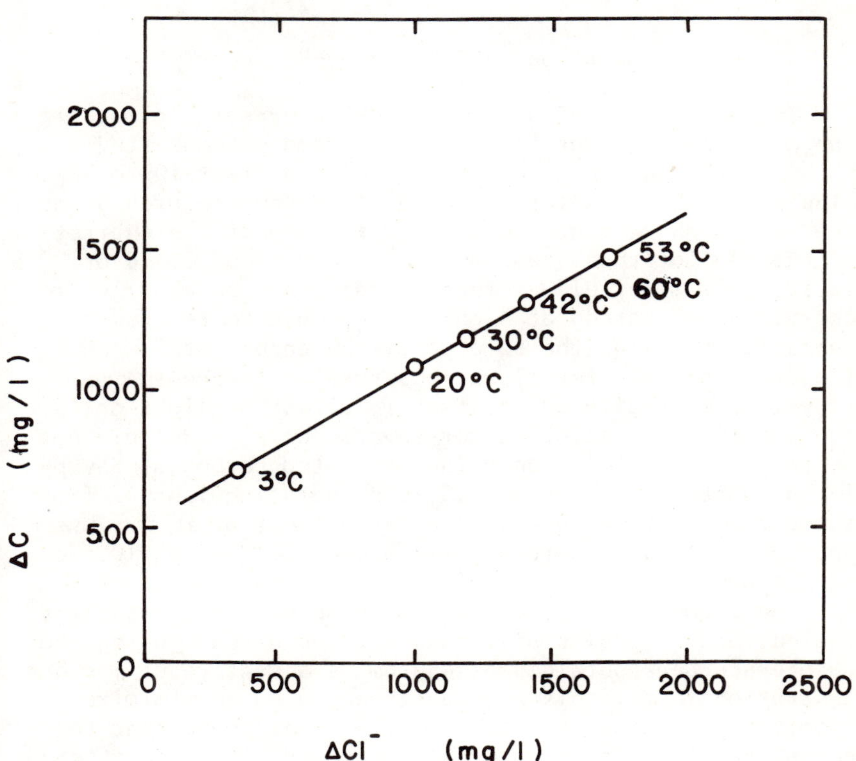

Figure 12.10. Initial chlorine uptake as a function of chloride production. 3.0 g/l Darco G-60.

Proton Production

The adsorption-reaction of aqueous chlorine with activated carbon produced protons as well as chlorides. The quantity of protons produced was determined by recording the volume of 1 *N* NaOH to keep the pH of the reaction at the desired level. If it is assumed that one proton is produced for every OCl^- molecule that reacts, it is possible to express the proton production in terms of mg/l of Cl. That was done, and Elovich plots (Figure 12.11) illustrate a two-part process just as in the disappearance of aqueous chlorine from solution (see Figure 12.6). The α_1 and α_2 values calculated for the various temperatures are shown in Table 12.VI. These α values are plotted as a function of the inverse of the temperature (Figure 12.12), and the lines of best fit for the data in Figure 12.7 are dotted in. The similarity of the results is striking.

Production of Colored By-Products

The production of a brown-black water-soluble material was observed to occur for both activated carbons studied. This material was not detectable for the first 10 to 20 minutes of the runs for those reaction temperatures greater than 20° C and was not detectable throughout the entirety of the run for those reaction temperatures of 20° C or less. This material was found to desorb more readily from the activated carbon at high pH and could be removed from the solution using the same activated carbon at low pH. The brown-black material was observed to change to green or yellow when allowed to stand in direct sunlight or in aqueous chlorine solution for several days, and could not be removed from solution using activated carbon at any pH. The products formed at pH 5.5 were indistinguishable from those formed at pH 10 with the exception that at the lower pH the non-sorbable product was green while at pH 10 the non-sorbable product was yellow.

The brown-black solution was scanned in the visible-UV region, and several small peaks were observed. Using the absorbance peak obtained at 450 nm as a measure of its concentration (note: this wavelength was used to minimize absorbance of other soluble ions), it was found that the concentration of this material increased with time after 20 minutes, and in some cases, decreased with increasing time after that, as shown in Figure 12.13.

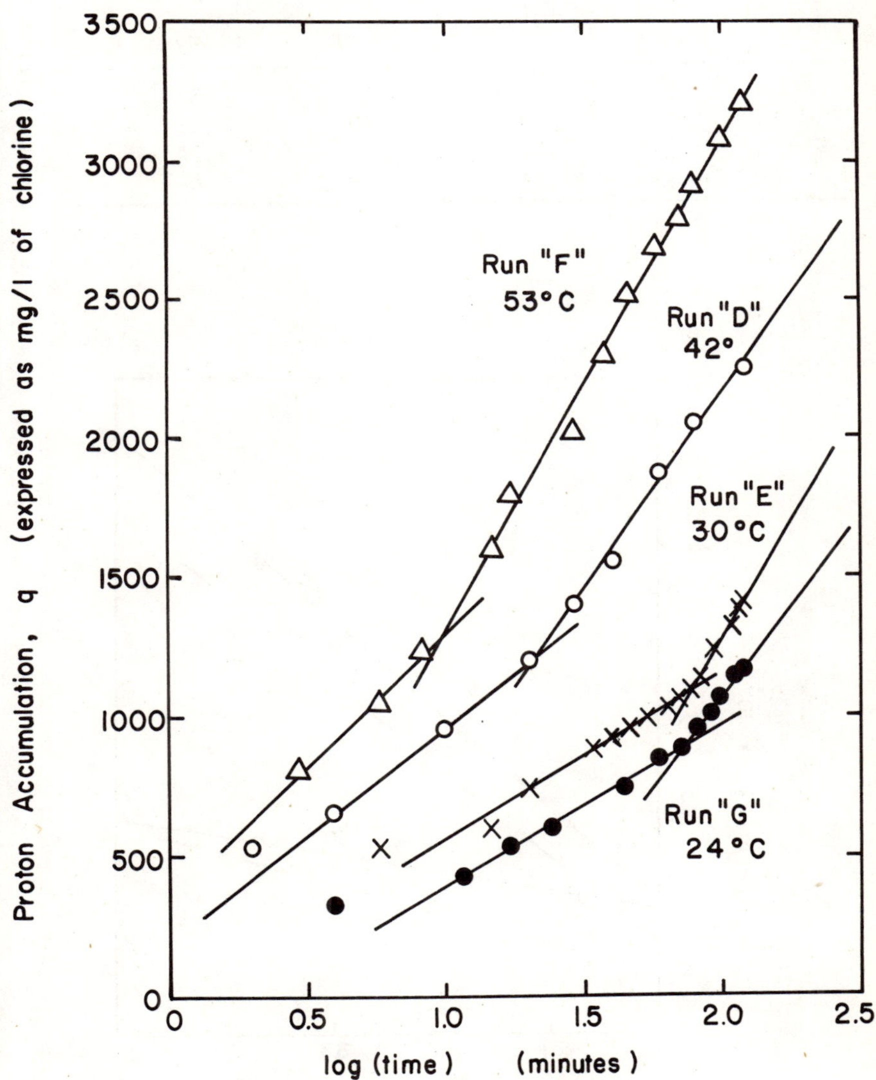

Figure 12.11. Elovich plots for proton production. 3.0 g/l Darco G-60 at pH 10.0.

Table 12.VI

Elovich Data for Proton Production. Darco G-60

Run #	Temp °C	$\alpha_1' \times 10^3$ (l/mg)	$\alpha_1' \times 10^3$ (l/mg)
G	24	3.37	6.32
E	30	3.30	5.55
D	42	3.17	5.24
F	53	3.07	4.44

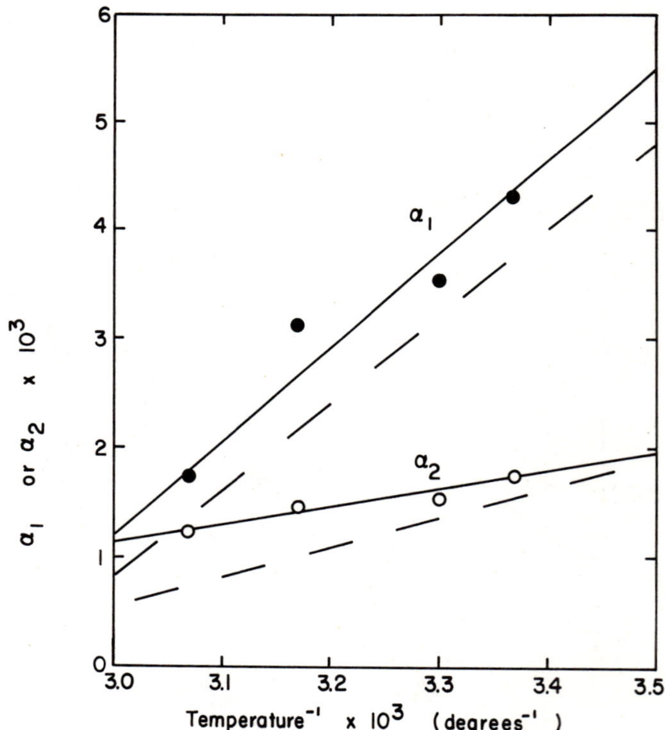

Figure 12.12. *Elovich data for proton production showing the variation of α_1 and α_2 with temperature. Similar data for chlorine removal dashed in.*

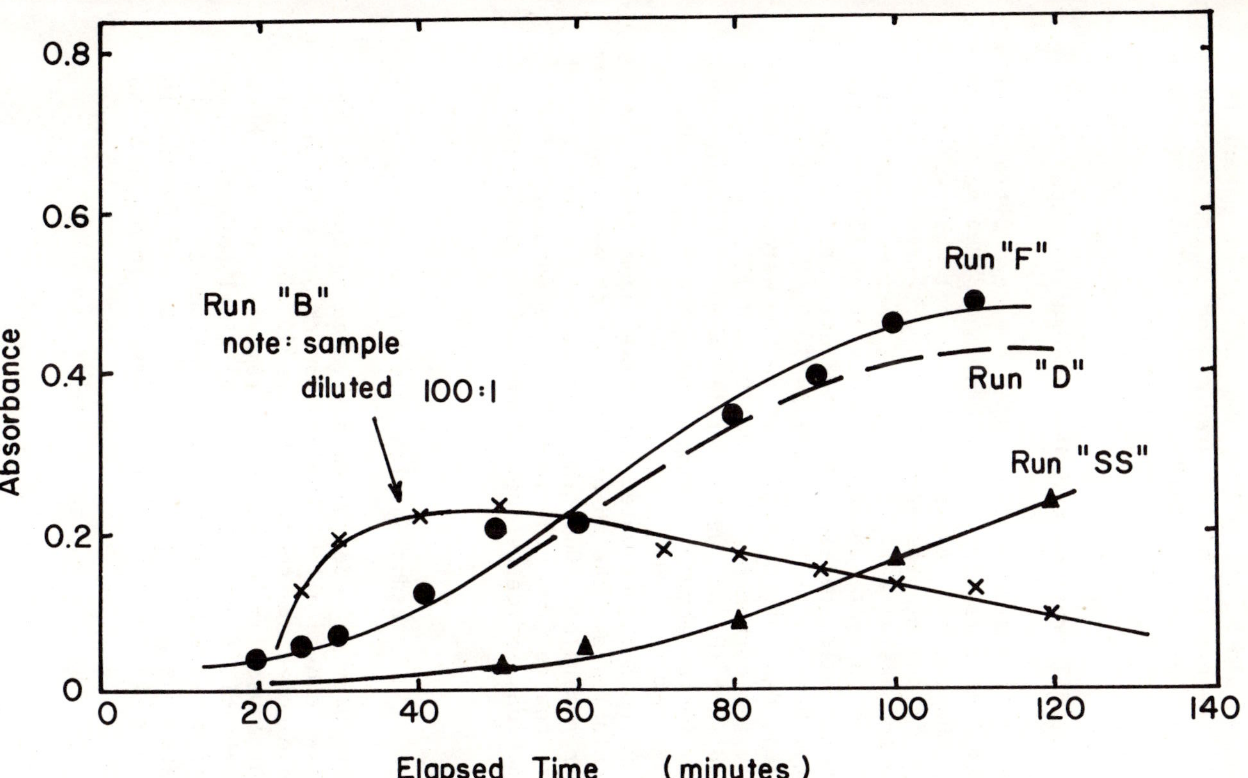

Figure 12.13. Absorbance at 450 nm against elapsed time. Darco G-60.

The brown-black, the green, and the yellow materials were also scanned in the infrared region of the spectrum and several peaks were observed. A discussion of the peaks and their significance is presented later.

Filtrate Titrations and Base Adsorption Capacity

Several filtrate samples from a run were titrated with 0.1 *N* HCl. At time zero, the titration curve characteristic of the carbon dioxide system was not observed to be present. At ten minutes and at two hours it was still not possible to observe a titration curve that is characteristic of such a system. This run was at pH 10.0 so that any carbon dioxide released from the activated carbon would have been entirely solubilized. Base adsorption capacities, BAC, of the activated carbon were measured on the activated carbon for time 0, 10, and 120 minutes. It was found that the BAC was initially about 0.06 meq/g; after 10 minutes about 1.3 meq/g; and after 120 minutes about 0.20 meq/g.

The Darco G-60 activated carbon was immersed in an acidic solution and the pH of the solution measured before and after immersion. One gram of activated carbon caused the pH to change from 4.40 to 4.20. The adsorption of hydroxide ions indicates that Darco G-60 is an "L" type carbon.[37]

Redosing with Chlorine

The results of redosing an activated carbon that had previously removed all of the aqueous chlorine to which it had been exposed are shown in Figure 12.12. The redosing (Run 14A) causes an initial rapid removal, followed by a semilogarithmic removal as in the original dosing. The amount of aqueous chlorine removed influences the first order rate constant in both cases; k_1 is less for Run 14 than for Run 14A; hence, k_1 is inversely proportional to ΔC.

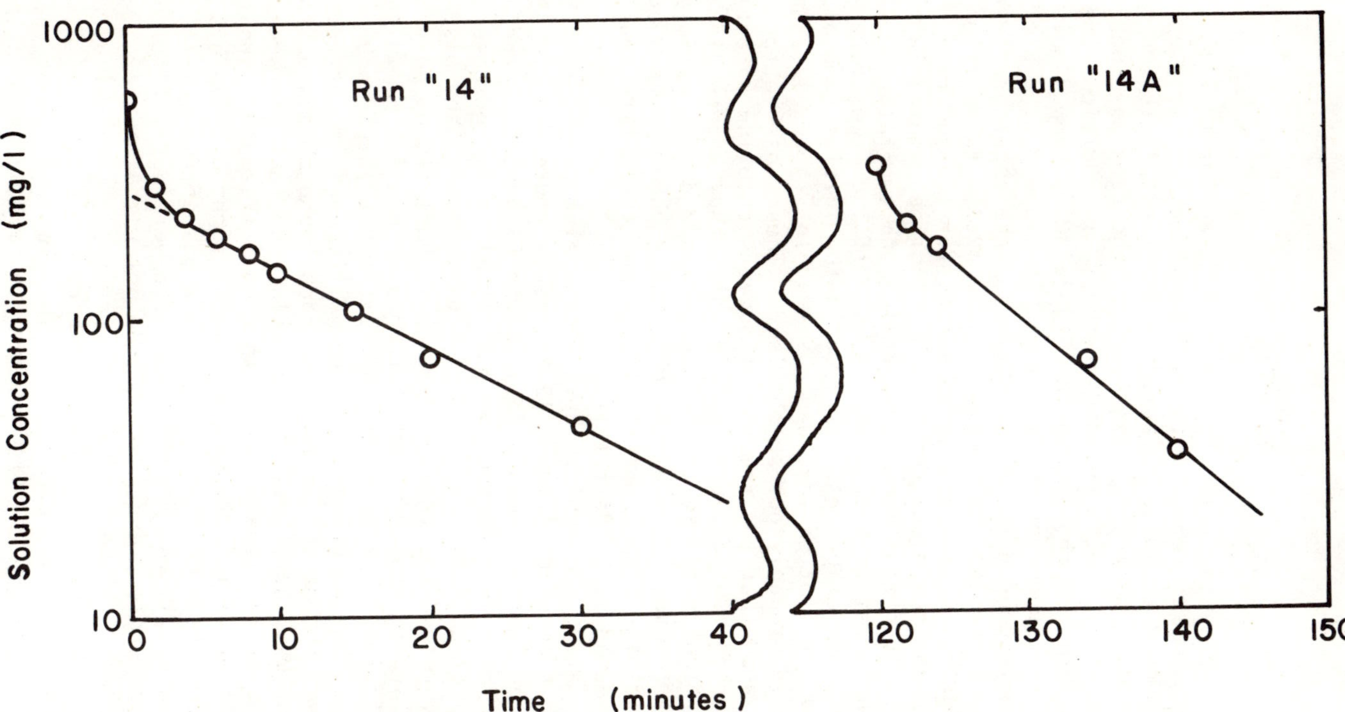

Figure 12.14. Solution concentration as a function of time. 1.0 g/l Nuchar C-1000N.

DISCUSSION

Kinetics

The data represented by Figure 12.1 appear to illustrate a dual-rate phenomenon. It was first thought that a zero order removal could be detected in the first 10 to 20 minutes of most of the runs; however, the data were not distinct enough to be certain of that. In addition, the Elovich plots of the same data (Figure 12.6) showed only a two-rate phenomenon. The changes in slope of the Elovich plots corresponded to the inception of the apparent first order phenomena illustrated in Figure 12.1. In addition the relationship between the apparent first order rate constant and the amount of aqueous chlorine removed initially yielded additional support for the two-rate phenomenon, rather than three rates. If there was more than one rate or process involved in the initial removal, the good correlation that was observed in Figures 12.2 and 12.3 would probably not be obtained. These observations seem to show that the removal of aqueous chlorine on activated carbon of the types used in this study occurs via two different processes that have correspondingly different rates of removal. The amount of aqueous chlorine removed in the initial step (Part I) affects the rate of removal in the second step (Part II).

The rate of removal in Part II can also be observed to be a function of pH (Figure 12.2), temperature (Figure 12.3), and the activated carbon used (compare Figures 12.2 and 12.3). The effect of the lower pH (5.5) is to increase the quantity of aqueous chlorine removed by Part I, and also to markedly retard the rate of removal in Part II, as evidenced by the larger negative slope for pH 5.5 than for pH 10 in Figure 12.2. The effect of a different activated carbon appears to be due primarily to the difference in surface area, even though the activated carbons were obtained from different manufacturers. For example, the ratio between the slope of the line at pH 10 in Figure 12.2 and a supposedly similar line in Figure 12.3 is about 0.79. This corresponds to the ratio of surface areas of the same carbons of 0.71. Admittedly, the data were not sufficient to validate this trend, but they do point the way for future study. If one looks at the ratios of the values of (x/m) apparent extrapolated to zero k_1, a ratio of 0.4 is obtained for the two carbons. This indicates that the Nuchar C-1000N removed significantly more aqueous chlorine than the Darco G-60 in Part I, and at a faster rate in Part II. These trends merit further study, particularly in light of the work by Puri, *et al.*[19] in which they showed that the chlorine-fixing capacity of carbon in a gaseous system is related to the initial hydrogen content of the activated carbon. This work was for gaseous systems, but based on certain similarities pointed out in another paper by Puri, *et al.*,[8] the effect of initial hydrogen content of the activated carbon could be expected to be important in the aqueous system also.

Temperature Effects

The effect of temperature on this system appears to be more important than for most adsorptive materials. The data in Figures 12.3, 12.4, and 12.5 indicate that for temperatures of 30° C or less, one phenomenon is occurring, while at temperatures of 40° C or greater, another occurs. The difference in calculated energy of activation (4.56 versus 7.40 Kcal/mole) for ΔC/AC of 0.1 is significant and would tend to indicate two different types of reactions. These energies of activation are in the same range as reported by others for gaseous systems.[20,36] In addition, the fact that no brown-black water-soluble material was produced for those temperatures of 20° C or less for the duration of the runs tested, while such a material was produced for those temperatures of 30° C or greater, would also indicate that two different types of reactions are occurring. However, the rate constants (α_1 and α_2) for the Elovich plots appear to vary linearly with the inverse of the temperature over the entire temperature range (see Figures 12.7 and 12.2). This is expected behavior according to Low,[37] and may be an indication that characterization by the Elovich equation is more appropriate. This will be discussed more completely later.

The effect of temperature on the production of chlorides, which seems to be associated only with Part I, is shown in Figure 12.10. Because it was not possible to correlate temperature and chloride production directly, though it was possible to correlate ΔC and ΔCl^-, it was felt that the effect of temperature was either complex or of secondary importance in the production of chlorides.

Data Plotted According to Elovich Equation

The relationship between C_o and α_2 (Figure 12.8) indicates that α_2 may be related to the initial concentration. The data were not adequate to fully support this trend. It was also found that there was very little relationship between α_1 and α_2. It appears that the removal in Part I is dependent only on the characteristics of the activated carbon because of the invarient nature of α_1. Because the values of α_1 for the removal of aqueous chlorine and for the production of chlorides are similar at 30° C, it can be assumed that they represent the same process. In addition, the α_1 value for proton production at 30° C is about the same as the other two α values. Consequently, the disappearance of aqueous chlorine and the production of chlorides and protons probably all occur via the same reaction.

Chlorides

The relationship between ΔC and ΔCl^- (Figure 12.10) illustrates that for low temperatures more aqueous chlorine is removed from solution than is converted to chlorides. This could mean that the portion that is not converted to chlorides is held at the surface by physical absorption, capillary condensation, or non-dissociative chemisorption. Of these three possible mechanisms, the non-dissociative chemisorption seems to be a logical precursor to dissociative chemisorption in which chlorides are produced. A more thorough discussion of this follows later.

It is interesting to note that for higher temperatures (Figure 12.10 and Table 12.V) more chlorides are produced than chlorine removed in Part I. This is probably an artifact of the way in which ΔC is calculated. Extrapolation of the apparent first order process causes the value of ΔC to be smaller than if the concentration was determined just at the inception of the apparent first order process. The chloride production at 53° C and 60° C is about the same. This could indicate that maximum chloride production had been obtained, or that the rate for Part II had become sufficiently larger than the rate for Part I that fewer chlorides were produced.

Products

The brown-black water-soluble material was produced 10-20 minutes into the run, and then only for those temperatures greater than 20° C (Figure 12.13). However, it was also observed for one run at approximately 25° C that this brown-black material could be leached from the filter cake for a sample at 10 minutes even though none of the material was evident in the filtrate. This leaching was accomplished using distilled water.

Whenever the brown-black material was observed in the filtrate, it was also noticed that the pressure drop across the filter cake was higher than usual. It was not determined if the difference in pressure drop was due to a change in the activated carbon cake on the filter or to the interaction of the water soluble brown-black material with the filter or the glass septum below it. It was observed that the glass septum required thorough rinsing between samples to obtain reproducible results with respect to production of the brown-black material as measured by UV spectroscopy.

The fact that the brown-black solutions changed to a green or a yellow solution in an oxidizing medium would indicate that the green and yellow solutions contained oxidation products of the brown-black solutions. Apparently, the green compound is more highly oxidized than the yellow, because the former was produced preferentially at pH 5.5, a stronger oxidizing medium than pH 10, while the latter was produced preferentially at pH 10.0. This conclusion is consistent with IR scans.

The fact that the adsorption of the brown-black material on activated carbon is pH-dependent tends to indicate that a dissociable acid group is present, either on the activated carbon or in the water-soluble material, or both. Visual observations illustrated no difference in solubility of the brown-black material at various pH values, indicating that its solubility was not due entirely to acid dissociation.

Ultraviolet Spectra

The spectral scan of the brown-black water-soluble material in the visible-UV region shows three distinct peaks: one at about 270 nm, another at about 390 nm, and another at less than 240 nm. The peak at less than 240 nm is very strong and may be due to absorption from ions in solution not compensated for in the reference cuvette.

The peaks at 270 nm and 390 nm correspond quite well to the spectral scan of a single crystal of graphite observed by Ergun and McCartney.[41] They observed a weak peak at about 260 nm (ε = 1.9) and a very weak peak at about 410 nm (ε = 1.10). The UV visible scan of the brown-black material seems to indicate that it is a graphite-like compound.

Infrared Spectra

The IR spectra may be interpreted according to spectra for similar systems, or in terms of expected peaks. Both techniques were used. The broad peak observed at 3400 cm^{-1} is in the region characteristic of hydrogen stretching in the hydroxyl group.[42] This peak would normally be assigned to water, but a study of the IR spectra of graphitic oxide by Hadzi and Novak[43] clearly establishes that a strong peak at 3430 cm^{-1} is associated with water molecules. Because of the similarity between the visible-UV spectra for the products observed in this study and the graphite in Ergun and McCartney's study[41] and also because activated carbon is graphite-like in its makeup,[22,44] the peak at 3460 cm^{-1} is probably due to hydroxyl groups on the graphite-like material. A large number of hydroxyl groups on the activated carbon, and then perhaps on its water-soluble products, is to be expected because of the fact that the Darco G-60 was observed to be an "L" type of activated carbon as evidenced by an increase in hydrogen ion concentration when the carbon was wetted. The presence of hydroxyl groups is also consistent with the increased concentration of hydrogen for activated carbon observed by Studebaker[45] when activated carbon is wetted with water.

The very weak peaks observed at 2830 and 2900 cm^{-1} are indicative of alkane-type groups[41] and appear similar to peaks for the spectra of a coal reported by Friedel.[46] These peaks appear in all three scans, but are stronger for the brown-black material. The peak which appeared at 2470 cm^{-1} for the brown-black material could not be assigned, and no comparable peaks could be found in the literature.

The strong peaks found at 1580 and 1405 cm^{-1} are normally assigned to aromatic and carbonyl groups, respectively.[41] The very weak shoulder found at 1750 cm^{-1} is also associated with the carbonyl group,[47] but because this group is almost always accompanied by a strong peak in the 1700 to 1740 cm^{-1} range, the 1405 cm^{-1} peak is probably not representative of the carbonyl group. The combination of the two peaks at 1580 and 1405 cm^{-1} is characteristic of the carboxylate anion[41] and the combination of the two peaks seen at 1405 and 1130 cm^{-1} is characteristic of alcohols and phenols.[41] The shoulder at about 1600 cm^{-1} on the 1580 cm^{-1} peak for the brown-black material corresponds to the peak at 1600 cm^{-1} for the yellow material and the shoulder for the green material. The 1600 cm^{-1} peak is characteristic of aromatic materials.[41] The smaller peak at this frequency for the green material would be consistent with the previously drawn conclusion that the green material is more highly oxidized than the yellow material. The more highly oxidized material has its aromatic ring structure broken, and hence, appears green rather than yellow. From these observations, it was concluded that the peak at 1405 cm^{-1} was made up of contributions from both the carboxylate anion and alcohol structures. The yellow material probably contains neither of these, the green material probably contains predominantly the alcohol structures, and the brown-black material probably contains both alcohol structures and the carboxylate anion.

The strong band observed at 1130 cm^{-1} may also be due to a mixture (because of its breadth) of esters and ethers[41] and similar spectra have been reported by Matson[27] for some internal reflectance spectroscopy on sugar carbons and cellulose derivatives. The sharp strong peak at 875 cm^{-1} can be assigned to any of a number of structures, including some inorganic structures such as $CO_3^{=}$.[41] No assignment of this peak to structure was made. The peaks at 620 and 690 cm^{-1} correspond to the C-Cl aliphatic peak and were assumed to be due to the C-Cl interaction of the "fixed" chlorine.

The infrared spectrum, therefore, indicates that the brown-black material has similarities to spectra observed for activated carbon, graphite, and graphitic oxide. This would indicate that the water-soluble material is composed of portions of the surface that have a small enough molecular weight to be soluble. The observance of the IR peak characteristic of the C-Cl interaction, along with the fact that there is virtually no chloride production while the brown-black material is being produced, illustrates that the aqueous chlorine is combined with the activated carbon, perhaps by substitution at the unsaturated sites as suggested by Puri, *et al.*[11] This brown-black material can then be oxidized further to produce the green or yellow materials. These conclusions are consistent with other observations. For example, the decrease in the concentration of the brown-black material as Run B progresses

(Figure 12.13) is probably due to the oxidation of the material to produce a compound that does not absorb at 450 nm. This was probably only observed for this run because of the higher temperature and higher solution concentration of the brown-black material.

Crank Model

The system studied herein involved an initial removal followed by a chemical reaction. Crank[48] has developed a model that accounts for adsorption and simultaneous first order reaction. His formulation is of the form of a second order differential equation.

$$\frac{\partial C}{\partial t} = D\,\frac{\partial^2 C}{\partial x^2} - k_1 C \qquad (6)$$

where C, t, D, x, and k_1 are the solution concentration, time, diffusion coefficient, thickness of the diffusion layer, and the first order rate constant, respectively.

The results from Run SS were compared to the model with the diffusion coefficient based on the coefficients reported for the sodium cation. This comparison was made because it was anticipated that diffusion into the pore would maintain electroneutrality and that the aqueous chlorine ion would diffuse as a cation-anion pair with the sodium.[49] Mysels[50] has reported a diffusion coefficient of 1.2×10^{-5} cm^2/sec for Na^+ in water at 20° C. Since the system in Run SS was maintained at 30° C, this diffusion coefficient was adjusted, using the Nernst-Einstein approximation for diffusion in liquids[51] to a value of 10.8×10^{-4} cm^2/min. The first order rate constant for this run had been determined, using the method of least squares, to be 0.00455 min^{-1}. When the initial chlorine concentration of 4744 mg/l was used in the Crank model, the first order removal rate was similar to the experimentally observed rate constant (Figure 12.15). However, using this initial concentration, the amount of chlorine in solution calculated using the Crank model was found to be above that observed experimentally by an amount approximately equivalent to that removed in Part I. When the apparent initial concentration (C_o') was substituted into the Crank model for the initial solute concentration (3569 mg/l), good agreement was observed between the model and the experimental data. The model, thus modified, proved very accurate during the first order removal rate portion of the uptake but less

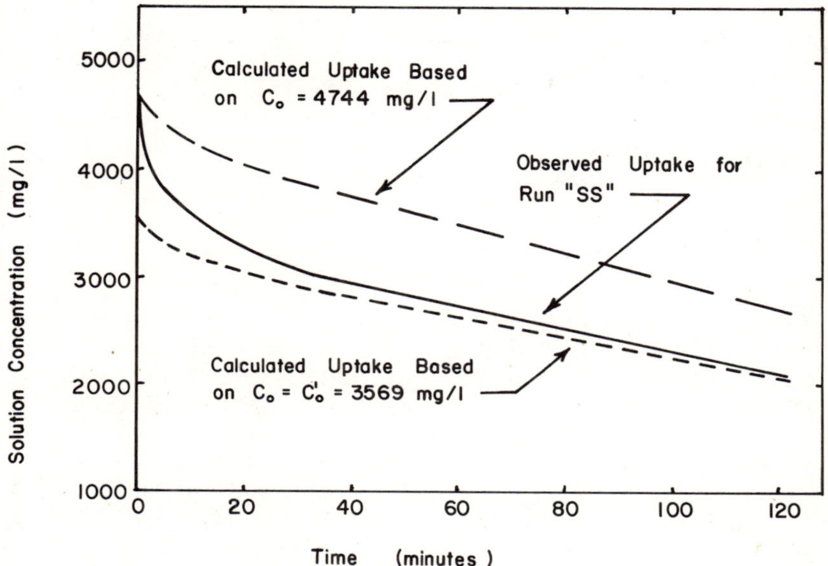

Figure 12.15. Crank model comparison to observed chlorine removal for run "SS."

reliable during the initial uptake phases. This was not wholly unexpected since the Crank model is designed to characterize a diffusion-reaction process and makes no provision to account for the case in which the initial adsorption removal is not diffusion controlled.

Rate Control

The data presented in Figures 12.4 and 12.5 were originally collected in an attempt to determine if Part II of the removal process was reaction controlled or diffusion controlled. Reaction control should be indicated by a linear Arrhenius plot[50] and if the diffusion coefficient can be properly expressed by the Nernst-Einstein equation, diffusion control should be indicated by a linear plot of k_1 versus 1/T. For a series of runs in which the activated carbon (Darco G-60) was at 3.0 g/l, the Arrhenius plot was linear and the plot of k_1 versus 1/T was not. However, for the plots in Figures 12.4 and 12.5, two regions of linearity were noticed. This tends to support the conclusion that different products may have been formed. Based on the linearity of the Arrhenius plot (the fact that the Crank model seemed to fit the data only when the chemical reaction was term controlled) and the characteristics of the system (small particle size, rapid mixing, and high concentration of aqueous chlorine), it can be concluded that Part II of the removal is reaction-controlled.

Mechanism

There can be no doubt that the amount of aqueous chlorine removed in Part I affects the rate of removal of aqueous chlorine in Part II (Figure 12.3). It is interesting to speculate as to the reason for this relationship. The increase in the base adsorption capacity from 0.06 to 1.3 meq/g at ten minutes is expected although it is greater than the two-fold increase observed by Puri, *et al.*[9] This indicates that the surface oxide groups, specifically those that evolve CO_2 on heating, are increasing in concentration. The lower value for the BAC at 120 minutes into the run means that fewer of these surface oxides are present at that time. The most obvious explanation for the decrease in the rate of removal for Part II is that the formation of surface oxides with concurrent chloride production in Part I "seals" the surface for the reaction that is present in Part II. However, two observations seem to be inconsistent with that interpretation. The first is that the BAC decreases with time, and one would expect that once the surface "sealing" was accomplished and prevented reaction at that point, any decrease in BAC would be accompanied by an increase in the removal rate for Part II. The other observation is embodied in Figure 12.14. The redosing of the activated carbon suspension with aqueous chlorine was accompanied by an initial sudden decrease followed by a semilogarithmic disappearance as in the initial dosing. However, the first order rate constant for the redosing was greater than for the original treatment, a relationship not expected if the rate was controlled by the presence of surface oxides. One would expect the concentration of surface oxides to remain the same or increase on redosing. Thus, the value of the first order rate constant for Run 14A should be equal to or smaller than that for Run 14. It was, however, about two times larger.

An alternate explanation for the inverse correlation between the amount of material removed in Part I and the rate of removal in Part II has to do with an observation on the alteration of physical characteristics with adsorption of gaseous chlorine. Puri and Bansal[52] investigated the variation of the adsorption of water vapor on activated carbons that have been treated with gaseous chlorine at various temperatures. They observed marked changes in the adsorption isotherms for water and concluded that the "fixed" chlorine appeared to alter the pore structure and size distribution of the capillary pores. The number of smaller capillaries increased as is consistent with this behavior.[53] It was not determined if this increase in small capillaries originated from constriction of larger ones or from formation of new ones. Because of the decrease in total adsorption capacity, the increase probably resulted from constriction of larger capillaries. The alteration of the pore structure upon treatment with gaseous chlorine may also occur in solution.

In the present study, for temperatures below 50° C the amount of chlorine removed from solution in Part I was observed to be larger than the chloride production. This remaining, unaccounted-for removal could be physically adsorbed or "fixed" at the surface. This adsorption or "fixation" may have been accompanied by a constriction of the pores, even to the point of blocking some of the pores so that solution diffusion of reactants in and products out was greatly hindered. This explanation is consistent with Runs 14 and 14A in that when the concentration of aqueous chlorine in solution for Run 14 became low, desorption occurred with subsequent adsorption in Run 14A. It was also consistent with the difference in BAC, because the rate control does not seem to be related to BAC, but rather to that portion of the removal in Part II that is not associated with an increase in BAC. This observed adsorption or "fixation" may be a form of capillary condensation, though such a conclusion is pure speculation.

An attempt was made to formulate a model for the removal of aqueous chlorine on activated carbon using the Elovich parameters, but the data were inadequate to allow such formulation. That approach does appear to hold promise for future studies.

Stoichiometry

The Elovich plots for proton production aid in the determination of stoichiometry of the reactions involved in Parts I and II. Comparing α_1 for the disappearance of aqueous chlorine and the production of protons in Part I, it can be seen that the variation of α_1 with temperature is very nearly the same (Figure 12.12). In addition, a comparison of the chloride production and the total proton production at the end of Part I (see Figure 12.11) shows that one chloride was produced for every proton produced in Part I. Consequently, there appears to be one proton produced for every chloride ion that is produced. The proton must originate either from the activated carbon surface or from the water because the aqueous chlorine is almost entirely OCl^- at pH 10.0. The increase in the BAC indicates that a CO_2 evolving group ($-COO^-$) must have been formed.[36] One such reaction might be

$$\geq C\text{-}OH + OCl^- \rightleftarrows -COO^- + Cl + H^+ \qquad (7)$$

If the chloride production was not used exclusively to form CO_2 evolving groups, an alternate reaction might be

$$\geq C\text{-}H + OCl^- \rightleftarrows \geq C{=}O + Cl^- + H^+ \qquad (8)$$

It was not found convenient to write any reaction in which water acted as the source for the protons without postulating some exotic redox reactions.

The slope of the α_2 curves (Figure 12.12) may indicate that one proton was produced for every two hypochlorite molecules that were removed from solution. However, the data are not conclusive. In Part II, the aqueous chlorine was "fixed" by the activated carbon, and a redox reaction in which chlorides were released into solution did not occur. The exact stoichiometry depends on the reaction products, and they could not be identified.

Relating to Other Studies

Most of the data in this study were collected at aqueous chlorine concentrations far in excess of those used in water treatment. For those runs at less than 10 mg/l, the data are consistent with the runs at high concentrations. Consequently, one could expect to observe a two-part removal of aqueous chlorine in water treatment plants using powdered activated carbon. The removal of aqueous chlorine using granular activated carbon in filters may be controlled by the kinetics of diffusion, and hence, removal patterns would be different, though the same principles and mechanisms should apply.

The results of this study appear to disagree with the observations of Baylis.[2] He observed the formation of carbon oxide and the complete conversion of aqueous chlorine to chlorides. However, his system was markedly different than the one used in this study, and the time required for complete recovery of the chlorides in his study was about 24 hours. The portion of chlorides that were not released until his filter had set overnight may correspond to the "fixation" of chlorine in this study with subsequent oxidation and release of the chlorides back into solution. Carbon dioxide production was not observed in this study, but that may have been due to the use of a different type of activated carbon. Different products for different carbon can be expected, particularly when approximately 40 years separate the technology of production of the carbon used in Baylis' study and this study.

The observation that activated carbons appear to recover some of their adsorptive capacity for chlorine if allowed to stand for a short time[2,7,32] can also be explained by the results of this study. If time is allowed for Part II of the removal process to proceed until the concentration in solution reaches zero, Part I can be utilized again as shown by Runs 14 and 14A, and the activated carbon appears to regain its adsorption capacity. Based on the fact that the removal process is a dynamic one that involves more than just equilibrium adsorption, it should be possible to design a removal system that has a tremendously high removal capacity (utilizing virtually all of the activated carbon) and a very long life.

The study also shows that activated carbon is solubilized; consequently, if dechlorination using activated carbon is practiced, one must expect to get soluble organics in the product water. Admittedly, the concentrations are very low, but if the water is to be used where no organic contamination is allowed, chlorination followed by dechlorination with activated carbon should not be practiced.

CONCLUSIONS

The removal of aqueous chlorine from solution using activated carbon is a two-part process, each having its characteristic rate of removal. Chloride production occurs only in Part I, while in Part II, the chlorine is "fixed" or combined with the activated carbon. Proton production occurred in both Parts I and II, but the largest portion of protons was produced in Part I. Concurrent with the production of chlorides and protons in Part I, surface oxides are formed on the activated carbon surface. The chlorine is "fixed" or combined with the activated carbon in Part I, and the activated carbon is also broken down and solubilized under certain conditions. These brown-black water-soluble products appear from UV-visible and IR spectroscopy to be graphitic oxides. If these graphitic oxides are allowed to remain in an oxidizing environment, they are further oxidized to green or yellow materials and apparently the "fixed" chlorine is released. They are not released into solution for reaction temperatures of 20° C or less for the time span of the experiments conducted in this investigation.

The rate of removal of aqueous chlorine using activated carbon in Part II of the removal process is inversely proportional to the amount of aqueous chlorine removed in Part I. The reason for this has been concluded to be due either to a "sealing" of the pores of the activated carbon, or a "filling" of these pores. The removal of aqueous chlorine in Part II was concluded to be reaction rate controlled rather than diffusion rate controlled. The stoichiometry of Part I could not be proven conclusively, but the following reactions seemed plausible:

$$\gtrdot C\text{-}OH + OCl^- \rightleftarrows -COO^- + Cl^- + H^+$$

$$\gtrdot C\text{-}H + OCl^- \rightleftarrows \, >C{=}O + Cl^- + H^+$$

The stoichiometry of Part II could not be determined.

The trends and mechanisms enumerated by this study were based primarily on observations for aqueous chlorine concentrations far in excess of those used in water treatment. However, the data collected at concentrations usually encountered in water treatment practices illustrate that the trends and mechanisms are present at both high and low concentrations. With further study, these trends should allow the formulation of reliable predictive removal equations for aqueous chlorine using activated carbon. The understanding of the mechanism of the removal process appears to yield a good explanation for both the time lag of chloride production observed by Baylis[2] and also the recovery of the adsorptive capacity of activated carbon with respect to aqueous chlorine observed by several authors.[16,22,33]

It was also concluded from this study that the data for the removal of aqueous chlorine using activated carbon could be fitted to the Elovich equation. In addition, the data for proton and chloride production also fit the Elovich equation. An attempt was made to formulate a comprehensive model for the system using the Elovich parameters and including other variables such as temperature and initial chlorine concentration, but the data were inadequate to allow this formulation.

REFERENCES

1. Baylis, J. R. *Elimination of Taste and Odor in Water* McGraw-Hill Book Co., New York, 1935.
2. Baylis, J. *J. Amer. Water Works Assoc., 21,* 787 (1929).
3. Fair, G. M., and J. C. Geyer. *Water Supply and Wastewater Disposal*. John Wiley and Sons, Inc., New York, 1954.
4. Fair, G. M., J. C. Geyer, and D. A. Okun. *Water and Wastewater Engineering,* Vol. 2, John Wiley and Sons, Inc., New York, 1968.
5. American Water Works Association. *Water Quality and Treatment*. McGraw-Hill Book Co., New York, 1971.
6. Babbitt, H. E., J. J. Doland, and J. L. Cleasby. *Water Supply Engineering*. McGraw-Hill Book Co., New York, 1955.
7. Hassler, J. W. *Activated Carbon*. Chemical Publishing Co., New York, 1963.
8. Puri, B. R., D. D. Singh, T. Chander, and L. R. Sharma. *J. Indian Chem. Soc., 35,* 181 (1958).
9. Puri, B. R., O. P. Mahajan, and D. D. Singh. *J. Indian Chem. Soc., 37,* 171 (1960).
10. Puri, B. R., and S. K. Sharma. *J. Indian Chem. Soc., 45,* 115 (1968).
11. Puri, B. R., A. K. Dhingra, and K. C. Sehgal. *Indian J. Chem., 7,* 174 (1969).
12. Donnet, J. B. *Chem. Abstracts, 54,* 23283 (1960).
13. Johnson, R. L., F. J. Lowes, Jr., R. M. Smith, and T. J. Powers. "Evaluation of the Use of Activated Carbons and Chemical Regenerants in Treatment of Waste Water," Advanced Works Treatment Research Publications (AWRR)-11 Report No. 999-WP-11, U. S. Dept. of Health, Education and Welfare, 1964.
14. Magee, V. *Proc. Soc. Water Treat. Exam., 5,* 17 (1956).
15. Kovach, J. L. *Ind. Water Eng., 8,* 30 (19).

16. Puri, B. R., and R. C. Bansal. *Indian J. Chem., 5,* 381 (1967).
17. Puri, B. R., S. L. Malhotra, and R. C. Bansal. *J. Indian Chem. Soc., 40,* 180 (1963).
18. Puri, B. R., and N. K. Sandle. *Indian J. Chem., 6,* 267 (1968).
19. Puri, B. R., S. S. Tulsi, and R. C. Bansal. *Indian J. Chem., 4,* 7 (1966).
20. Reyerson, L. H., and A. W. Wishart. *J. Phys. Chem., 42,* 679 (1938).
21. Boehm, H. P., E. Diehl, W. Heck, and R. Sappok. *Ang. Chem., 3,* 669 (1964).
22. Coughlin, R. W. "Effect of Surface Groups on Adsorption of Pollutants," Report to Water Quality Office, Environmental Protective Agency Grant #WP-00969-01, Program $17020-06/70, 1970.
23. Coughlin, R. W., and F. S. Ezra. *Environ. Sci. Technol., 2,* 291 (1968).
24. Coughlin, R. W., F. S. Ezra, and R. N. Ton. *J. Colloid Interface Sci., 28,* 386 (1968).
25. Anderson, R. B., and P. H. Emmett. *J. Phys. Chem., 56,* 753 (1952).
26. Anderson, R. B., and P. H. Emmett. *J. Phys. Chem., 56,* 756 (1952).
27. Mattson, J. S., and H. B. Mark, Jr. *Activated Carbon.* Marcel Dekker, New York, 1971.
28. Puri, B. R. In *Chemistry and Physics of Carbon*, Vol. 6, Marcel Dekker, New York, 1970, p. 191.
29. Puri, B. R. In *Proceedings of the Fifth Conference on Carbon,* Vol. 1, The Macmillan Co., New York, 1962.
30. Puri, B. R., S. Singh, and O. P. Mahajan. *J. Indian Chem. Soc., 42,* 427 (1965).
31. Studebaker, M. L., E. W. D. Huffman, A. C. Wolfe, and L. G. Nabors. *Ind. Eng. Chem., 48,* 162 (1956).
32. Baumann, E. "Chlorine Removal Capacity of Precoat Carbon Filters," Progress Report of the Iowa Engineering Station, 1963.
33. Bansal, R. C., F. J. Vastola, and P. L. Walker, Jr. *J. Colloid Interface Sci., 32,* 187 (1970).
34. Deitz, V. R., and E. F. McFarlene. In *Proceedings of the Fifth Conference on Carbons,* Vol. 2, The Macmillan Co., New York, 1963.
35. Newman, J. O. H., L. Stanley, P. L. Evans, A. J. T. Coldrick, and T. J. Kempton. *Nature, 213,* 280 (1967).
36. Puri, B. R., D. D. Singh, and K. C. Sehgal. *J. Indian Chem. Soc., 48,* 513 (1971).

37. Low, M. J. D. *Chem. Rev., 60,* 267 (1960).
38. *Standard Methods for the Examination of Water and Wastewater,* 13th ed., American Public Health Association, New York, 1972.
39. Puri, B. R., and D. D. Singh. *J. Indian Chem. Soc., 37,* 401 (1960).
40. Nelson, F. M., and F. T. Eggertsen. *Anal. Chem., 30,* 1387 (1958).
41. Ergun, S., and J. T. McCartney. In *Proceedings of the Fifth Conference on Carbon,* Vol. 2, The Macmillan Co., New York, 1963.
42. Reilley, C. N., and D. T. Sawyer. *Experiments for Instrumental Methods,* McGraw-Hill Book Co., New York 1961.
43. Hadzi, D., and A. Novak. *Trans. Faraday Soc., 51,* 1614 (1955).
44. Snoeyink, V. L., and W. J. Weber, Jr. *Environ. Sci. and Technol., 1,* 228 (1967).
45. Studebaker, M. L. In *Proceedings of the Third Conference on Carbon,* Pergamon Press, New York, 1959.
46. Friedel, R. A. In *Proceedings of the Fourth Conference on Carbon,* Pergamon Press, 1960.
47. Friedel, R. A., and L. J. E. Hofer. *J. Phys. Chem., 74,* 2921 (1970).
48. Crank, J. *The Mathematics of Diffusion.* Clarendon Press, London, 1965.
49. Binning, C. D., Jr. "The Kinetics of Aqueous Chlorine Interaction with Activated Carbon," Thesis, University of Delaware, 1973.
50. Mysels, K. J. *Introduction to Colloid Chemistry.* Interscience Publishers, New York, 1967.
51. Bird, R. B., W. E. Stewart, and E. N. Lightfoot. *Transport Phenomena. Part I. Material and Energy Balances.* John Wiley and Sons, Inc., New York, 1954.
52. Puri, B. R., and R. C. Bansal. *Indian J. Chem., 5,* 556 (1967).
53. Smisek, M., and S. Cerny. *Active Carbon.* Elsevier Publishing Co., New York, 1970.

CHAPTER 13

DYNAMICS OF BREAKPOINT CHLORINATION

Irvine W. Wei and J. Carrell Morris*

INTRODUCTION

Elucidation of the breakpoint chlorination, the oxidation of ammonia by chlorine in dilute aqueous solution, has generally been attributed to the work of Calvert[1] and of Griffin and Chamberlin[2] in the late 1930's. It should be noted, however, that both Holwerda[3] and Chapin[4] described the phenomena associated with this reaction at an earlier date, even though the terminology and conventional form for presentation of the data were not used. The major contribution of Calvert and of Griffin and Chamberlin was their recognition that this complex reaction provided a means for the achievement of free-residual chlorination, with attendant benefits to water quality, without the problem of excessive available chlorine in the distributed water. The importance of this finding has increased constantly over the past thirty years as the quality of raw water supplies has tended everywhere to be degraded and as the potential hazard of waterborne pathogenic viruses, which are quite insensitive to combined chlorine, has been recognized.

The stoichiometric pattern of the reaction, which develops after an extended period of reaction, is an odd and characteristic one that has given rise to the

*I. W. Wei, Department of Civil Engineering, Northeastern University, Boston, Massachusetts. J. C. Morris, Division of Engineering and Applied Physics, Harvard University, Cambridge, Massachusetts.

appellation "breakpoint." It was first described in detail by Griffin;[5] an extensive contribution to the understanding of the stoichiometry was later made by Griffin and Chamberlin[2] and by Palin.[6]

A diagrammatic representation of the standard pattern, which has been derived primarily from Palin's data, is shown in Figure 13.1. Such a pattern develops most clearly

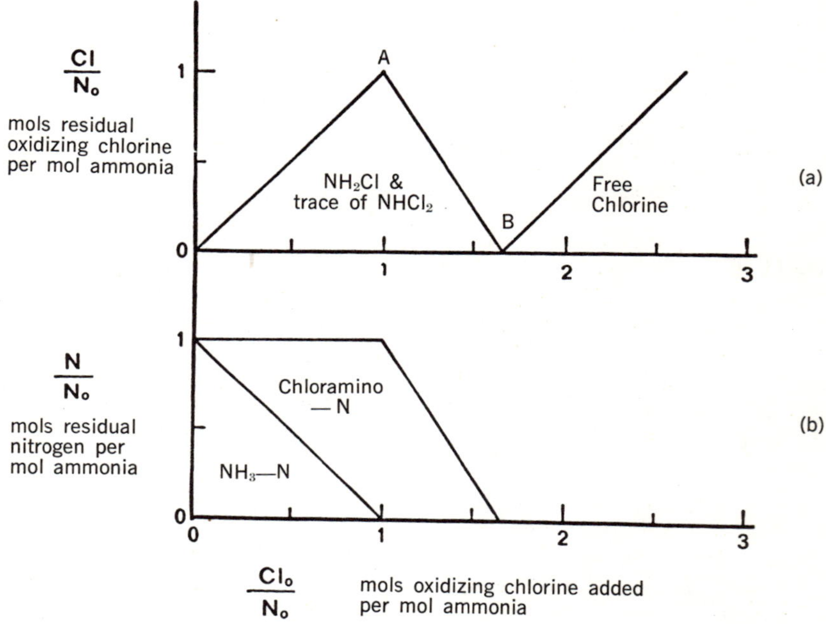

Figure 13.1. Schematic diagram of breakpoint chlorination.

and typically when the pH is near neutrality and in water containing no chlorine-reducing substances and no nitrogenous matter other than ammonia.

When successively increasing concentrations of aqueous chlorine are added to separate portions of water containing ammonia, and the mixed solutions allowed to sit for from several hours to a day or more, the following characteristics are noted. The residual oxidizing chlorine found increases proportionately with the dosage of chlorine until a molar ratio of added chlorine to initial ammonia equal to unity is reached, as shown in the segment of Figure 13.1(a) up to point A. Also, as shown in Figure 13.1(b), there is essentially no oxidation of combined nitrogen over this range. However, once the ratio of oxidizing chlorine to combined nitrogen is in excess of unity, a decrease in residual oxidizing chlorine is found with an increased dosage of chlorine. At point B, the "breakpoint," essentially all the oxidizing chlorine is reduced after an appropriate period of contact and all the ammonia-N is oxidized. At ratios in excess of B there remains a residuum of free chlorine as HOCl or OCl^-. Also shown by Figure 13.1 is the fact that in the area under point A, the so-called "hump," the residual chlorine is present exclusively as chloramines.

Essentially, the development of the breakpoint results from redox reaction between aqueous chlorine and ammonia, with the location of the breakpoint indicating the stoichiometry of the reaction. The molar ratio of reduced chlorine to oxidized nitrogen, equal to about 1.65 at the breakpoint, shows that the major oxidation of the ammonia is to gaseous nitrogen, which requires a ratio equal to 1.50. The excess over 1.50 is accounted for by a small side formation of nitrate as shown by Palin.[6]

Stoichiometrically, then, the breakpoint process, except for the nitrate formation, can be accounted for by two reactions

$$HOCl + NH_3 = NH_2Cl + H_2O \tag{1}$$

and

$$2NH_2Cl + HOCl = N_2 + H_2O + 3H^+ + 3Cl^-. \tag{2}$$

The second reaction is presumed to occur whenever excess free chlorine remains after the formation of NH_2Cl.

These equations should not be interpreted as representing the mechanism of the breakpoint process. Although the first reaction undoubtedly occurs as written, the second certainly does not occur in that way. Additional facts about the breakpoint, which must be accounted for in any mechanistic set of reactions, are the maximum rate of development of the breakpoint in the pH region near neutrality, and the intermediate formation of considerable $NHCl_2$ only on the acid side of the pH region of maximum rate.

In order to understand properly the dynamics and the mechanism of the breakpoint process, it is essential to have detailed information about its time course, that is, about the rate of reduction of chlorine (loss of oxidizing chlorine) and about the formation and reaction of various chloramines. Previous studies have provided only very limited data of this sort. Although the studies of Griffin and Chamberlin[2] and of Palin[6] covered broad ranges of pH and of chlorine to nitrogen ratios, they dealt only scantily with the time factor. Griffin and Chamberlin carried out analyses only for contact times of 20 minutes, 2 hours, and 24 hours; Palin made analytical measurements only at 10 minutes, 2 hours, and 24 hours.

It is the first hour of reaction that is most significant in terms of practical application of the breakpoint process, for disinfection of water supplies must almost always be accomplished within that period. Therefore, knowledge of the species and concentrations of possible germicides present during the first hour is important. The same early period, and particularly the first few minutes of it, is also of greatest theoretical interest for elucidation of the fundamental dynamics and mechanism. At neutral pH about half the oxidizing chlorine has been reduced after the first ten minutes.

In an attempt to elucidate major aspects of kinetics and mechanism, the work now being reported has focused on the dynamics of the process during the first hour or so. No attempt has been made to cover the broad pH ranges and the chlorine to nitrogen ratios dealt with by others. Only pH values near neutrality and the maximum rate of the breakpoint process, and chlorine to nitrogen ratios close to that of the breakpoint itself have been investigated. On the other hand, some factors, notably temperature and the total concentrations of reactants that are relevant to the dynamics of the reaction and have not been studied extensively by others, have been varied.

EXPERIMENTAL METHODS

The essential requirements for a successful study of the dynamics of the initial stages of breakpoint chlorination are twofold. First, appropriate analytical methods must be available for determination of the concentrations of as many pertinent chemical species as possible. Second, sufficient determinations must be obtained within a relatively short period of time in order for a delineation of the dynamic pattern of the determined constituents to be made.

A comparative study was made of various methods of determining the forms of oxidizing chlorine available, including amperometric titration, NOT-FAS method (neutral orthotolidine as colorimetric indicator), and DPD-FAS method (diethyl - p - phenylenediamine as indicator). The

DPD-FAS method, using ferrous ammonium sulfate as titrant and diethyl - p - phenylenediamine as indicator with differing conditions of titration, was finally adopted as the analytical means for differentiation of free chlorine and components of chloramines in these kinetic experiments. This method was selected on the basis of the following considerations:

1. DPD-FAS can determine free chlorine without interference from NCl_3. Unambiguous determinations of free chlorine were considered highly important because it is the most potent species available for disinfection.
2. Operationally, DPD-FAS is the simplest of the three methods evaluated. This is an absolute advantage when as many samples as possible are to be determined in a relatively short period of time.
3. Determinations by the DPD-FAS method are all done at a relatively neutral pH of 6.4, whereas the amperometric and NOT-FAS methods require adjustments of pH to about 4 during the procedure.
4. DPD-FAS method was extensively tested and found to give adequately accurate and discriminatory results.
5. A recent comprehensive study[7] of nine analytical methods presently available for determining free and total residual chlorine rated DPD-FAS as the most accurate and precise method.

DPD-FAS Method

The DPD-FAS method[8,9] is essentially a modification of NOT-FAS, with diethyl - p - phenylenediamine (DPD) replacing neutral orthotolidine as the indicator. A standard solution of ferrous ammonium sulfate is used as the titrant for both methods.

The procedure involves three successive steps of titration: In the absence of iodide ion, free chlorine reacts instantly with DPD at pH 6.4 to produce a red color. Titration with the ferrous solution results in a disappearance of the red color at the end-point. The subsequent addition of a small amount of iodide acts catalytically to cause NH_2Cl to reform the red color. After titration with ferrous solution to a second disappearance of the red color, further addition of iodide to excess evokes a third development of the red color corresponding to the $NHCl_2$. A third titration with FAS yields concentrations of this substance.

Unlike the reaction with neutral orthotolidine, any NCl_3 present is included with the $NHCl_2$ and does not react like free chlorine. Addition of iodide before the DPD, however, causes a proportion of the NCl_3 to appear with the free chlorine. A supplementary procedure based upon alteration of the order of addition of reagents to a separate sample permits the estimation of NCl_3 when this result is desired.

In our experience, the colors produced are stable, few reagents are required and a full response is obtained from $NHCl_2$. In the titrimetric procedure, decolorization by standard ferrous ammonium sulfate solution is instantaneous, thus enabling each step to be performed rapidly.

The accuracy of the DPD-FAS method for differentiation of free chlorine, NH_2Cl, and $NHCl_2$, was examined in preliminary tests. Pure solutions of free chlorine, NH_2Cl and $NHCl_2$ were prepared with concentrations on the order of 10^{-4} *M*. Then the concentrations were determined by DPD-FAS, standard iodometric, and UV spectrophotometric methods. A comparison of the results indicated that free chlorine and $NHCl_2$ could be determined by the DPD-FAS method with a discrepancy of less than 2 per cent, while the determination of NH_2Cl was within about 6.5 per cent. Closer agreement was obtained when measurements of total available chlorine were compared. With pure NH_2Cl solutions, about 3.3 per cent of the total chlorine appeared to be free chlorine with the DPD-FAS method, and about 2 per cent was measured as $NHCl_2$. Dissolution of atmospheric oxygen into the sample during mixing seemed to obscure the end points of the titrations to a certain extent. However, this interference was greatly minimized through sufficient experience with the titration.

General Laboratory Operations

Experiments on the rate of breakpoint reaction are essentially kinetic studies of the oxidation of ammonia by chlorine in dilute aqueous solution. Because of the low concentrations of ammonia and chlorine solutions involved in normal occurrence of the breakpoint process and used in these studies, extreme care had to be taken to eliminate any extraneous reducing substances and any extraneous ammonia in the preparation of reagents or in the dilution process. As in most chemical kinetic studies, the temperature and pH of the reacting system had to be maintained as nearly constant as possible.

Glassware and Reagents

All glassware underwent a standard cleaning procedure before each use. The procedure involved soaking the glassware overnight in a chlorine solution of about 10 mg/l, followed by rinsing, first with distilled water and then with chlorine-demand-free water.

Chlorine-demand-free (CDF) water, used in the preparation of most reagents and all dilutions, was prepared from glass-distilled water by adding approximately 10 mg/l of chlorine, allowing it to stand stoppered for at least one week in the laboratory, and then exposing it to direct sunlight or to ultraviolet lamps until the DPD-FAS method showed the absence of residual chlorine. The chlorine demand of the distilled water was generally rather low. One test showed the presence of about 9 mg/l chlorine after the distilled water dosed with 10 mg/l of chlorine had been allowed to stand overnight. The use of ultraviolet lamps was almost as efficient as sunlight in decomposing aqueous chlorine, and was a convenient procedure to use during winter months or when strong sunlight was not available.

Stock solutions of chlorine were prepared by bubbling chlorine gas (Matheson lecture cylinder) through a sintered-glass diffuser into doubly-distilled water until a faint yellow color was apparent. Potassium hydroxide solution (ca. 1 *M*) was then added to raise the pH from approximately 1.5 to 4.2. In this way equilibrium was shifted toward the right in the equation:

$$Cl_2 + H_2O = HOCl + H^+ + Cl^-$$

and loss of chlorine by volatilization was decreased. The solutions were stored in a refrigerator (ca. 0° C) in a low-actinic Pyrex bottle to avoid loss of chlorine through any of the reactions,

$$2\ HOCl \xrightarrow{\text{light}} 2H^+ + 2Cl^- + O_2$$

$$HOCl + OCl^- \longrightarrow ClO_2^- + H^+ + Cl^-$$

$$HOCl + ClO_2^- \longrightarrow ClO_3^- + H^+ + Cl^-.$$

All working chlorine solutions were prepared by dilution of stock solution with CDF water. Stock and working chlorine solutions were standardized by titration of the iodine, liberated from iodide, with thiosulfate, using starch as an indicator. The concentrations of chlorine stock solutions were usually on the order of 10^{-2} *M*.

Ammonium chloride was used as a source of ammonia for the breakpoint process. Its stock solutions (0.01 *M*) were prepared by dissolving a specific amount of predried reagent-grade ammonium chloride in CDF water, and its working solutions by dilution of portions of stock solution with CDF water.

The pH values of the reaction systems studied ranged from 6.7 to 7.2. Mixtures of $H_2PO_4^-$ and $HPO_4^=$ provided adequate buffer capacity and a total phosphate concentration was maintained at 0.01 *M* throughout all the experiments.

The buffer solutions were also required to be free of chlorine demand. Individual stock solutions of Na_2HPO_4 and NaH_2PO_4 were prepared by dissolving the proper amount of each chemical (Mallinckrodt reagent grade) in a few hundred milliliters of glass-distilled water. Ten mg/l of chlorine were added, and dilution continued until the concentration of phosphate reached 0.1 *M*. These stock solutions then underwent the same procedure for dechlorination as was used in the preparation of CDF water. Any subsequent dilutions were made with CDF water so that no chlorine demand would be introduced.

Working buffer solutions at a specific pH were prepared by mixing proper volumes of 0.02 *M* Na_2HPO_4 and NaH_2PO_4 solutions to the desired pH, as measured by a pH meter. The pH of the reacting systems, buffered at the total phosphate concentration of 0.01 *M*, could be maintained within ±0.05 pH unit during the course of reaction.

Another buffer system was also used in some experiments to more closely simulate natural water conditions. This was accomplished by adding $NaHCO_3$ stock solution to the solutions of reactants so that a total carbonate concentration of about 0.01 *M* could be maintained in the final reaction system. The $NaHCO_3$ (Merck reagent grade) stock solution had also undergone the same procedure of chlorination and dechlorination as was used in the preparation of CDF water. The pH of this buffer solution was adjusted with strong acid or base to the desired value shortly before the reaction was initiated. There were more variations in pH during the period of reaction with this buffer system, generally about ±0.2 unit, depending on the concentrations of reactants.

All of the analytical reagents for the DPD-FAS determinations were prepared according to Palin's directions.[8] Sources of the solid chemicals were: DPD Oxalate-Eastman #7102; $NaHPO_4$ and KH_2PO_4 Mallinckrodt, analytical reagent; KI-granular, Merck, reagent; and, ferrous ammonium sulfate-Merck, reagent.

All pH values were measured with a Beckman Model 76A Expandomatic pH meter. The electrodes consisted of a glass electrode with a pH range of 0 to 14 and a calomel reference electrode. All pH values were measured at a room temperature of 20° C. The pH values of buffer solutions for low-temperature experiments at 15, 10, and 5° C were adjusted to the desired figure of 20° C before the temperature of the buffer solution was decreased. No attempt was made to accurately measure pH at the low temperatures. The buffer systems used vary only slightly in pH with temperature.

Because reaction rates are usually strongly dependent upon temperature, constant, controlled temperatures are essential for kinetic studies. For these experiments the room temperature was controlled at 20° C by a central air-conditioning system, the temperature of reacting solutions was controlled by the use of a constant-temperature water-bath, manufactured by Wilkens-Anderson Compnny: and heating or cooling was controlled by a thermo-regulator with a tested sensitivity of $\pm 0.05°$ C.

Procedures for Kinetic Experiments

A standard procedure was developed for all the experiments with phosphate buffer, as outlined in the following directions:

1. Mix proper amounts of Na_2HPO_4 and NaH_2PO_4 solutions of 0.02 *M* to give a volume of 400 ml at the desired pH level, measured by the pH meter.
2. Prepare 200 ml of an ammonium chloride solution with four times the desired final nitrogen concentration.
3. Mix equal volumes, normally 200 ml each, of solutions 1 and 2 in a 500-ml Erlenmeyer flask, and place the flask in the constant-temperature water-bath.
4. Prepare 200 ml of a chlorine solution with four times the desired final chlorine concentration.

5. Mix equal volumes of solutions 1 and 4 in a 500-ml Erlenmeyer flask, and place the flask in the constant-temperature water-bath.
6. After the contents of the flasks have reached the temperature of the water-bath (up to one hour of contact) mix them rapidly at time zero by pouring simultaneously into a one-liter beaker with a magnetic stirring bar spinning inside.
7. Immediately withdraw the first sample from the beaker, using a Nessler tube (50- or 100-ml) for volume measurement and then quickly return the beaker (reaction vessel) to the water bath.
8. Record as the reaction time the time when the sample was poured into a 150-ml beaker containing the reagents for free chlorine determination by the DPD-FAS method.
9. Determine free chlorine by titration with standard ferrous ammonium sulfate according to the DPD-FAS method.
10. Proceed to determine NH_2Cl and $NHCl_2$ content of the same sample according to the DPD-FAS method.
11. Continue sampling and immediate determinations at regular, frequent times until sufficient samples have been analyzed; ordinarily the times are to be spaced at least 6 minutes apart.
12. Repeat step 3 and steps 5 through 11 with the same starting solutions, staggering the reaction times at which samples are taken so as to fill the intervals between the sample times of the previous run. In most instances, it is necessary to perform two such repetitions in order to obtain sample points spaced approximately every two minutes when the data from the three identical runs are consolidated.

For those experiments with bicarbonate buffer, essentially the same procedure was followed. The only differences were minor modifications in the preparation of the buffered solutions of aqueous chlorine and ammonia as described previously.

KINETIC STUDIES

Design of Experiments

The kinetic experiments were designed to provide information about the pattern of the breakpoint process, particularly during the initial stages of reaction. Emphasis was placed on the effects of the following parameters:

pH

The rate of the breakpoint process at 15° C is greatest near pH 7.0 to 7.2.[6] Studies of the kinetic pattern near these pH values, where all of the characteristic features of the breakpoint process are manifested during a short period, were considered to be especially significant. Three pH values, 7.2, 7.0, and 6.7, were chosen for experimentation.

Temperature

Most previous studies of the breakpoint process were performed at ambient temperature, whatever it might have been. Since reaction rates are normally strongly dependent upon temperature, experiments at various controlled temperatures are needed to evaluate the importance of this parameter. In the present study, four temperatures, 20, 15, 10 and 5° C, were employed. This range of temperature was chosen to cover seasonal variations of water temperature in water treatment plants, and to provide a more complete picture of the phenomena occurring in breakpoint chlorination, especially during cold weather.

Initial Total Reactant Concentrations

Some indication of the effect of total concentration of reactants on the rate of breakpoint chlorination was obtained by varying the initial ammonia concentration, N_o, from 0.25 to 1.5 mg/l. Since, as will be described below, the chlorine ammonia ratio remained fixed, the aqueous chlorine concentration was varied in the same manner. Morris and Wei[10,11] on the basis of a proposed mechanism, has suggested that the pattern of intermediate species formation should vary characteristically and significantly with N_o.

Buffer System

A few experiments were done with bicarbonate instead of the phosphate buffer that was utilized in most of the experiments. Bicarbonate buffering is found in the natural condition, and the effect of the buffer, if any, may be detected by a comparison of the relative reaction patterns in the two buffers.

Molar Ratio of Chlorine to Ammonia

A molar Cl:N ratio of 1.8 was maintained for all of the experiments with phosphate buffers. Since the molar ratio at the breakpoint has been found to be near 1.65, a molar ratio of 1.8 should leave a detectable free chlorine residuum after the completion of the breakpoint process.

Experimental Limitations

As a result of unavoidable constraints, the kinetic experiments had certain limitations that kept the experimental results from being ideal.

Since it usually took approximately six minutes to completely determine the free chlorine, monochloramine, and dichloramine of a single sample, two or three identical repeated experiments with properly staggered sampling times were necessary to accumulate enough data to define the kinetic pattern satisfactorily. This technique resulted in greater scattering of results among the successive, identical experiments than is usually found for the successive points in a single run. As shown in the figures which follow, the scattering in free chlorine and monochloramine, due to the technique used, was very slight; however, in dichloramine it was rather noticeable. Nevertheless, the general consistency of the results from the successive, identical experiments serves to indicate a satisfactory degree of reproducibility for these kinetic experiments.

No determination of NCl_3 was made in any sample. All kinetic experiments were at pH values between 6.7 and 7.2, at which range the formation of NCl_3, according to Palin's observations,[6] was expected to be rather insignificant. Moreover, as the DPD-FAS method requires a separate sample for the estimation of NCl_3, the work involved would have required more time than was available if the schedule for the other determinations was to be maintained.

Also, no attempt was made to determine probable reaction products, *i.e.*, nitrogen or nitrate. Determinations of nitrogen would have required degasification of all solutions and exclusion of air. Determinations of nitrate are not especially reliable at 10^{-5} *M* concentration and have little significance for explication of the major reaction mechanism.

Results of Experiments

The results of all the kinetic experiments are presented in a series of figures which display the variations of free chlorine, NH_2Cl, $NHCl_2$, and total chlorine, with reaction time for the reaction conditions investigated. The total chlorine has been calculated as the sum of the free chlorine, NH_2Cl, and $NHCl_2$, as determined by the DPD-FAS method. To facilitate the evaluation of reaction stoichiometry and kinetic characteristics, the concentration of each species is expressed as molar ratio to initial ammonia concentration and has been plotted logarithmically. This means that the initial point for the total chlorine in phosphate-buffered systems is always 1.8.

Effect of Temperature

A central series of experiments was conducted with an initial ammonia-N concentration of 1.0 mg/l and initial chlorine to ammonia molar ratio of 1.8 at four different temperatures -- 20, 15, 10, and 5° C -- each at three different pH values. Figures 13.2 and 13.3 give results for pH 6.7; Figures 13.4 and 13.5 for pH 7.0; and Figures 13.6 and 13.7 for pH 7.2. Figures 13.8 and 13.9 show a supplementary series of experiments with an initial ammonia concentration of 1.5 mg/l, the same molar ratio of chlorine to ammonia of 1.8, a pH value of 7.0 and the four different temperatures.

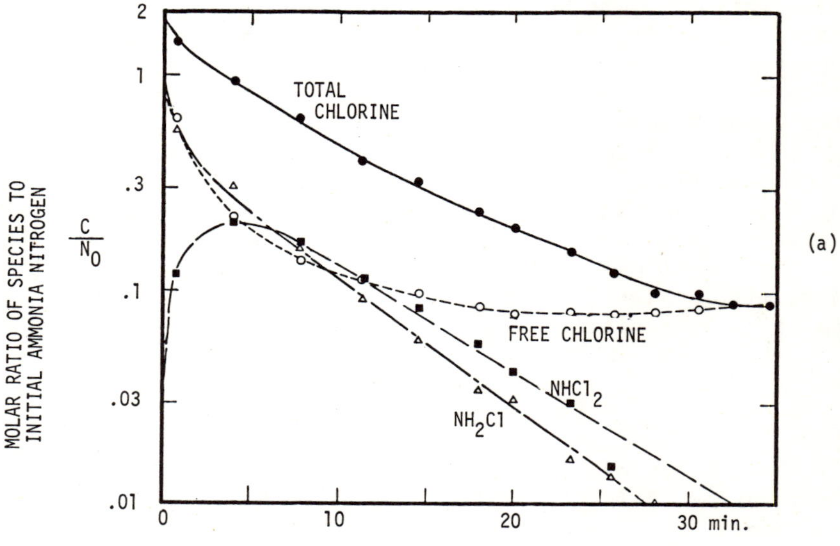

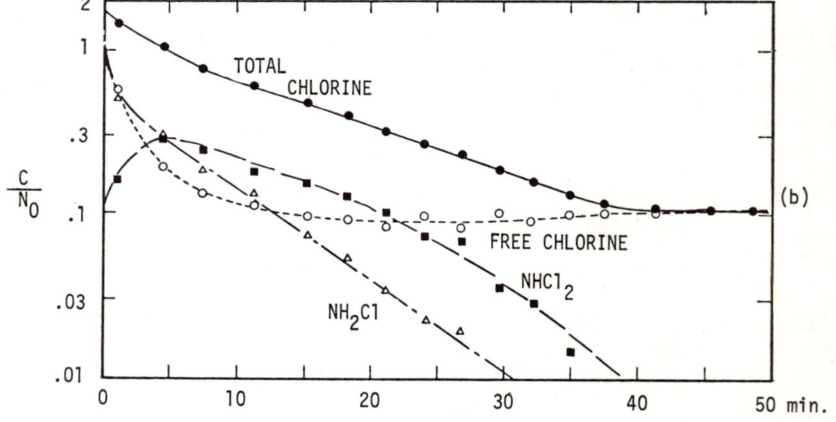

Figure 13.2. Chlorination of ammonia at pH 6.7, N_0 1 mg/l N, C_0/N_0 1.8, P_t 0.01 M, (a) 20° C, (b) 15° C.

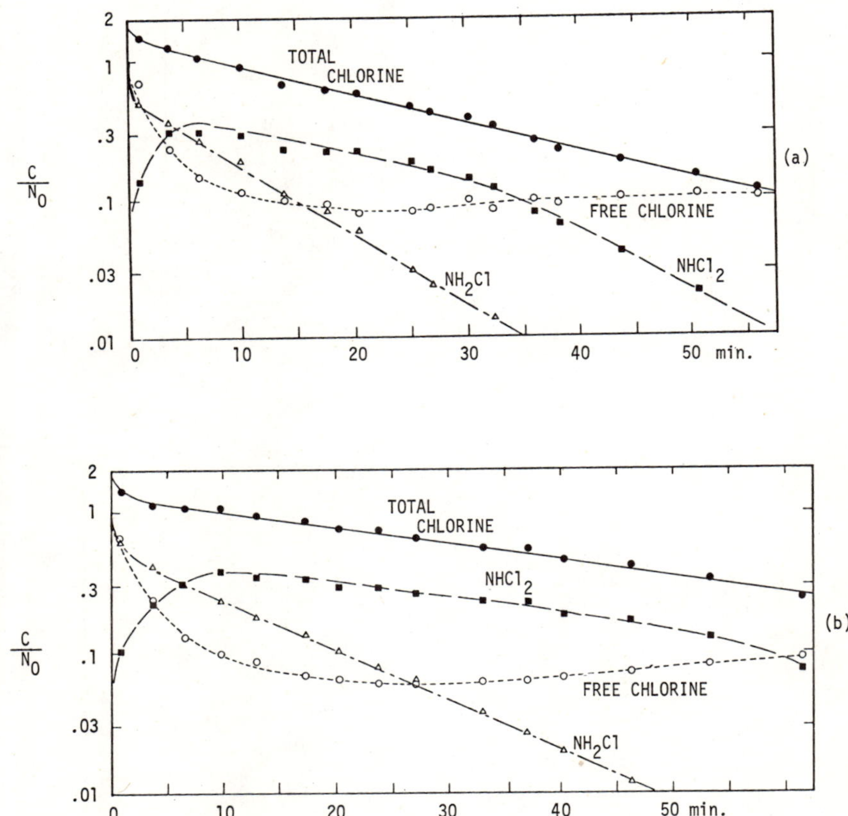

Figure 13.3. Chlorination of ammonia at pH 6.7, N_0 1 mg/l N, C_0/N_0 1.8, P_t 0.01 M, (a) 10° C, (b) 5° C.

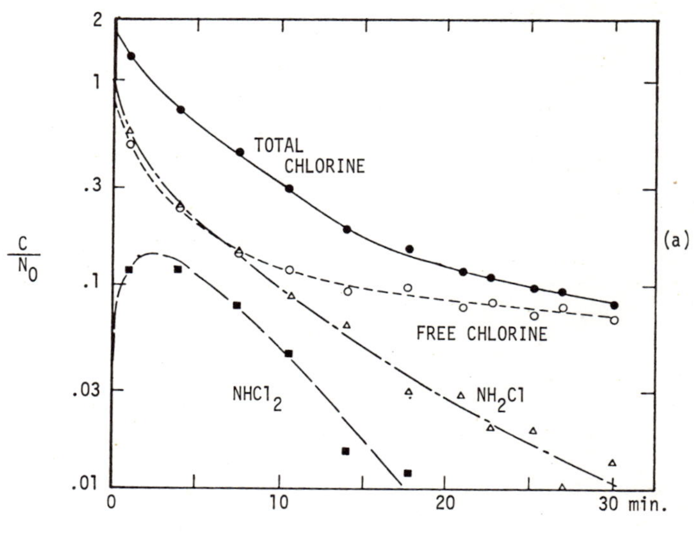

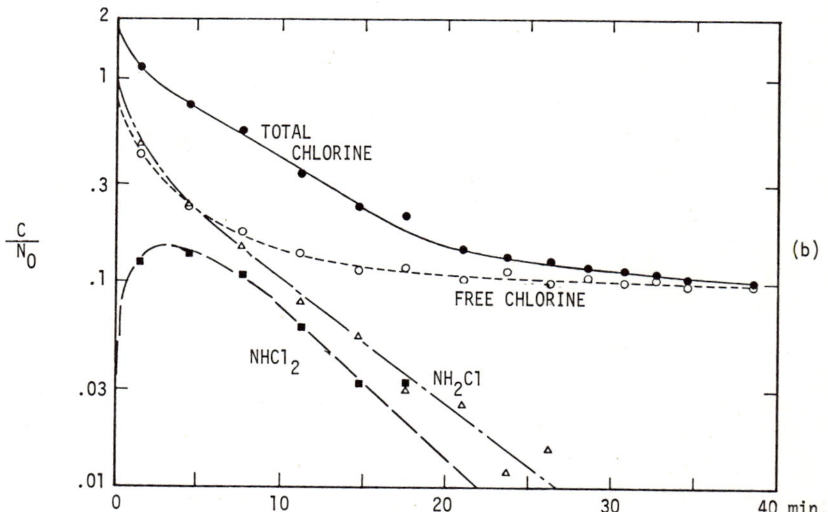

Figure 13.4. Chlorination of ammonia at pH 7.0, N_0 1 mg/l N, C_0/N_0 1.8, P_t 0.01 M, (a) 20° C, (b) 15° C.

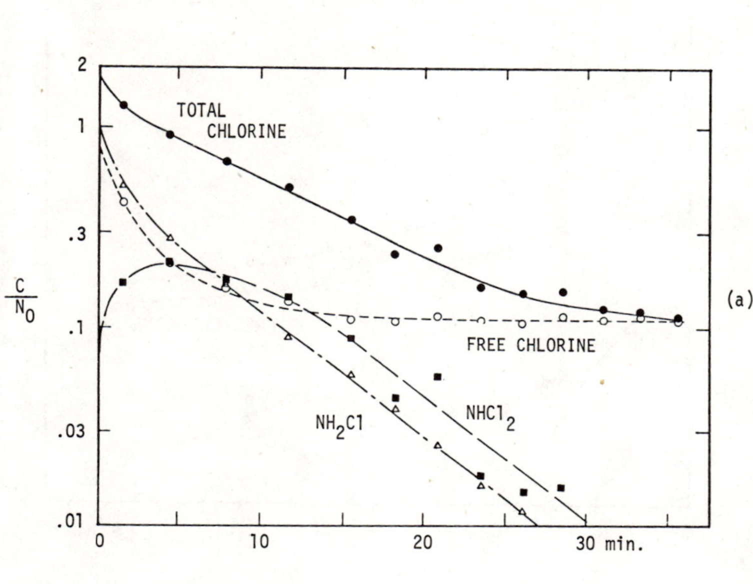

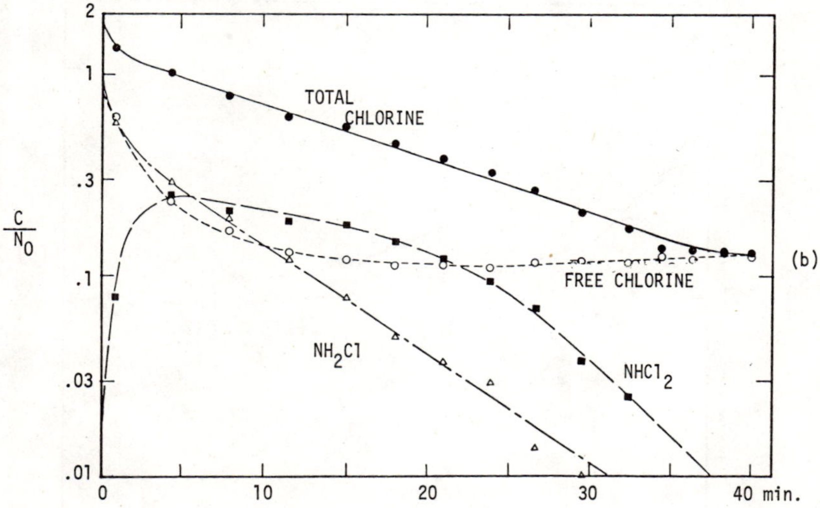

Figure 13.5. Chlorination of ammonia at pH 7.0, N_0 1 mg/l N, C_0/N_0 1.8, P_t 0.01 M, (a) 10° C, (b) 5° C.

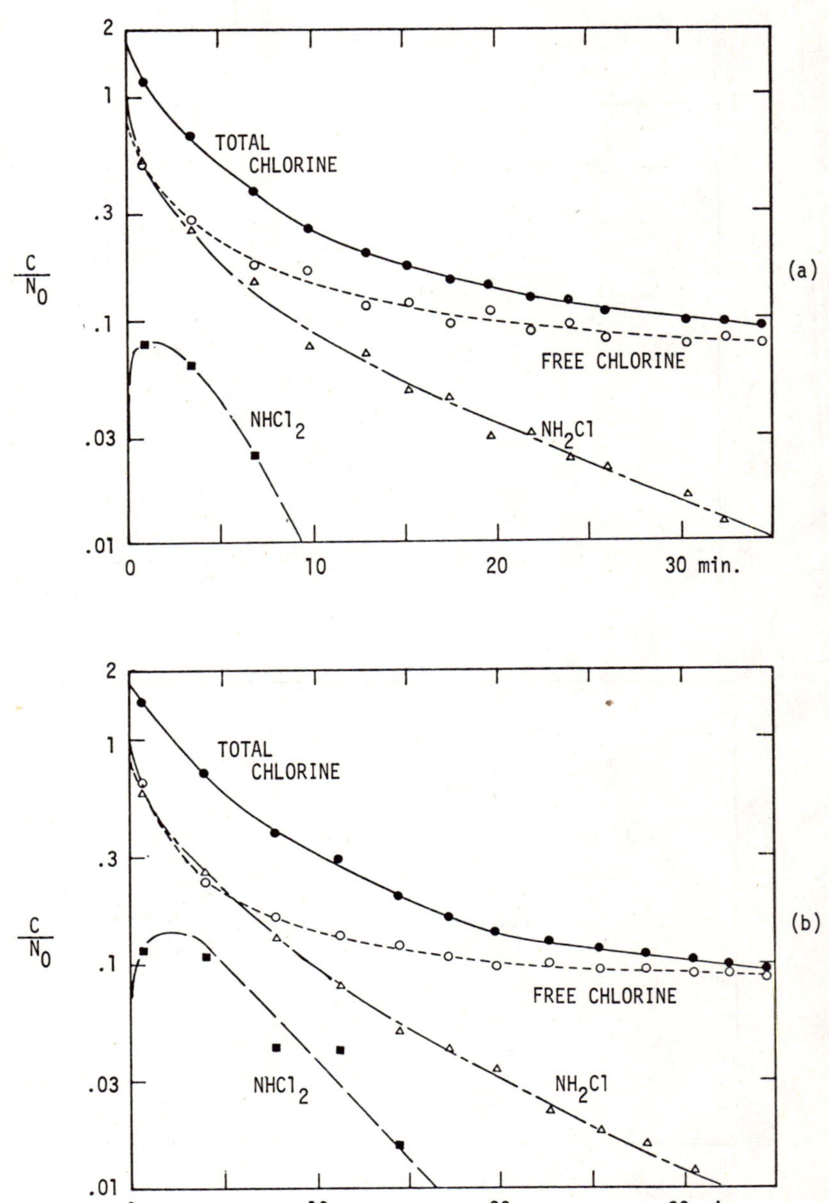

Figure 13.6. Chlorination of ammonia at pH 7.2, N_0 1 mg/l N, C_0/N_0 1.8, P_t 0.01 M, (a) 20° C, (b) 15° C.

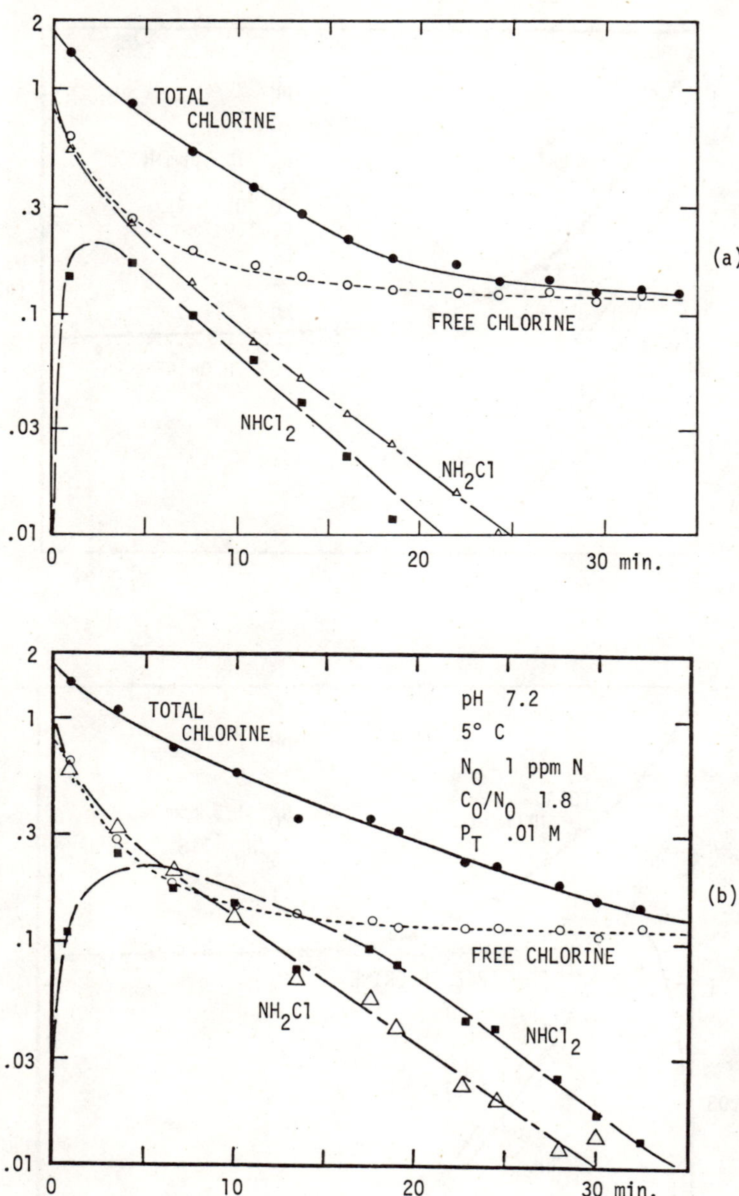

Figure 13.7. Chlorination of ammonia at pH 7.2, N_0 1 mg/l N, C_0/N_0 1.8, P_t 0.01 M, (a) 10° C, (b) 5° C.

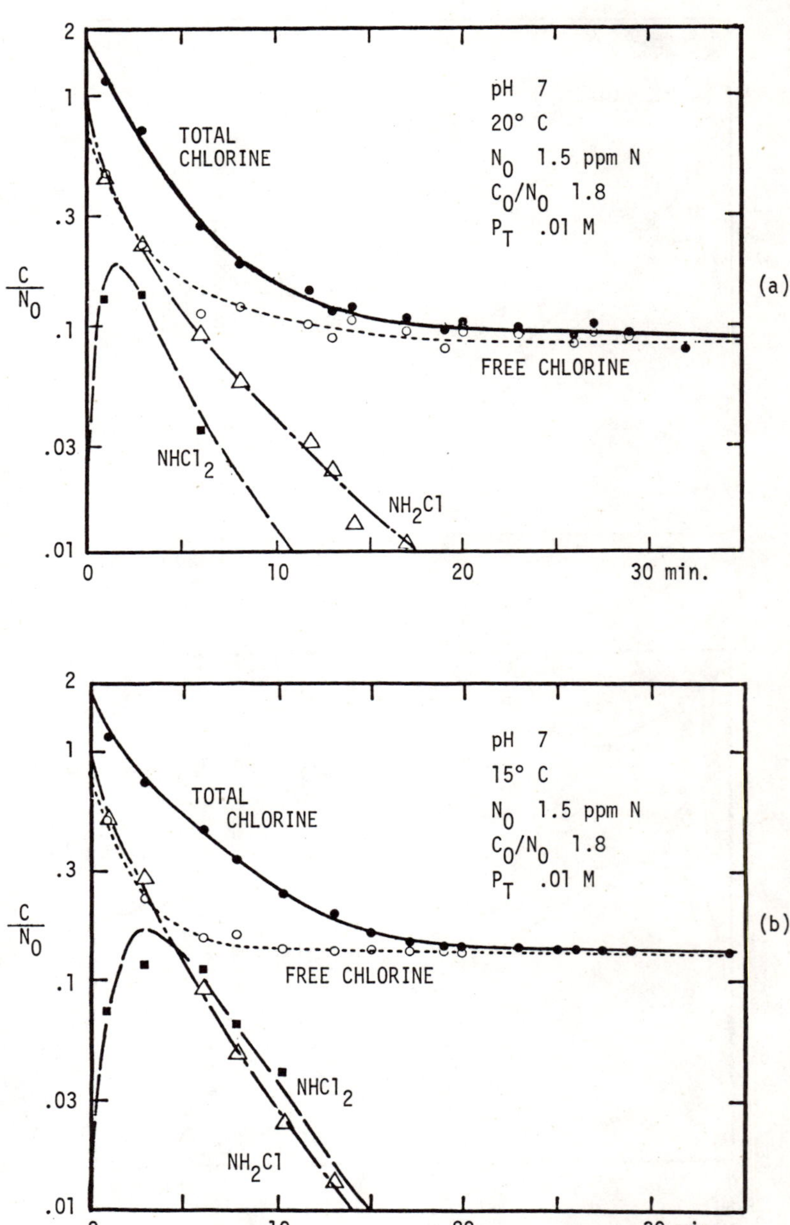

Figure 13.8. Chlorination of ammonia at pH 7, N_0 1.5 mg/l N, C_0/N_0 1.8, P_t 0.01 M, (a) 20° C, (b) 15° C.

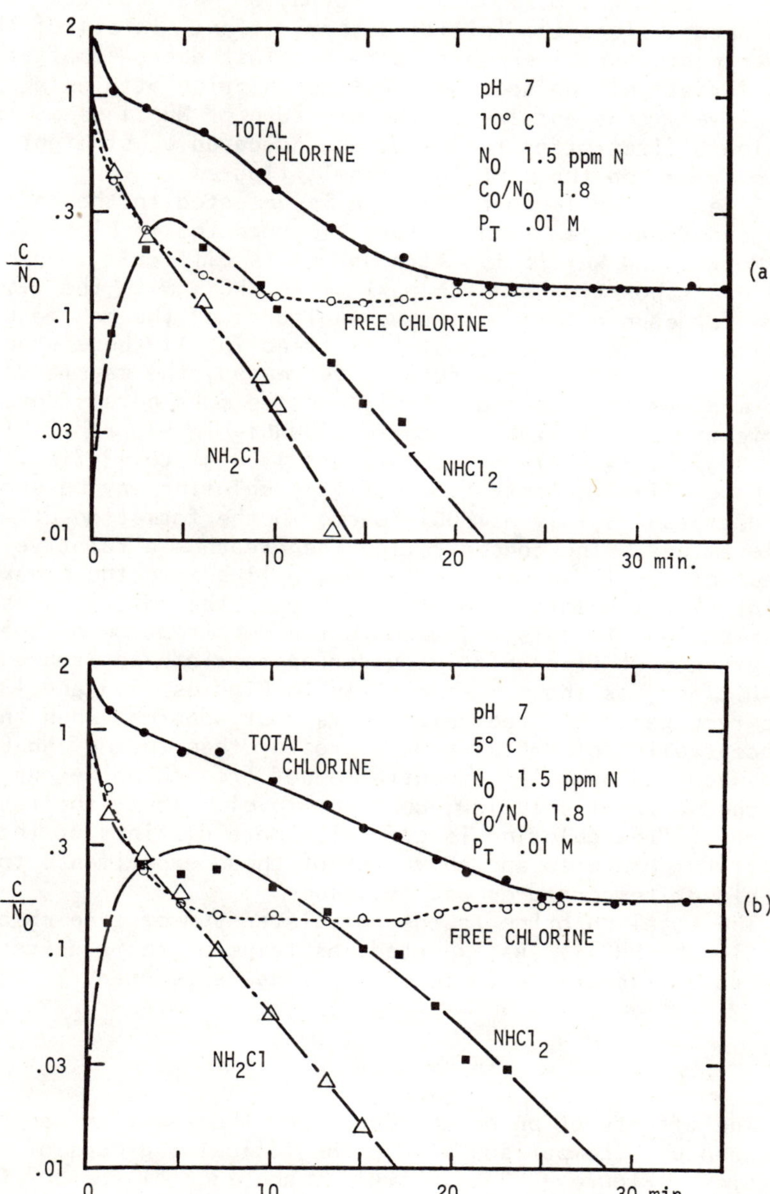

Figure 13.9. Chlorination of ammonia at pH 7, N_0 1.5 mg/l N, C_0/N_0 1.8, P_t 0.01 M, (a) 10° C, (b) 5° C.

The formation of NH_2Cl is very rapid in all these experiments. Indeed, maximum NH_2Cl formation has occurred in instances before the initial determination. In all of the experiments, NH_2Cl exhibits a rather fast decrease after its formation. As soon as the free chlorine attains a relatively constant level, the decrease of NH_2Cl seems to follow a first-order rate law, as indicated by straight line traces on the semi-logarithmic figures.

A maximum $NHCl_2$ concentration is detected in the initial stages of each reaction mixture, in much the fashion predicted by Morris and Wei.[10,11] Although it is difficult to pinpoint the exact position of each maximum in $NHCl_2$ with the limited data for each experiment, the magnitudes of the maxima do follow a definite and consistent trend in all these experiments. When the temperature is decreased, the magnitude of the maximum increases and $NHCl_2$ becomes more persistent, as indicated by the widening curve of $NHCl_2$ in Figure 13.2(a) to 13.3(b), 13.4(a) to 13.5(b), and 13.6(a) to 13.7(b).

Generally, concentrations of free chlorine may be seen to decrease rapidly as HOCl reacts in the formation of chloramines. The concentration then reaches a relatively constant level, in most instances, well before the breakpoint process nears completion. One of the most interesting observations in this research is the confirmation of the resurgence of free chlorine under appropriate environmental conditions, as shown most clearly in Figures 13.3 and 13.9. The resurgence of free chlorine is most apparent when the concentration of $NHCl_2$ is much greater than that of NH_2Cl, and when the overall concentration of free chlorine has reached a relatively low, constant level. Thus, the resurgence of free chlorine is generally more distinct at the lower temperatures and pH values of these experiments than at higher temperatures and pH values.

The total chlorine is the calculated sum of free chlorine, NH_2Cl, and $NHCl_2$. As expected, as temperature is decreased, the rate of decrease in total chlorine is reduced.

Effect of pH

The effects of pH on the course of the reaction can best be seen by a comparison within the following groups of figures: Figure 13.2(a), 13.4(a), and 13.6(a) for 20° C; Figures 13.2(b), 13.4(b), and 13.6(b) for 15° C; Figures 13.3(a), 13.5(a), and 13.7(a) for 10° C; and Figures 13.3(b), 13.5(b), and 13.7(b) for 5° C.

The most important effect of pH in this range seems to be on the transient concentration of $NHCl_2$. As the pH is decreased from 7.2 to 6.7, the trace for $NHCl_2$ concentration as a function of time undergoes three pronounced changes. These changes are an increase in the magnitude of the maximum, a delay in the time of occurrence of the maximum, and an increase in the persistence of $NHCl_2$. These effects run parallel to the observations that the resurgence of free chlorine becomes more distinct at lower pH values, and that this trend is especially evident at lower temperatures. The rate of the overall breakpoint reaction, as measured by the rate of decrease of total chlorine, decreases somewhat as the pH is reduced from 7.2 to 6.7. This is apparently related to the slower decomposition of $NHCl_2$ at lower pH values.

Effect of Initial Reactant Concentration

It was predicted by Morris and Wei[10,11] that variations in initial reactant concentrations will have important effects on the reaction pattern, especially on the trace for the concentration of $NHCl_2$. Consequently, a series of experiments was conducted, in which smaller initial reactant concentrations were used, to provide a test of the predicted effect. Results are shown in Figure 13.10 for pH 6.7, Figure 13.12 for pH 7.0, and Figure 13.14 for pH 7.2. Summaries of significant data from these figures and from corresponding figures previously given are shown in Figures 13.11, 13.13, and 13.15. For clarity only the curves for the concentrations of $NHCl_2$ and for total chlorine have been shown in the summary figures. The organization of data may be better illustrated in Table 13.I.

According to the mechanism proposed by Morris and Wei,[10,11] the formation of $NHCl_2$ is a second-order reaction while its decomposition is first-order. Thus, the maximum $NHCl_2$ concentration should occur sooner and have a larger magnitude when the initial reactant concentrations are increased. Consequently, the rate of decrease in both $NHCl_2$ and total chlorine should also be increased. It is appropriate to point out that all of the magnitudes in the above reasoning refer to the molar ratio of respective species to initial ammonia concentration. As shown in Figures 13.11, 13.13, and 13.15, the experimental results are indeed in accord with the results expected on the basis of the proposed mechanism.

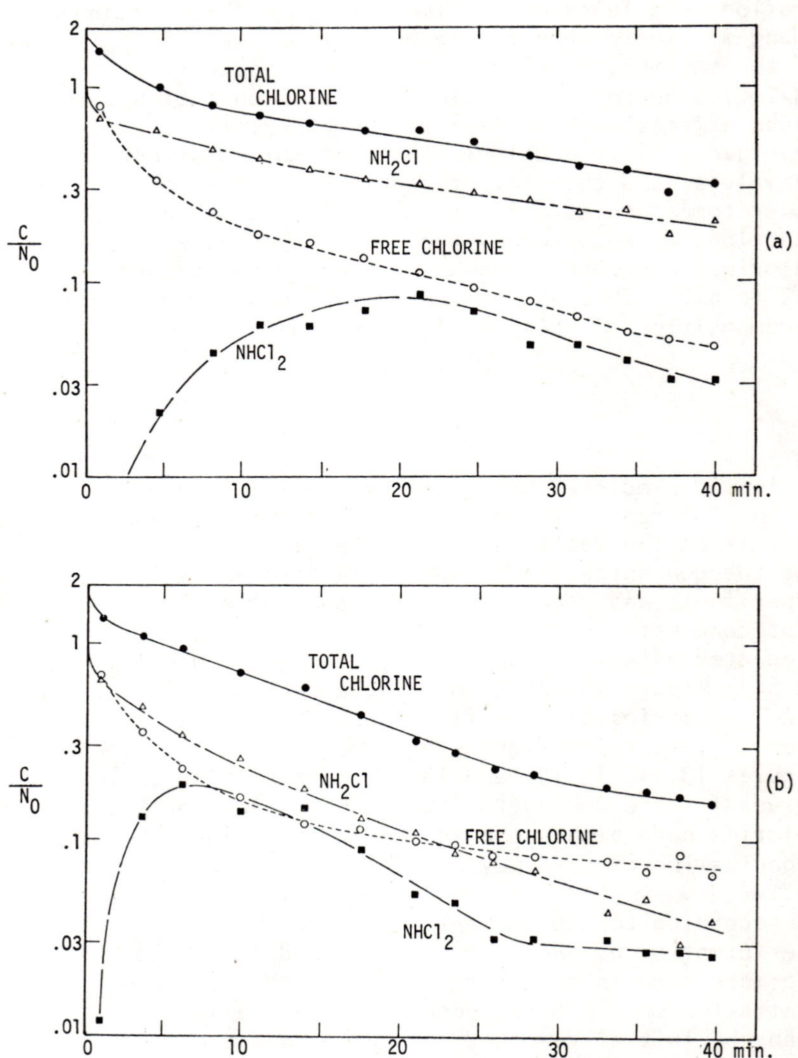

Figure 13.10. Clorination of ammonia at pH 6.7, 20° C, C_0/N_0 1.8, P_t 0.01 M, & N_0 at (a) 0.25 (b) 0.5 mg/l N.

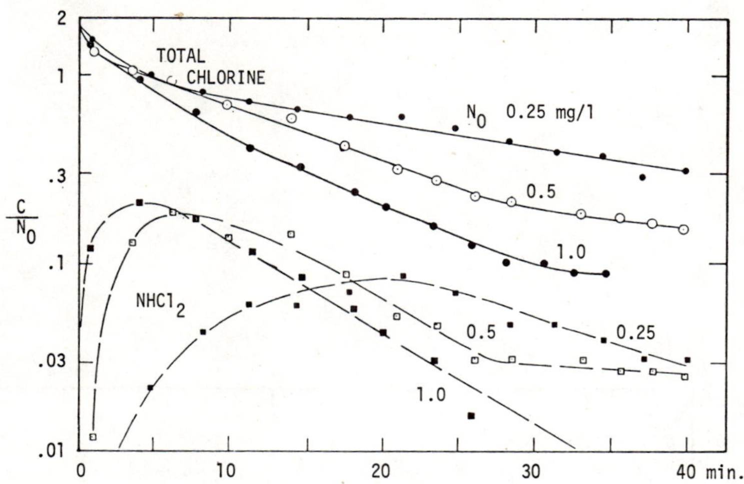

Figure 13.11. Effect of initial ammonia concentration on chlorination of ammonia at pH 6.7, 20° C, C_0/N_0 1.8, P_t 0.01 M.

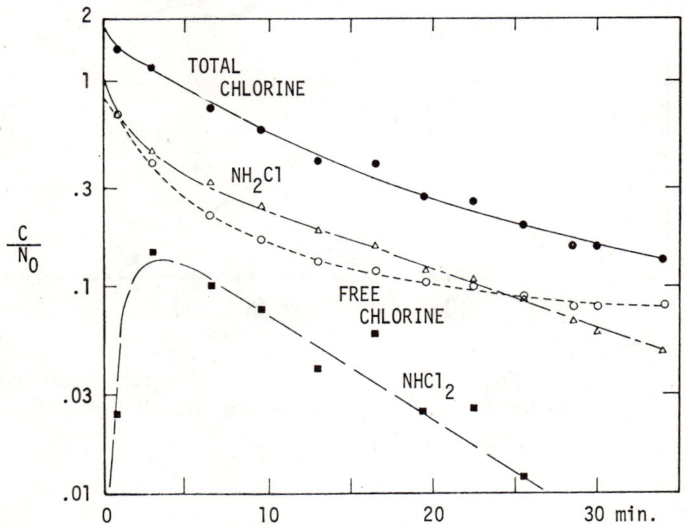

Figure 13.12. Chlorination of ammonia at pH 7.0, 20° C, N_0 0.5 mg/l N, C_0/N_0 1.8, P_t 0.01 M.

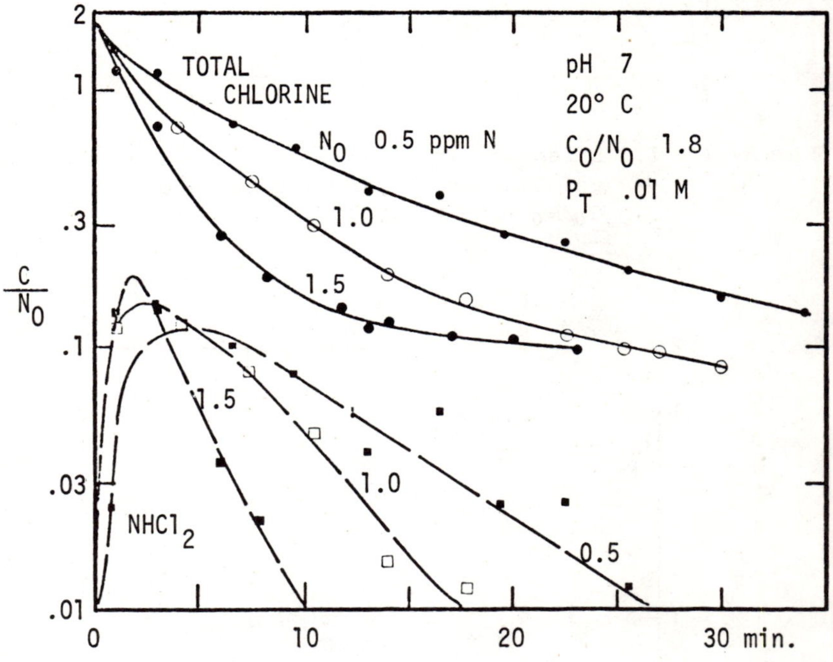

Figure 13.13. Effect of initial ammonia concentration on chlorination of ammonia at pH 7, 20° C, C_0/N_0 1.8, P_t 0.01 M.

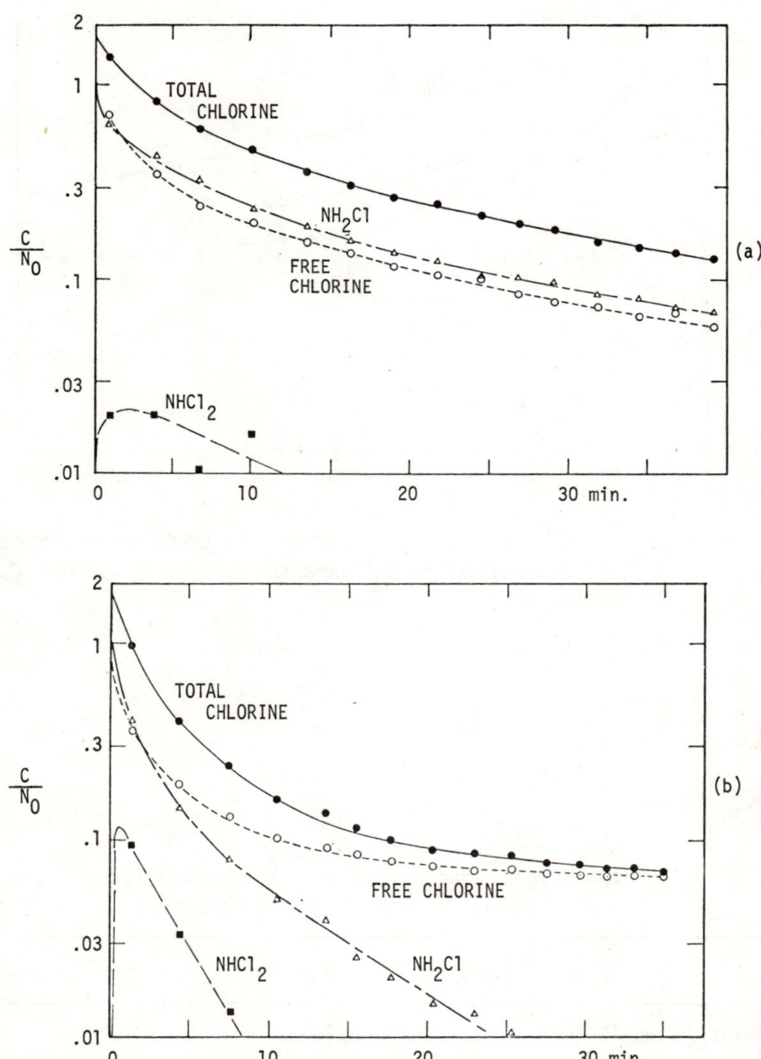

Figure 13.14. Chlorination of ammonia at pH 7.2, 20° C, C_0/N_0 1.8, P_t 0.01 M, & N_0 at (a) 0.5 (b) 1.5 mg/l N.

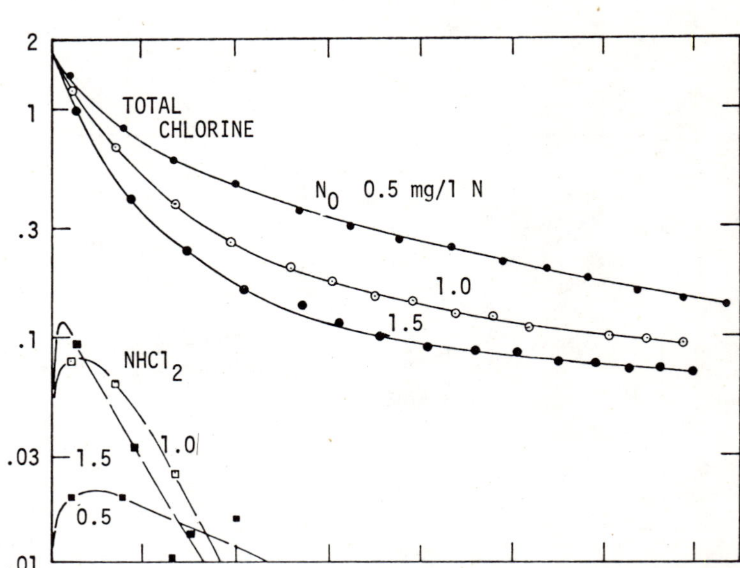

Figure 13.15. Effect of initial ammonia concentration on chlorination of ammonia at pH 7.2, 20° C, C_0/N_0 1.8, P_t 0.01 M.

Table 13.I

Organization of Figures for the Effect of Initial Reactant Concentration

pH =	*6.7*	*7.0*	*7.2*
N_o = 0.25 mg/l N	13.10(a)		
0.5	13.10(b)	13.12	13.14(a)
1.0	13.2(a)	13.4(a)	13.6(a)
1.5		13.8(a)	13.14(b)
Comparison	13.11	13.13	13.15

Effect of Buffer System

In an attempt to simulate natural water conditions, three experiments using $NaHCO_3$ as a buffer were also conducted. The results are shown in Figures 13.16, 13.17 and 13.18. Because of the relatively poor buffer capacity of the bicarbonate system near pH 7, there was more variation in pH during the course of reaction than in the reactions with the phosphate buffer. The range of pH is shown in the figure titles, with the smaller value indicating the initial stage of reaction. The gradual increases in pH in the latter stages of reaction are probably due to the loss of CO_2 from the reacting system during the process of equilibration with the atmosphere.

Because of differences in the reaction conditions, no rigorous comparison between the reactions of the phosphate buffer and the bicarbonate buffer is possible. Nevertheless, a comparison of Figures 13.17 and 13.18 and their closest counterparts in phosphate buffer, Figures 13.14(a) and 13.6(a), respectively, shows a number of major differences.

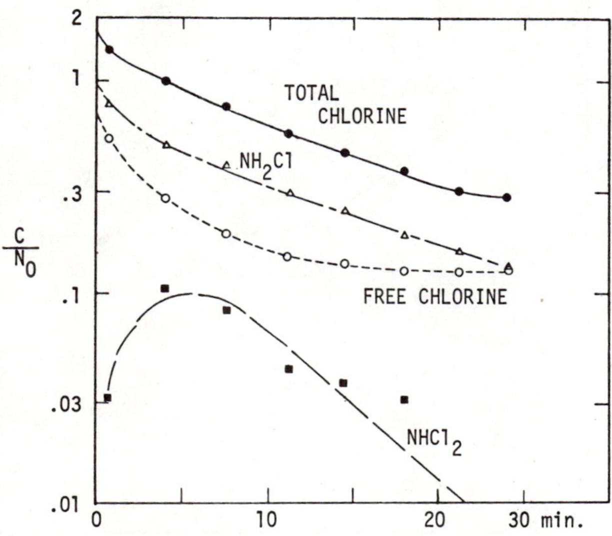

Figure 13.16. Chlorination of ammonia at pH 7.04 to 7.24, 20° C, N_0 0.25 mg/l N, C_0/N_0 1.7, $(HCO_3^-)_t$ 0.005 M.

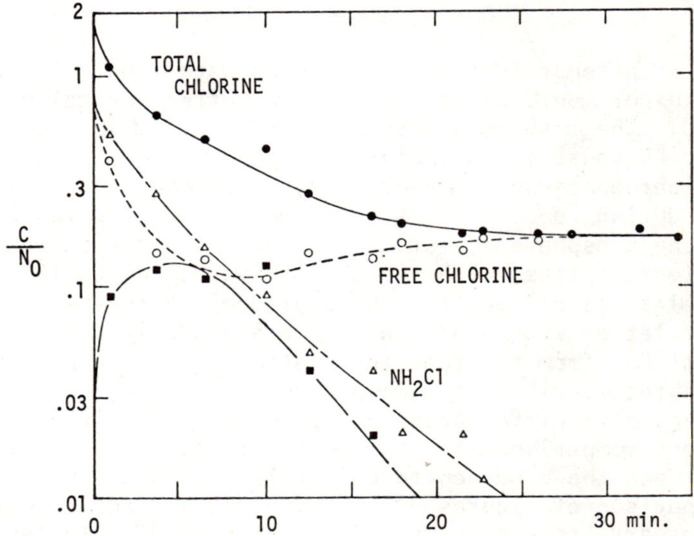

Figure 13.17. Chlorination of ammonia at pH 7.05 to 7.25, 20° C, N_0 0.5 mg/l N, C_0/N_0 1.7, $(HCO_3^-)_t$ 0.01 M.

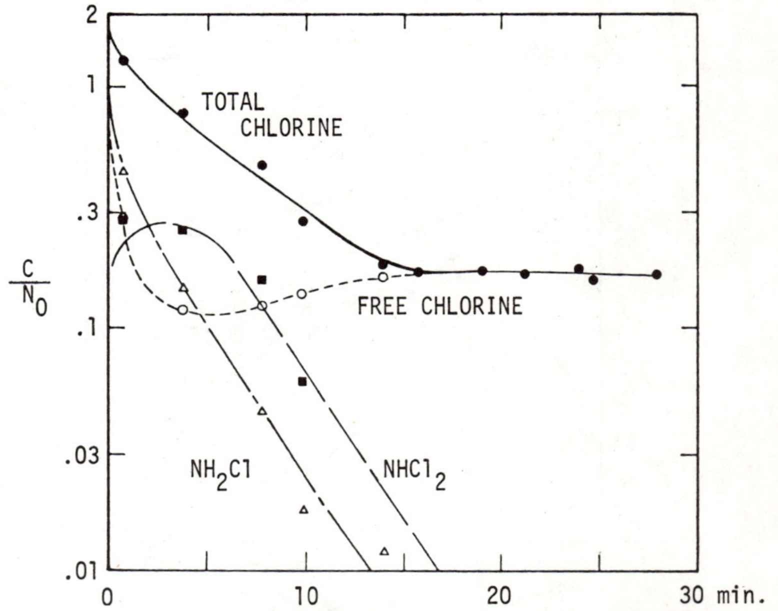

Figure 13.18. Chlorination of ammonia at pH 7.16 to 7.22, 20° C, N_0 1.0 mg/l N, C_0/N_0 1.7, $(HCO_3^-)_t$ 0.01 M.

In the bicarbonate system the following were observed: 1) NH_2Cl decreased faster; 2) $NHCl_2$ had larger maxima in concentration and decomposed more rapidly; 3) a distinct resurgence of free chlorine was evident, which was not detectable in the corresponding phosphate buffer. Generally, these differences may be attributed to the variation of pH in the bicarbonate system; therefore, the results from phosphate systems may be taken as generally valid for constant pH conditions.

A comparison of Figures 13.16, 13.17 and 13.18 shows that the same conclusions on the effect of initial reactant concentrations can be drawn from the bicarbonate system as were previously drawn from the phosphate system.

DISCUSSION

Stoichiometric Ratio

The molar ratio of chlorine reduced to ammonia oxidized during the breakpoint process constitutes the overall stoichiometry of the breakpoint process. If chlorine is applied in a molar ratio larger than the stoichiometric ratio, the excess chlorine will remain in the form of free chlorine when the reaction is completed. Therefore, the stoichiometric ratio may be calculated by substracting the remaining free chlorine after completion of the breakpoint process from the initial chlorine dose, both as a molar ratio to initial ammonia. A detailed analysis of the stoichiometric ratios from the experiments is shown in Table 13.II. It must be noted, however, that this calculation makes the assumption that all the nitrogen has been oxidized. The experimental values are high if any nitrogen remains in the form of unmeasured NCl_3.

In the limited range of pH studied, -- 6.7, 7.0, and 7.2 -- there seems to be no significant dependence of the stoichiometric ratio on pH. This agrees with observations in previous studies.[2,4]

A careful look at Table 13.II reveals several consistent trends in the stoichiometric ratio. The ratio decreases as temperature decreases. The means of the ratios at each temperature for the first sixteen experiments in Table 13.II give the average stoichiometric ratio at various temperatures as summarized in Table 13.III.

Table 13.II

Comparison of Stoichiometric Ratios

Figure	*pH*	*Temp.* ^{o}C	N_o mg/l N	C_o/N_o[a]	C_f/N_o[b]	R[c]
13.2(a)	6.7	20	1	1.8	0.09	1.71
13.2(b)	6.7	15	1	1.8	0.10	1.70
13.3(a)	6.7	10	1	1.8	0.11	1.69
13.3(b)	6.7	5	1	1.8	*	*
13.4(a)	7.0	20	1	1.8	0.07	1.73
13.4(b)	7.0	15	1	1.8	0.10	1.70
13.5(a)	7.0	10	1	1.8	0.11	1.69
13.5(b)	7.0	5	1	1.8	0.13	1.67
13.6(a)	7.2	20	1	1.8	0.07	1.73
13.6(b)	7.2	15	1	1.8	0.08	1.72
13.7(a)	7.2	10	1	1.8	0.12	1.68
13.7(b)	7.2	5	1	1.8	0.12	1.68
13.8(a)	7.0	20	1.5	1.8	0.09	1.71
13.8(b)	7.0	15	1.5	1.8	0.13	1.67
13.9(a)	7.0	10	1.5	1.8	0.14	1.66
13.9(b)	7.0	5	1.5	1.8	0.16	1.64
13.10(a)	6.7	20	0.25	1.8	*	*
13.10(b)	6.7	20	0.5	1.8	*	*
13.12	7.0	20	0.5	1.8	*	*
13.14(a)	7.2	20	0.5	1.8	*	*
13.14(b)	7.2	20	1.5	1.8	0.07(-)	1.73(+)
13.16	7.04-7.24	20	0.25	1.7	*	*
13.17	7.05-7.25	20	0.5	1.7	0.17	1.53
13.18	7.16-7.22	20	1	1.7	0.17	1.53

* Data are not available because the reaction had not reached completion at the time the experiment was terminated.

[a] C_o/N_o = Molar ratio of initial chlorine to initial ammonia.

[b] C_f/N_o = Molar ratio of final free chlorine to initial ammonia.

[c] R = Stoichiometric ratio as a molar ratio of chlorine reduced to ammonia oxidized.

$$R = \frac{C_o}{N_o} - \frac{C_f}{N_o}$$

Table 13.III

Temperature Dependence of the Stoichiometric Ratio

Temperature, °C	*Average Stoichiometric Ratio from Experiments*
20	1.72
15	1.70
10	1.68
5	1.66

As means of all his experiments between pH 6.5 and 8.0, Palin[6] reported molar stoichiometric ratios ranging from 1.54 to 1.88 for initial molar ratios of chlorine to nitrogen ranging from 1.46 to 3.12. At an initial molar ratio equal to 1.82 the average stoichiometric ratio was 1.64, a figure only slightly lower than those shown in Table 13.III.

The effect of initial reactant concentration upon the stoichiometric ratio can be best illustrated by a comparison of Figures 13.4 and 13.5 and Figures 13.8 and 13.9. As the initial reactant concentration is increased from 1.0 to 1.5 mg/l N, the experimental stoichiometric ratio decreases by 0.02 to 0.03 unit at each of the four different temperatures studied. This finding is in agreement with the early observation of Chapin,[4] who reported that the stoichiometric ratio tends to decrease toward 1.5 as the initial ammonia concentration is increased. Apparently, the reaction pathway leading to a stoichiometric ratio of 1.5 is favored by an increase in initial ammonia concentration, or, conversely, nitrate formation is favored at low reactant concentrations.

The Maximum in $NHCl_2$ Concentration

One of the important and interesting findings of this study of the breakpoint reaction is the recognition and detection of the intermediate maximum in $NHCl_2$ concentration. The analyses for $NHCl_2$ showed that the concentration of this substance increases during the early stages of the breakpoint process, reaches a maximum after several minutes, and then declines as the reduction of chlorine proceeds. Though a number of previous studies of the breakpoint process had assumed, on stoichiometric grounds, that dichloramine was a type of intermediate, this study is the first unequivocal demonstration that the kinetic pattern of its growth and decay is that of a true intermediate substance.

Some aspects of the proposed mechanism by Morris and Wei[10,11] have received important confirmation from trends in characteristic features of the kinetic pattern rather than from numerical accord. One such feature is the maximum in $NHCl_2$ concentration, both as to magnitude and time of occurrence. The proposed mechanism has the maximum in $NHCl_2$ as a result of the competition between formation of $NHCl_2$ according to

$$NH_2Cl + HOCl \longrightarrow NHCl_2 + H_2O \tag{3}$$

which is described as a second-order reaction and the decomposition of $NHCl_2$ according to

$$NHCl_2 + H_2O \longrightarrow NOH + 2H^+ + 2Cl^- \tag{4}$$

which is described as a first-order reaction. If these hypotheses are valid, because the rate of a second-order reaction is favored relative to that of a first-order reaction as the total concentration is increased, the magnitude of the maximum in $NHCl_2$ concentration should be increased and should come earlier with increased total concentrations of reactants.

Both of these predictions were found to be valid in the kinetic experiments, as shown in Figures 13.11, 13.13, and 13.15, thus establishing the relative orders of the reactions by which $NHCl_2$ is formed and decomposed, regardless of what the specific processes may be. Moreover, since the evidence of Weil and Morris[12] is reasonably good in establishing Reaction 3 as a second-order reaction, the present study then affirms the first-order character of Reaction 4.

Resurgence of Free Chlorine

One of the initial reactants in the breakpoint process, HOCl, is partially regenerated at a later stage of the reaction and is a very unusual circumstance. This phenomenon, that had been occasionally observed by Palin,[6] was confirmed as a regular feature of the overall breakpoint process for a certain range of conditions. Not only is the confirmed existence of this phenomenon of general kinetic interest, but also it serves to establish strongly that the major overall reaction leading to the breakpoint is

$$2NHCl_2 + H_2O \longrightarrow N_2 + 3H^+ + 3Cl^- + HOCl. \qquad (5)$$

It is difficult to conceive any alternate reaction that will satisfy the requirements of the stoichiometric ratio and, at the same time, give a return of free chlorine.

Temperature Dependence

The temperature dependence of the overall breakpoint process has no direct relation to problems of mechanism. On the other hand, the shifting concentrations of intermediate substances as a function of temperature reflect both the mechanism and the temperature coefficients of individual elementary reactions in the mechanism. It should be noted, however, that this is the first investigation of the breakpoint process in which the effect of temperature has been studied. Previous studies have been carried out at various temperatures ranging from 10° C to 25° C, but no one study has been done at more than one temperature.

The kinetic parameter used to evaluate the temperature dependence is the half-reaction time of total chlorine. The total chlorine for the kinetic experiments was calculated as the sum of free chlorine, NH_2Cl, and $NHCl_2$, assuming NCl_3 in negligible concentration. If NCl_3 is present in significant concentration, the calculated total residual chlorine should always be less than the actual value.

The major finding of this study with regard to the effect of temperature was that it is unexpectedly great. The apparent activation energy at constant pH based on half-reaction times ranged from 7.8 kcal/mole at pH 7.0 to 12.85 kcal/mole at pH 6.7. There are two points of major significance connected with this finding. The first is that no great amount of confidence can be placed in the quantitative degrees of reaction observed at intermediate times by other investigators who worked at uncontrolled temperatures. The second, a practical point, is that what can be expected of breakpoint chlorination by the water-plant operator depends quite strongly on the temperature of the water that is being treated, as well as on the more controllable characteristics such as pH.

ACKNOWLEDGEMENTS

This investigation was conducted during 1969-1971 at Harvard University, and was supported by Training Grant 5-T01-ES00020 from the National Institute of Environmental Health Sciences, National Institute of Health, U. S. Department of Health, Education and Welfare.

REFERENCES

1. Calvert, C. K. *J. Amer. Water Works Assoc., 32,* 1155 (1940).
2. Griffin, A. E., and N. S. Chamberlin. *J. New England Water Works Assoc., 55,* 371 (1941).
3. Holwerda. *Mededeelingen van den Dienst der Voklsgezondheid in Nederlandschindie, 17,* 251 (1928); *19,* 325 (1930).
4. Chapin, R. M. *J. Amer. Chem. Soc., 53,* 912 (1931).
5. Griffin, A. E. *J. Amer. Water Works Assoc., 31,* 2121 (1939).
6. Palin, A. T. *Water and Water Eng., 54,* 151, 189, 248 (1950).
7. "Water Chlorine (Residual) No. 1," Public Health Service, U.S. Dept. of Health, Education and Welfare, Analytical Reference Service, Study No. 35, 1969.
8. Palin, A. T. *J. Amer. Water Works Assoc., 49,* 873 (1957).
9. Palin, A. T. *J. Inst. Water Eng., 21,* 537 (1967).
10. Morris, J. C., and I. W. Wei. Presented before the Division of Water, Air, and Waste Chemistry at the 157th National Meeting of American Chemical Society, Minneapolis, Minnesota, April 1969.
11. Wei, I. W. Ph.D. Thesis, Harvard University, Cambridge, Massachusetts, 1972.
12. Weil, I., and J. C. Morris. Presented at the 117th National Meeting of the American Chemical Society, Detroit, Michigan, April 1950.

CHAPTER 14

EQUILIBRIA IN AQUEOUS SOLUTIONS OF CHLORINATED ISOCYANURATE

J. E. O'Brien, J. C. Morris, and J. N. Butler*

INTRODUCTION

Cyanuric acid and cyanurates have been used for many years in outdoor swimming pools as stabilizers for active chlorine to prevent its photolytic decomposition. In the absence of such stabilizers, usual concentrations of active chlorine can be completely dissipated in 1-2 hours on a sunny day, whereas in the presence of as little as 25 mg/l (0.0002 *M*) cyanuric acid, residual chlorine may persist for several days as a result of the formation of chlorinated cyanurates. Moreover, chlorinated cyanurates, in addition to acting as stabilizers, exhibit a limited degree of hydrolysis to yield a relatively constant level of germicidally potent, free chlorine. In other words, chlorinated cyanurates may be considered as analogous to a protected reservoir which liberates a small but relatively constant level of free chlorine in accordance with clearly defined principles of chemical equilibrium.

Because of the importance of chlorinated cyanurates and because the equilibria in aqueous chlorine-cyanurate mixtures constitute an interesting and challenging physicochemical system in its own right, studies to elucidate quantitatively the equilibrium situation were felt desirable. Additional information beyond that in the literature was needed to predict accurately the concentrations of free chlorine available for germicidal action as a function of reservoir chlorine, total cyanurate, and pH.

---

*J. E. O'Brien, Lawrence Experiment Station, Lawrence, Massachusetts. J. C. Morris and J. N. Butler, Division of Engineering and Applied Physics, Harvard University, Cambridge, Massachusetts.

Thus far two terms have been used, "reservoir chlorine" and "free chlorine." Although their meanings may be apparent, it is preferable that precise definitions be provided. These are as follows: (1) Free chlorine is the sum of hypochlorous acid (HOCl) and hypochlorite ion (OCl^-), and (2) Reservoir chlorine is free chlorine plus all chlorinated cyanurate species.

Ordinarily, most of the reservoir chlorine consists of chlorinated cyanurates. Although these provide a readily available source of active chlorine, Andersen[1] has submitted evidence that chlorinated cyanurates, as such, are not particularly germicidal. As a result, the germicidal activity must be borne by the relatively small fraction of free chlorine present at any given time. Since, in general, increase in cyanurate concentration results in decreased free chlorine, the use of large cyanurate concentrations to achieve maximum stability will tend to give inadequate germicidal activity. For the same reason the continual addition of chlorinated cyanurates as a source of chlorine is not recommended since this will lead to build up of cyanurate concentration and consequent repression of the concentration of free chlorine below that necessary for effective germicidal activity.

Analysis of Chlorinated Cyanurate Solutions

Although standard analytical techniques tend to show that there are substantial fractions of free chlorine present in solutions containing excess cyanurate, the actual concentration of free chlorine in chlorinated cyanurate systems is usually only a small percentage of the total reservoir chlorine. Whittle[2] has demonstrated that "wet" chemical methods for estimating free chlorine concentrations include some chlorinated cyanurates as well due to the rapid interconversions of species that occur during analysis.

Because of this ambiguity, reliable estimation of equilibrium concentrations of free chlorine, or, indeed, of any individual species in solutions of chlorine plus cyanurate, either requires a physical measurement that does not disturb the equilibrium or must be computed from known equilibrium constants. Computer programs have been developed that permit computation of all the individual species as a function of reservoir chlorine, total cyanurate, and pH value based on knowledge of the individual acidity and hydrolysis constants in solutions of chlorine plus cyanurate.

The Aqueous Chlorine-Cyanurate System

A member of the symmetrical triazine family, cyanuric acid or isocyanuric acid is systematically designated 2,4,6-trihydroxytriazine. It occurs in both the enol and keto tantomeric forms shown in Figure 14.1. Since there is some uncertainty as to which form predominates in solution, the term "Cy" has been used to represent the cyanurate structure. Thus, H_3Cy is cyanuric acid itself, H_2Cy^- its first ionization product, and H_2ClCy its monochlor derivative.

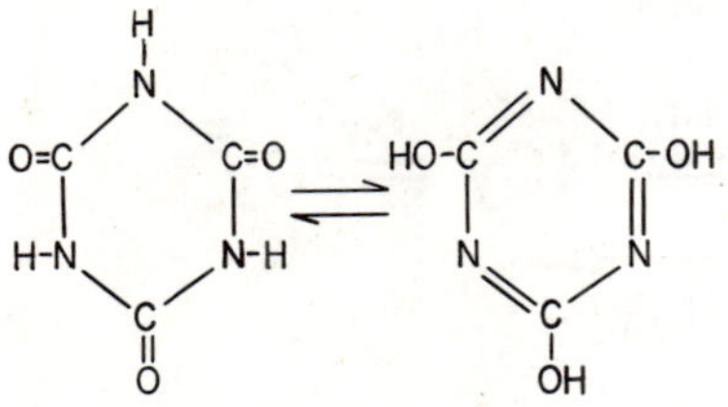

Keto form ⇌ Enol form

Figure 14.1. Cyanuric acid structure.

Cyanuric acid is a weak tribasic acid that ionizes progressively to H_2Cy^-, $HCy^=$, and $Cy^{\equiv}$. Also, the hydrogens may be successively replaced by unipositive chlorine to give H_2ClCy, HCl_2Cy, and Cl_3Cy. The first two of these compounds also dissociate as acids. As a result solutions of chlorinated cyanurate form a rather complex equilibrium system in which there may be as many as 10 species, regulated in concentration by nine independent equilibrium constants exclusive of the species HOCl, OCl^-, and H_2O, and constants K_{HOCl} and K_w. Some idea of the complexity of aqueous solutions of chlorinated cyanurates can be obtained from Figure 14.2. This is essentially the same illustrative diagram given by Brady, Sancier and Sirine.[3] Assignment of numbers to the equilibrium constants begins with the hydrolysis of trichlorcyanuric acid in the upper left. Ionization processes proceed from left to right and have even-numbered K values (K_2, K_4, etc.). Hydrolytic processes proceed downward and have odd-numbered K values (K_1, K_3, etc.).

In a solution of chlorinated cyanurate, total chlorine concentration is given by the following equation

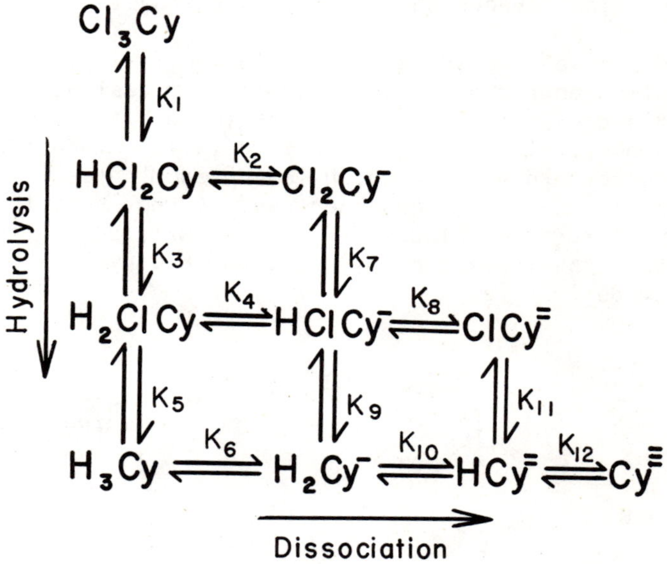

Cl_3Cy

$\rightleftharpoons K_1$

$HCl_2Cy \overset{K_2}{\rightleftharpoons} Cl_2Cy^-$

$\rightleftharpoons K_3$ $\quad$ $\rightleftharpoons K_7$

$H_2ClCy \overset{K_4}{\rightleftharpoons} HClCy^- \overset{K_8}{\rightleftharpoons} ClCy^=$

$\rightleftharpoons K_5$ $\quad$ $\rightleftharpoons K_9$ $\quad$ $\rightleftharpoons K_{11}$

$H_3Cy \overset{K_6}{\rightleftharpoons} H_2Cy^- \overset{K_{10}}{\rightleftharpoons} HCy^= \overset{K_{12}}{\rightleftharpoons} Cy^{\equiv}$

Hydrolysis

Dissociation

Figure 14.2. Equilibria among cyanuric acid and its chlorinated derivatives.

$$Cl_T = [H_2ClCy] + [HClCy^-] + [ClCy^=] + [HOCl] + [OCl^-] + 2[HCl_2Cy] + 2[Cl_2Cy^-] + 3[Cl_3Cy] \quad (1)$$

Similarly total cyanurate, Cy_T, is made up as follows

$$Cy_T = [H_3Cy] + [H_2Cy^-] + [HCy^=] + [Cy^{\equiv}] = [H_2ClCy] + [HClCy^-] + [ClCy^=] + [HCl_2Cy] + [Cl_2Cy^-] = [Cl_3Cy] \quad (2)$$

The concentrations of each species in these equations may be expressed as a function of $[Cy^{\equiv}]$, $[HOCl]$, and $[H^+]$ and the appropriate equilibrium constants shown in

Figure 14.2.* Expressions for the cyanurate species in these terms are:

$$[HCy^{=}] = \frac{[H^+][Cy^{\equiv}]}{K_{12}}$$

$$[H_2Cy^-] = \frac{[H^+]\ [Cy^{\equiv}]}{K_{12}K_{10}}$$

$$[H_3Cy] = \frac{[H^+]^3[Cy^{\equiv}]}{K_{12}K_{10}K_6}$$

$$[ClCy^{=}] = \frac{[H^+][HOCl][Cy^{\equiv}]}{K_{12}K_{11a}}$$

$$[HClCy^-] = \frac{[H^+]^2[HOCl][Cy^{\equiv}]}{K_{12}K_{11a}K_8} = \frac{[H^+]^2[HOCl][Cy^{\equiv}]}{K_{12}K_{10}K_{9a}}$$

$$[H_2ClCy] = \frac{[H^+]^3[HOCl][Cy^{\equiv}]}{K_{12}K_{11a}K_8K_4} = \frac{[H^+]^3[HOCl][Cy^{\equiv}]}{K_{12}K_{10}K_6K_{5a}}$$

$$[Cl_2Cy^-] = \frac{[H^+]^2[HOCl]^2[Cy^{\equiv}]}{K_{12}K_{11a}K_8K_{7a}} = \frac{[H^+]^2[HOCl]^2[Cy^{\equiv}]}{K_{12}K_{10}K_{9a}K_{7a}}$$

$$[HCl_2Cy] = \frac{[H^+]^3[HOCl]^2[Cy^{\equiv}]}{K_{12}K_{11a}K_8K_{7a}K_2} = \frac{[H^+]^3[HOCl]^2[Cy^{\equiv}]}{K_{12}K_{10}K_6K_{5a}K_{3a}}$$

$$[Cl_3Cy] = \frac{[H^+]^3[HOCl]^3[Cy^{\equiv}]}{K_{12}K_{11a}K_8K_{7a}K_2K_{1a}} = \frac{[H^+]^3[HOCl]^3[Cy^{\equiv}]}{K_{12}K_{10}K_6K_{5a}K_{3a}K_{1a}}$$

The molar concentration of any species may be represented as follows

$$[H_nCl_mCy] = [H^+]^{n+m}[HOCl]^m[Cy^{\equiv}]\beta_{nm} \qquad (3)$$

Both m and the sum of n + m have values running from 0 to 3. The factor, β_{nm}, is equal to the reciprocal of the product of the appropriate equilibrium constants for the species in question. Logarithms of the "beta" values for any species are equal to the sums of the appropriate pK

*When hydrolytic reactions are considered in the form $XCl + H_2O \rightleftarrows XH + HOCl$ instead of the form $XCl + OH^- \rightleftarrows XH + OCl^-$, then the same numerical suffix given in Figure 14.2 is used for the constant, but the symbol "a" is attached.

values. The expressions for the individual β_{nm} and values for n and m are tabulated in Table 14.I.

Table 14.I

Cyanurate Species Expressed as a Function of [$Cy^{=}$], Number of H Atoms (n), Number of Cl Atoms (m), and the Sum of the Appropriate pK values (log β_{nm})

Species	*n*	*m*	*Log* β_{nm}
[$Cy^{\equiv}$]	0	0	0
[$HCy^{=}$]	1	0	pK_{12}
[H_2Cy^{-}]	2	0	$pK_{12} + pK_{10}$
[H_3Cy]	3	0	$pK_{12} + pK_{10} + pK_6$
[$ClCy^{-}$]	0	1	$pK_{12} + pK_{11a}$
[$HClCy^{-}$]	1	1	$pK_{12} + pK_{11a} + pK_8$ or $pK_{12} + pK_{10} + pK_{9a}$
[H_2ClCy]	2	1	$pK_{12} + pK_{11a} + pK_8 + pK_4$ or $pK_{12} + pK_{10} + pK_6 + pK_{5a}$
[Cl_2Cy^{-}]	0	2	$pK_{12} + pK_{11a} + pK_8 + pK_{7a}$ or $pK_{12} + pK_{10} + pK_{9a} + pK_{7a}$
[HCl_2Cy]	1	2	$pK_{12} + pK_{11a} + pK_8 + pK_{7a} + pK_2$ or $pK_{12} + pK_{10} + pK_6 + pK_{5a} + pK_{3a}$
[Cl_3Cy]	0	3	$pK_{12} + pK_{11a} + pK_8 + pK_{7a} + pK_2 + pK_{1a}$ or $pK_{12} + pK_{10} + pK_6 + pK_{5a} + pK_{3a} + pK_{1a}$

One form of the working equation is obtained by substitution in the mass balance

$$Cl_T = Cy_T \frac{\Sigma\Sigma_m \beta_{nm} [HOCl]^m [H^+]^{n+m}}{\Sigma\Sigma\beta_{nm} [HOCl]^m [H^+]^{n+m}} + HOCl \left[1 + \frac{K_{HOCl}}{[H^+]} \right] \quad (4)$$

There are ten terms in the denominator of Equation 4 and six terms in the numerator. A computer program that permits calculations based on Equation 4 is given elsewhere.[5] Equation 4 can be used directly with the program to compute the reservoir chlorine required to give any desired level of free chlorine at a specific Cy_T and pH value. The computation of free chlorine concentration present for a given reservoir value (Cl_T) is more complex. It requires a convergence or iteration technique such as the Newton-Raphson Method.

So, provided the equilibrium K values are accurately known, it is possible to compute results and prepare simplified tables showing the necessary conditions for maintenance of adequate germicidal levels coupled with optimal stabilization of residual chlorine as a function of pH. Later sections of the paper are concerned with the accurate determination of the constants and with a presentation of the results obtained using them.

EXPERIMENTAL APPROACH

Experimental evaluations of K were based primarily on ultraviolet (U.V.) spectrophotometry, *i.e.*, techniques such as differential spectrophotometry and spectrometric titrimetry. Fundamentally the particular measurements performed exploited the often large spectral differences between "adjoining" species such as: (1) cyanuric acid and its successive ionization products; (2) chlorinated cyanuric acids and their successive ionization products; and (3) chlorinated cyanurates and their successive hydrolysis products.

When pH was controlled appropriately and concentrations and ratios of Cy_T and Cl_T were selected carefully, it was possible to "isolate" and determine the individual ultraviolet absorption spectrum of nearly every species. It was then possible to devise experiments that would permit determination of nine independent stability constants on the basis of observed spectra of solutions prepared under appropriate conditions and the known values of the molar absorptivities of the individual species. There are potentially ten cyanurate species present at equilibrium. Conditions were chosen, however, to limit the number of significant species present to two or three by selection of pH to control the degree of ionization and by choice of chlorine-cyanurate ratio to repress or encourage hydrolysis. The experimental observations were primarily measurements of absorbance changes accompanying systematic alterations in pH values, total concentration and ratio of cyanurate to chlorine. In addition, some constants were determined by direct electrometric acid-base titration.

An attempt was made to obtain experimental values not only for each of the independent constants, but for some of the redundant constants as well. This provided for independent check on the validity of some of the determinations. Complete details of the experimental measurements can be found elsewhere.[5] All the data used for computation of the molar absorptivities or the equilibrium constants is presented there.

The computer program was of great assistance to the experimental studies. Computed estimates of species' distributions obtained from preliminary values of the equilibrium constants were invaluable for establishment of the conditions and concentrations that would appropriately restrict the number of significant species in subsequent experiments. Reiteration of this procedure eventually permitted selection of conditions and measurement of individual equilibria without strong interference from extraneous processes. Thus, a more accurate determination of many of the constants was possible than had been achieved previously. In some instances errors as great as 0.5 in pK were found for previously determined values.

Importance of Chloride-Free HOCl

Chlorine hydrolyzes in water in accordance with the following equation

$$Cl_2 + H_2O \rightleftarrows HOCl + H^+ + Cl^- \qquad (5)$$

According to Connick and Chia[6] the hydrolysis constant, K, for this reaction is 3.94×10^{-4} at 25° C, where

$$K = \frac{[HOCl][H^+][Cl^-]}{[Cl_2]} \quad (6)$$

At neutral pH and concentrations of chloride ion as great as 10^{-3} *M*, the ratio of $HOCl:Cl_2$ is always greater than 10^6. So, molecular chlorine is essentially absent under such conditions for all practical purposes.

Some of the experimental work in the present study was performed at pH values less than 2, and chlorine concentrations, Cl_T, greater than 10^{-2} *M* were occasionally used. Under these conditions, unless chloride-free HOCl is employed, the ratio of $HOCl:Cl_2$ will be only about 4. Since this condition would have seriously affected equilibria and spectra, chloride-free HOCl was used for all experimental work.

PROCEDURE

The chloride-free HOCl was prepared by neutralizing commercial grade Chlorox with dihydrogen phosphate to a pH of approximately 6, and distilling HOCl under vacuum in a rotary distilling apparatus at 30-35° C. The chlorine concentration of the chloride-free condensate was 0.20-0.23 *M* and, when it was kept in low actinic glass under refrigeration, the solution was stable for several months.

Measurement of Cyanurate Spectra

Previous studies such as the electrometric titrations reported by Monsanto[4] indicated that the first and second ionization constants of H_3Cy and H_2ClCy (K_6 and K_{10}; K_4 and K_8) are separated by more than 4 orders of magnitude. These separations not only permit separation of the unchlorinated species from one another by pH control alone, but also by careful selection of Cy_T and Cl_T concentrations; $Cy_T:Cl_T$ ratios and pH allow isolation of the nonovalent species (H_2Cy^-, $HClCy^-$, and Cl_2Cy^-) and make possible reliable measurement of important portions of their individual spectra.

The chlorinated cyanuric acids can be studied in their undissociated forms at pH values less than 2, for the pK value of the strongest acid (that of HCl_2Cy) is 3.75. Because of the high molar absorptivities of these chlorinated species (1,000-10,000 cm^{-1} 1 mol^{-1}) 0.1-cm light paths and Cy_T concentrations of 10^{-3} to 10^{-2} *M* could be used for spectrophotometric measurements. In this concentration range, hydrolysis of many of the chlorinated species can be effectively repressed by an excess of Cy_T or Cl_T.

As an example, the H_2ClCy spectrum was obtained at pH 2 in the presence of a large excess of Cy_T. If pK_{5a} is approximately 4.0 as estimated by Brady *et al.*,[3] hydrolysis of 10^{-3} *M* H_2ClCy can be completely repressed by 10^{-2} *M* H_3Cy. Read against a blank of 10^{-2} *M* Cy_T, the differential absorbance of H_2ClCy is $A = (e_2 - e_1)bc$, where A is the differential absorbance of H_2ClCy; e_2 and e_1 are the molar absorptivities of H_2ClCy and H_3Cy, cm^{-1} 1 mol^{-1}; b is the light path, cm; and c is the concentration of Cl_T (and H_2ClCy).

The spectra of HCl_2Cy and Cl_3Cy were also obtained at pH 2 in systems with $Cl_T >> Cy_T$. Molar absorptivities at any given wavelength were obtained by plotting absorbance as a function of the reciprocal of the $Cl_T:Cy_T$ ratio and extrapolating to zero. Since Cl_3Cy began forming before HCl_2Cy formation was complete, two linear extrapolations were involved.

The spectrum of slightly hydrolyzed $HClCy^-$ was determined in the presence of a slight excess of H_2Cy^- at a pH value where H_2ClCy and $ClCy^=$ were less than one per cent of the $HClCy^-$ concentration. The total absorbance, read against a water blank, was $A = e_1 bc_1 + e_2 bc_2$, where e_1 and e_2 and c_1 and c_2 are, respectively, the molar absorptivities and concentrations of H_2Cy^- and $HClCy^-$ (H_3Cy has no absorbance at the wavelengths examined).

The interesting spectrum of Cl_2Cy^-, a rather widely employed bleach, was readily observed at pH 6 in the presence of excess HOCl.

Spectral determinations were made for eight of the ten cyanurate species (summarized later in Table 14.III). For complete experimental details, reference is made to O'Brien.[5] Conditions did not permit measurement of $Cy^{\equiv}$ and $ClCy^=$ spectra.

Determination of Equilibrium Constants

First Ionization Constant of H_3Cy, K_6

Exploitation of the rather large difference in spectral properties of adjacent or conjugate cyanurate species was one method successfully employed for determining equilibrium constants. A simple example was the spectrometric titration of cyanuric acid to determine its first ionization constant, K_6. Cyanuric acid, H_3Cy, exhibited almost no absorbance at wavelengths longer than 205 nm, whereas its first ion, H_2Cy^-, had an absorbance peak of 8800 cm^{-1} liters mol^{-1} at 214 nm. In order to determine the value for the constant as accurately as possible, one series of experiments measured absorbance as a function of pH at constant Cy_T while another measured absorbance as a function of Cy_T at constant pH.

Second Ionization Constant of H_3Cy, K_{10}

The species H_2Cy^- and $HCy^=$ have overlapping spectra and share an isobestic point at 218 nm. At wavelengths longer than 220 nm, e for $HCy^=$ is much greater than that for H_2Cy^-. Also, at these wavelengths and a light path of 0.1 cm, there is minimal interference from hydroxide ion. Therefore, absorbance as a function of pH at 225 and 230 nm was used to determine pK_{10}.

Third Ionization Constant of H_3Cy, K_{12}

As the pH exceeded 13, absorbance values at wavelengths less than 225 nm (when corrected for OH^- absorbance) showed a slight decrease. Unless this resulted from overcompensation for the large hydroxide absorbance, the third ionization of H_3Cy would appear to have a pK value between 13 and 14.

First Ionization Constant of H_2ClCy, K_4

The experimental determination of the first ionization constant of H_2ClCy was made by a differential spectrometric titration in the presence of excess Cy_T. Sample and blank each contained 0.001 *M* Cl_T and 0.008 *M* Cy_T at an initial pH of 1.8. All subsequent changes in absorbance resulted from increases in pH value. Spectra of the three absorbing species are shown in Figure 14.3 while typical absorbance data are shown in Figure 14.4. Differential absorbance was given by $A = c_1(e_1 - e_3) + e_2c_2$, where e_1, e_2, and e_3 are the molar absorbtivities of $HClCy^-$, H_2Cy^-, and H_2ClCy, respectively, and c_1 and c_2 are the molar concentrations of $HClCy^-$ and H_2Cy^-, respectively.

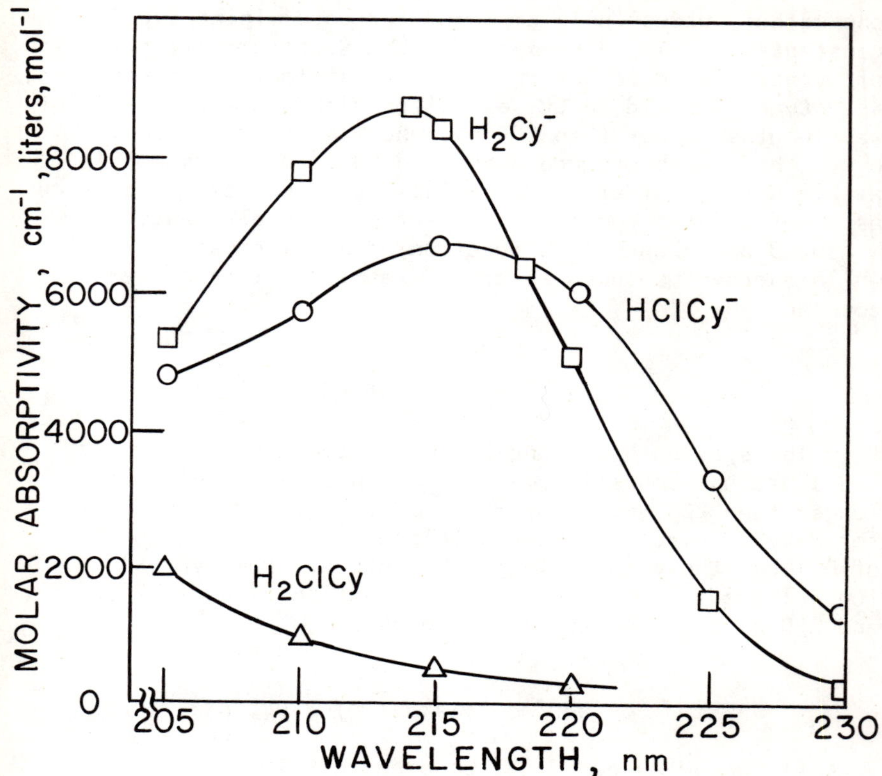

Figure 14.3. U.V. spectra of $HClCy^-$, H_2ClCy, *and* H_2Cy^-.

Absorbance data between pH values of 4.5 and 5.5 are considered quite significant because formation of $HClCy^-$ was more than half complete within this pH range while interference from H_2Cy^- was still minimal.

Second Ionization Constant of H_2ClCy, K_8

Alkaline hydrolysis of $HClCy^-$ (K_9) was quite extensive in the 10^{-3} *M* range at pH values above 9 and made spectrometric techniques impractical. The method of choice was the electrometric titration of as concentrated a solution as possible with 0.4 *M* NaOH as titrant.

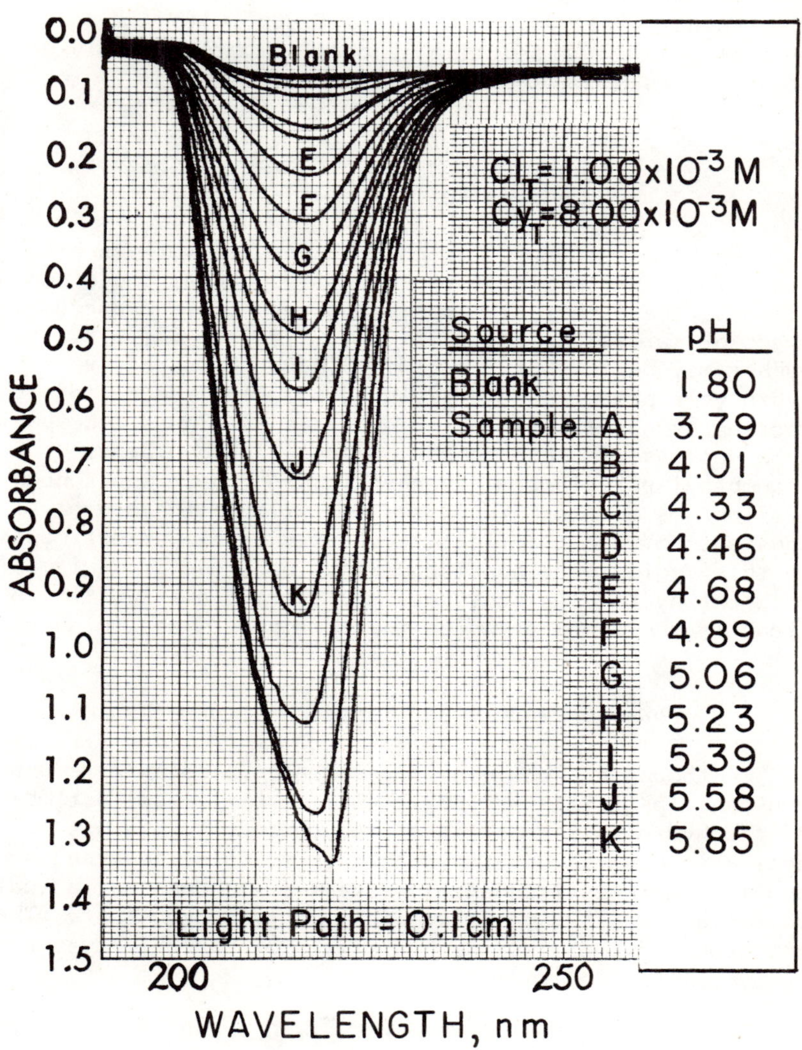

Figure 14.4. Differential absorbance of 10^{-3} M $HClCy^-$ as a function of pH.

Ionization Constant of HCl_2Cy, K_2

The experimental value of pK_2 was obtained by an electrometric titration of 10^{-1} *M* Cl_2Cy^- with 1 *M* acid titrant. A 10^{-1} *M* solution is only slightly hydrolyzed at pH values above the equivalence point.

Hydrolysis Constants of Unionized Chlorinated Cyanuric Acids, K_{1a}, K_{3a}, K_{5a}

Cyanuric acid and its chlorinated derivatives are essentially unionized at pH below 1.8. Under these conditions, the number of significant species is reduced to 4 and by careful selection of Cl_T, Cy_T, and Cl_T:Cy_T, this number may be further reduced. Since the successive hydrolysis products of Cl_3Cy have parallel spectra of decreasing absorptivities, hydrolysis is always accompanied by a decrease in absorbance. Changes in absorbance accompanying reciprocal changes in light path and concentration was one technique used to estimate the hydrolysis constants of the chlorinated cyanuric acids. Experiments in this region were handicapped by the close proximity of adjacent hydrolysis constants and limited availability of light in the range at which these compounds absorb energy.

Hydrolysis Constant of Monochlorcyanurate, K_{9a}

Although K_{9a} could have been calculated as a redundant constant ($pK_{9a} = pK_{5a} + pK_6 - pK_4$), experimental estimation of the hydrolysis constant of monochlorcyanurate was desirable because of its importance to chlorine-cyanurate equilibria in environmental systems. The constant actually determined was K_9, the alkaline hydrolysis constant. The reaction for the alkaline hydrolysis of monochlorcyanurate, $HClCy^-$, is as follows:

$$HClCy^- + OH^- = H_2Cy^- + OCl^- \qquad (7)$$

$$K = \frac{[H_2Cy^-][OCl^-]}{[HClCy^-][OH^-]} = K_{9a} \times \frac{K_{HOCl}}{K_w} .$$

The experimental method of choice involved measurement of the U.V. absorption of OCl^- as the pH was varied from

9.4 to 10.6 in a system containing 0.002 *M* Cl_T and 0.003 *M* Cy_T. Experimental determination of K_9 made possible the algebraic computation of K_{5a} and K_{11a} as redundant constants without the necessity of propagating errors through two "equilibrium boxes" (see Figure 14.2). The constant K_{5a} was also estimated experimentally (K_{11a} was not).

RESULTS AND DISCUSSION

Cyanurate Spectra and Stability Constants

Molar absorptivities of the cyanurate species for which spectral determinations were made are listed in Table 14.II.

Experimental values of the equilibrium constants are summarized in Table 14.III. In accordance with Figure 14.2, hydrolysis constants have odd numbers and ionization constants are even-numbered. The table also includes brief comments on experimental methods. When the respective values are placed in their proper location in Figure 14.2, some interesting observations may be made. The addition of each chlorine atom increases constants by 1.4-1.6 orders of magnitude and hydrolysis constants by 1.1 log units (*i.e.*, $HClCy^-$, the chlorinated cyanurate of importance in environmental equilibria, is 40 times less hydrolyzed than its acid). This pattern is consistently observed throughout each of the three "equilibrium boxes" of Figure 14.2.

A summary of results obtained by the Stanford Research Institute[3] and the Inorganic Research Department of Monsanto Chemical Co.[4] are given in Table 14.IV, which also includes those of the present study. In spite of the rather widespread use of cyanuric acid and its chlorinated derivatives, the only values found in the literature for hydrolysis constants were those of the S.R.I. study as reported by Brady, Sancier and Sirine.[3] The Monsanto study determined primarily ionization constants by electrometric titration.

Correlation with Bacteriological Studies

Anderson[1] found that at a pH of 9.0, the addition of 25 mg/l cyanuric acid (1.94×10^{-4} *M* Cy_T) increased the 99 per cent kill time for 0.24 mg/l chlorine (3.40×10^{-6} *M*

Table 14.II

Molar Absorptivities of Cyanurates and Their Chlorinated Derivatives

Species	*196 nm*	*200*	*205*	*210*	*215*	*220*	*225*	*230*
H_3Cy	1830	950	325	155	95	-	-	--
H_2ClCy	3110	2770	1980	960	510	-	-	-
HCl_2Cy	-	-	4030	2900	1710	1310	1060	-
Cl_3Cy	-	-	-	-	3800	3400	2900	2250
H_2Cy^-	-	(3400)	5300	7840	8410	5070	1530	260
$HClCy^-$	-	(4000)	4950	5770	6710	6060	3320	1330
Cl_2Cy^-	-	(8600)	6800	6060	6260	6350	4430	2100
$HCy^=$	-	-	(6000)	5330	5540	6470	5300	2300

Table 14.III

Equilibrium Constants of Cyanurates and Their Chlorinated Derivatives

pK Number	*Method of Determination*	*pK Value*
1a	UV Spectrum at pH<2, $Cl_T >> Cy_T$	1.8 ± 0.2
3a	UV Spectrum at pH<2, $Cl_T > Cy_T$	2.93 ± 0.07
5a	$pK_4 + pK_{9a} - pK_6$	4.07 ± 0.08
(5a)	Differential Spectrophotometry at pH<2, $Cy_T >> Cl_T$	(4.10 ± 0.12)
7a	$pK_{3a} + pK_4 - pK_2$	4.51 ± 0.09
9a*	OCl^- Spectrum during alkaline hydrolysis, $Cy_T > Cl_T$	5.62 ± 0.05
11a	$pK_{9a} + pK_{10} - pK_8$	6.90 ± 0.11
2	Electrometric Titration	3.75 ± 0.03
4	Differential Spectral Titration	5.33 ± 0.05
6	Spectral Titration	6.88 ± 0.04
8	Electrometric Titration	10.12 ± 0.02
10	Spectral Titration	11.40 ± 0.10
12	Spectral Changes at pH 13	13.5

*$pK_9 = -0.84 \pm 0.05$ at 25° C.

$pK_{9a} = pK_w - pK_{HOCl} + pK_9 = 14.00 - 7.54 - 0.84$

$pK_{9a} = 5.62 \pm 0.05$

Table 14.IV

Comparison of Results with Those of Other Investigators

Reaction	*pK No.*	*Present Study*	*Monsanto*[4]	*S.R.I.*[3]
$Cl_3Cy + H_2O = HCl_2Cy + HOCl$	1a	1.8 ± 0.2	--	0.3
$HCl_2Cy = H^+ + Cl_2Cy^-$	2	3.75 ± 0.03	3.95	4.0
$HCl_2Cy + H_2O = H_2ClCy + HOCl$	3a	2.93 ± 0.07	--	3.0
$H_2ClCy = H^+ + HClCy^-$	4	5.33 ± 0.05	5.31	5.7
$H_2ClCy + H_2O = H_3Cy + HOCl$	5a	4.07 ± 0.08	--	4.1
$H_3Cy = H^+ + H_2Cy^-$	6	6.88 ± 0.04	6.93	7.0
$Cl_2Cy^- + H_2O = HClCy^- + HOCl$	7a	4.51 ± 0.09	--	-1.7[a]
$HClCy^- = H^+ + ClCy^-$	8	10.12 ± 0.02	10.09	9.4
$HClCy^- + H_2O = H_2Cy^- + HOCl$	9a	5.62 ± 0.05	--	-1.1[a]
$H_2Cy^- = H^+ + HCy^-$	10	11.40 0.10	11.0	10.6
$ClCy^= + H_2O = HCy^= + HOCl$	11a	6.90 0.11	--	-0.1[a]
$HCy^= = H^+ + Cy^{\equiv}$	12	13.5		(13)

[a]These values are expressed as alkaline hydrolysis constants. For comparison with the present study a factor of 6.5 should be added to these values (*i.e.* $pK_w - pK_{HOCl}$).

Cl_T) from 3.5 minutes to 15.5 minutes. Utilizing the PPL computer program based on the experimentally determined pK values of the equilibrium model (in which the three master variables are pH, Cy_T, and Cl_T), the computed free chlorine concentration under these conditions was 8.6 x 10^{-7} *M* or 0.06 mg/l. The ratio of free chlorine in the presence and absence of cyanurate was 0.25:1.00. The ratio of 99 per cent kill-times was 1.00:0.23. Since inactivation-times were almost inversely proportional to free chlorine concentrations, it would appear probable that most, if not all, bactericidal activity resulted from the free chlorine fraction. Although these computations are based on a single comparison, it would seem advisable to regard the chlorinated cyanurate molecule as a free chlorine donor with little bactericidal potency of its own. Any error in this assumption would be in the direction of safety.

Distribution of Individual Species

Examples of computer output based on the equilibrium model are shown graphically in Figures 14.5 and 14.6. In each case the master variables were Cl_T and Cy_T. Molar concentrations of individual species were plotted as a function of pH. Figure 14.5 is intended to depict the distribution of species in a realistic environmental system, *i.e.*, $Cy_T = 2.0 \times 10^{-4}$ *M* (∿25 mg/l), $Cl_T = 2.0 \times 10^{-4}$ *M* (∿1.4 mg/l), and $Cy_T:Cl_T$ = 10:1. Figure 14.6 depicts the distribution of individual species in a 10^{-3} *M* solution of $NaCl_2Cy$, a commercial bleach. Between pH values of 3 and 10, 11 of the 12 possible species are present in significant concentrations.

Germicidal responsibility in chlorinated cyanurate solutions is borne principally by the free chlorine fraction. The distribution of free chlorine in three interrelated solutions is summarized graphically in Figure 14.7. All concentrations were computed from the equilibrium system. Symbols were used only for differentiation. Hydrolysis to chlorine is least at a pH of approximately 7.2. The fraction of free chlorine increases rather dramatically at both greater and smaller pH values. At pH values greater than 8, free chlorine exists primarily as the hypochlorite ion. At pH values below 7 free chlorine exists primarily in the form of hypochlorous acid (HOCl) which is approximately 80 times more effective as a bactericide than hypochlorite (OCl^-).

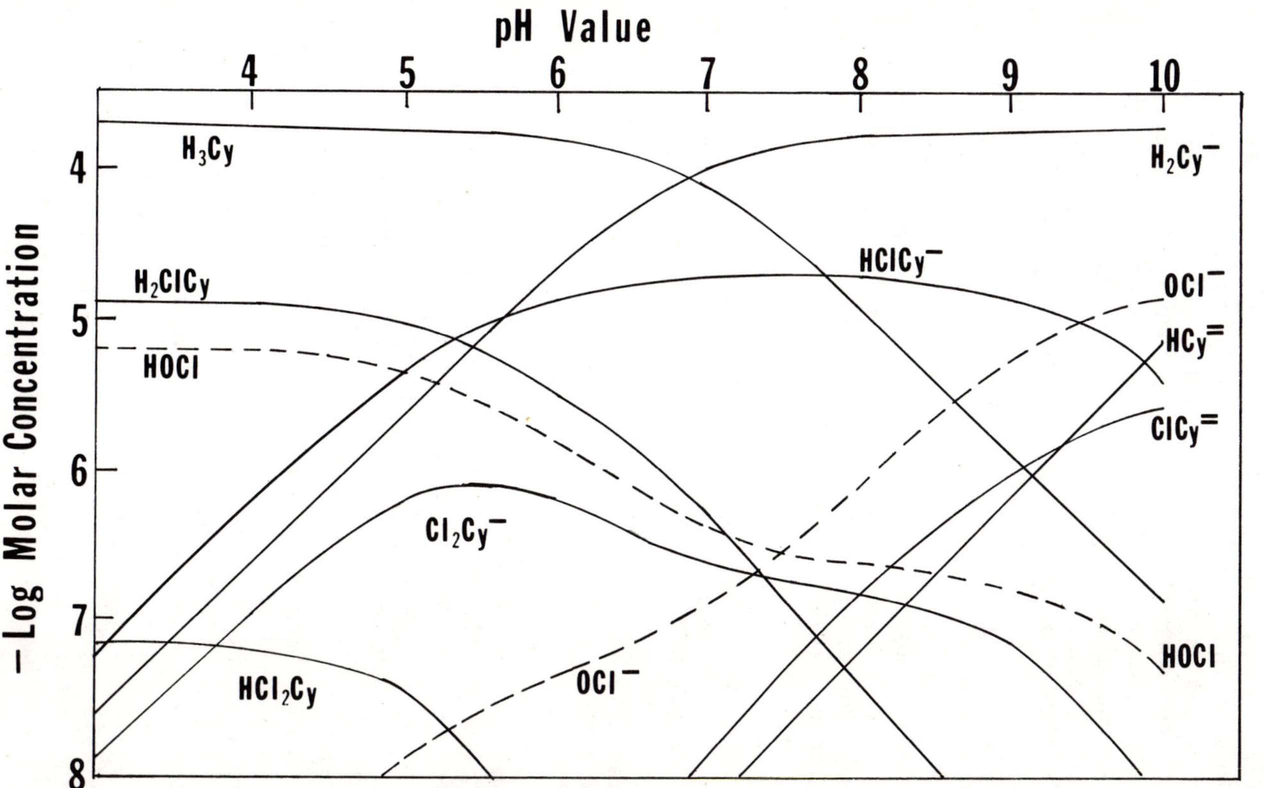

Figure 14.5. Distribution of aqueous chlorine and cyanurate as a function of pH. ($Cy_T:Cl_T$ = 10:1)

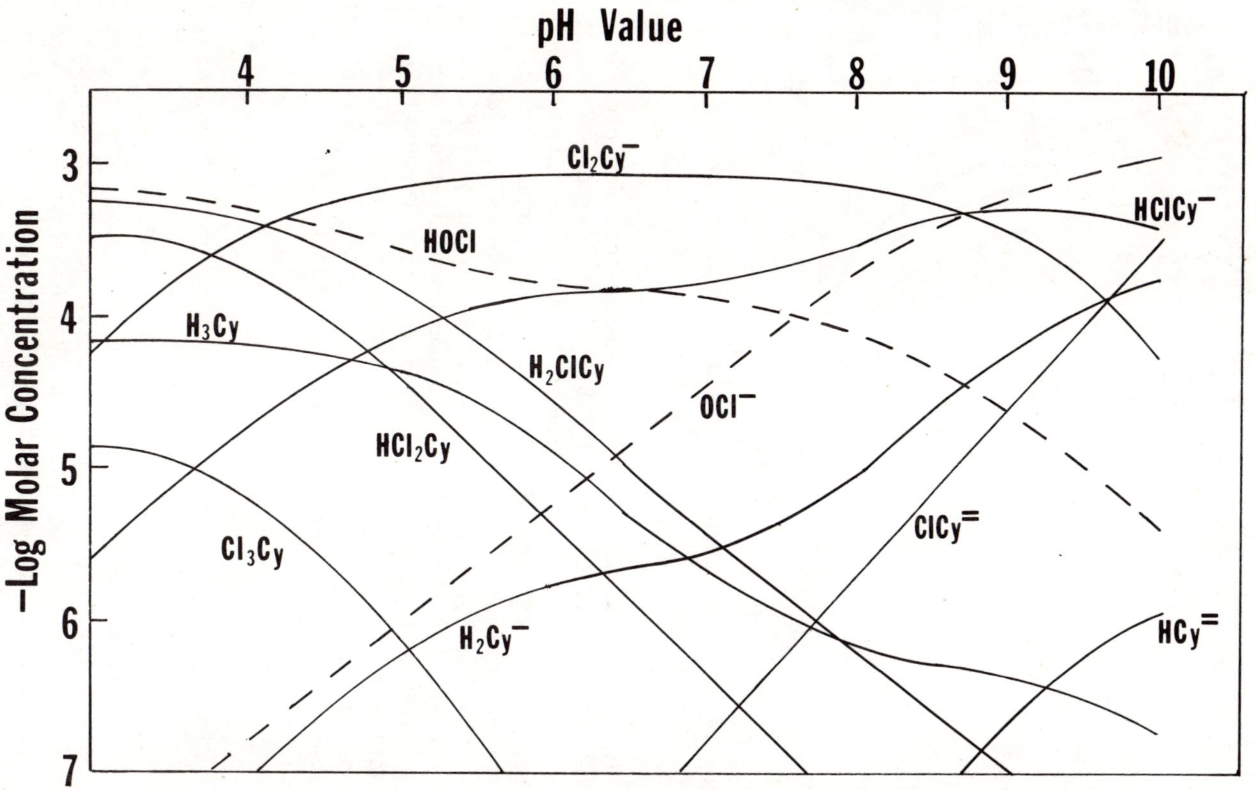

Figure 14.6. Distribution of species for solution of $NaCl_2Cy$ (10^{-3} M).

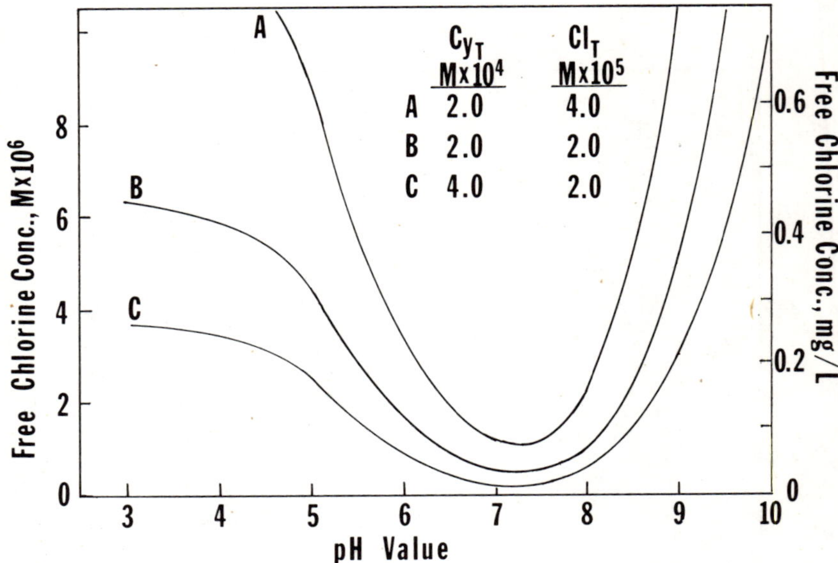

Figure 14.7. Free chlorine as a function of pool pH.

In "environmental" systems such as that depicted in Figure 14.7 (*i.e.*, Cl_T = 1.4-2.8 mg/l and Cy_T:Cl_T = 5.1 to 10:1) free chlorine varies inversely with the total concentration of cyanurate in the neutral pH range (6.0-8.5), indicating that equilibria in such systems is dominated largely by the hydrolysis of monochlor cyanurate ($HClCy^-$).

Figure 14.8 shows the Cl_T concentrations required for desired concentrations of free chlorine (or HOCl) as a function of pH in a system containing a constant Cy_T concentration. It represents the most practical and direct application of the basic computer program to environmental systems.

Photolytic Decomposition of Chlorinated Cyanurate

The effect of sunlight on solutions of cyanuric acid and reservoir chlorine was studied. It would appear that photo-stability is greatest at a pH of about 6.8 and decreases more rapidly at higher pH values than it does at lower pH values. The maximum stability at pH 6.8 accords rather well with the pH of minimum hydrolysis to free

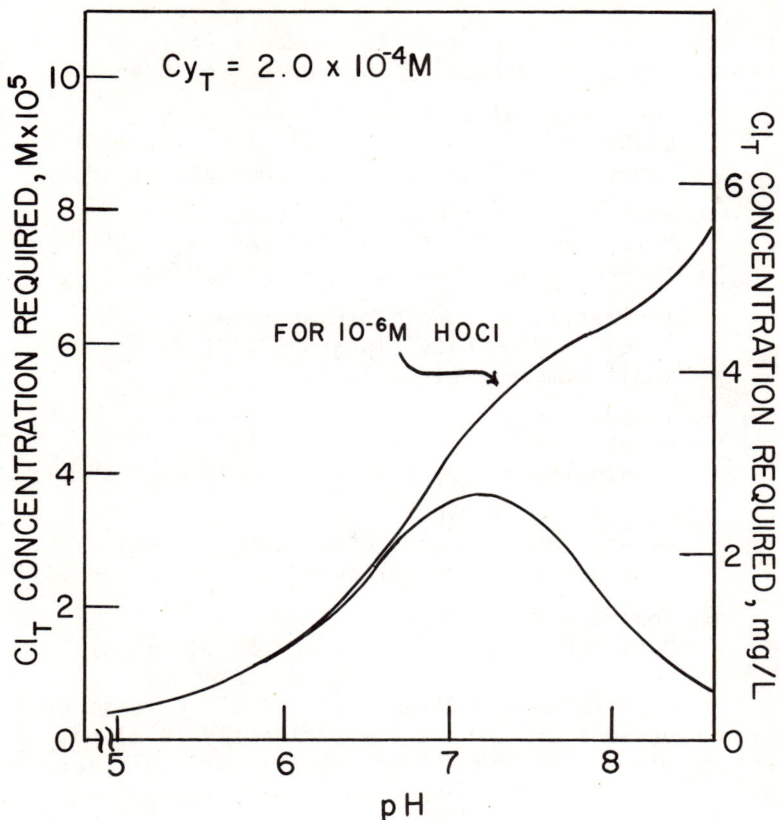

Figure 14.8. Cl_T required as a function of pH for an equilibrium concentration of 10^{-6} M HOCl or 10^{-6} M ($HOCl + OCl^-$).

chlorine; the rapidly increasing instability with increased pH indicates that HOCl has greater stability in sunlight than its ionization product, OCl^-, which has a moderately strong absorptivity in the energy range of sunlight reaching the earth's surface.

Evaluation of Equilibrium Constants

The myriad of changes occurring during alkaline hydrolysis of a chlorinated cyanurate system in which $Cy_T = 10^{-3}$ *M* and $Cl_T:Cy_T > 2$ made it useful as one method of evaluating the experimentally determined equilibrium constants. Consider the reactions occurring in such a system as the pH is gradually increased from 8 to 11:

1. Hydrolysis of dichlorcyanurate:
   $Cl_2Cy^- + OH^- \rightleftarrows HClCy^- + OCl^-$ ($pK_7 = -1.95$)
   (approximately 50 per cent complete at pH 9)

2. Hydrolysis of monochlorcyanurate:
   $HClCy^- + OH^- \rightleftarrows H_2Cy^- + OCl^-$ ($pK_9 = -0.84$)
   (approximately half complete at pH 10.2)

3. Ionization of monochlorcyanurate:
   $HClCy^- \rightleftarrows H^+ + ClCy^=$ ($pK_8 = 10.12$)
   (half complete at pH 10.12)

4. Ionization of H_2Cy^-:
   $H_2Cy^- \rightleftarrows H^+ + HCy^=$ ($pK_{10} = 11.40$
   (approximately 30 per cent complete at pH 11)

5. Hydrolysis of $ClCy^=$:
   $ClCy^= + OH^- \rightleftarrows HCy^= + OCl^-$ ($pK_{11} = 0.44$)
   (approximately 20 per cent complete at pH 11)

6. Ionization of HOCl:
   $HOCl \rightleftarrows H^+ + OCl^-$ ($pK_{HOCl} = 7.54$ at 25° C)

Five equilibria are directly involved in these overlapping reactions and others are indirectly involved. One effect resulting from reactions 1-6 is a gradual increase of hypochlorite ion concentration with increasing pH. It appeared that one method of determining reliability of the equilibrium system was to compare computed values of OCl^- concentrations with those found experimentally. Results of such a comparison are shown in Figure 14.9. Good agreement was attained.

SUMMARY

Wet chemical methods for estimating free chlorine concentrations include chlorinated cyanurates as well. Because of this ambiguity, reliable estimation of equilibrium concentrations of free chlorine, or indeed of any individual species in solutions of chlorine plus cyanurate, either requires a physical measurement that does not disturb the equilibrium or must be computed from known equilibrium constants. A computer program has been developed that permits computation of all individual species as a function of reservoir chlorine, total cyanurate, and pH value based on knowledge of individual

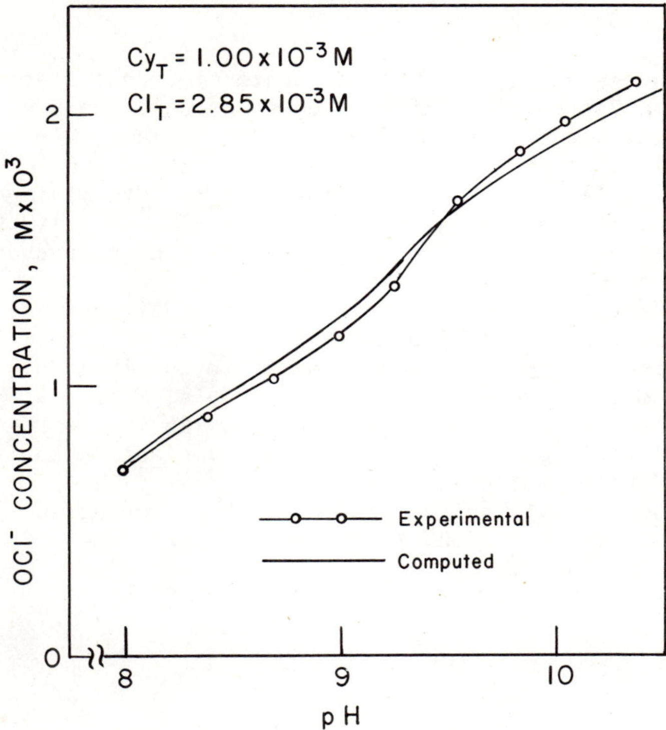

Figure 14.9. Comparison of experimental OCl^- concentrations with theoretical distribution line.

acidity and hydrolysis constants in solutions of chlorine plus cyanurate.

Although chlorinated cyanurates serve as a reservoir of free chlorine, bactericidal efficacy is more closely related to the relatively small fraction of free chlorine present at equilibrium. Therefore, the use of excessive cyanurate in an overly zealous attempt to reduce photolysis may repress free chlorine to the point of suppressing germicidal activity. For the same reason the continual addition of chlorinated cyanurates as a source of chlorine is not recommended because this will lead to build up of cyanurate concentration.

REFERENCES

1. Anderson, J. B. "The Influence of Cyanuric Acid on the Bactericidal Effectiveness of Chlorine," Ph.D. Thesis, University of Wisconsin, Madison, Wisconsin, 1963.
2. Whittle, G. P. "Recent Advances in Determining Free Chlorine," presented at the National Specialty Conference on Disinfection, University of Massachusetts, Amherst, Massachusetts, July, 1970.
3. Brady, A. P., K. M. Sancier, and G. Sirine. *J. Am. Chem. Soc., 85,* 3101 (1963).
4. Special Report No. 6862, Monsanto Chemical Co., Inorganic Research Dept., St. Louis, Missouri, prepared by G. D. Nelson, March, 1967.
5. O'Brien, J. E. "Hydrolytic and Ionization Equilibria of Chlorinated Isocyanurate in Water," Ph.D. Thesis, Harvard University, Cambridge, Massachusetts, 1972.
6. Connick, R. E. and Y. T. Chia. *J. Am. Chem. Soc., 81,* 1280 (1959).

CHAPTER 15

EFFECTIVENESS OF HYPOCHLOROUS ACID AND HYPOCHLORITE ION IN DESTRUCTION OF VIRUSES AND BACTERIA

Pasquale V. Scarpino, Merrilee Lucas, Daniel R. Dahling, Gerald Berg, and Shih Lu Chang*

INTRODUCTION

The disinfecting ability of chlorine and chlorine compounds in the treatment of water and wastewater has been known for many years. A considerable amount of information and expertise has been amassed concerning the destruction by chlorination of human enteric bacterial pathogens, and bacterial indicators of such bacterial pathogens. However, little precise information is presently available concerning the inactivation of animal viruses in water and wastewater by chlorine and its compounds. Human viruses are found in large numbers in sewage, sewage effluents, and in rivers and streams. Their importance lies, however, not in their numbers but in their great capacity to infect their human hosts. Today information of a precise nature dealing with the efficiency of applied disinfectants such as chlorine on human enteric pathogens is of increased significance as we turn to the use of renovated water for human needs. This information is especially needed for animal viruses of enteric origin, which are considerably more resistant than the bacteria.

---

*P. V. Scarpino and M. Lucas, Department of Civil and Environmental Engineering, University of Cincinnati, Cincinnati, Ohio. D. R. Dahling and G. Berg, Virology Research Section, National Environmental Research Center, Environmental Protection Agency, Cincinnati, Ohio. S. L. Chang, Division of Water Hygiene, Environmental Protection Agency, Cincinnati, Ohio.

This research was undertaken to determine the capability of chlorine to inactivate viruses in water. It is part of the cooperative comparative disinfection study coordinated by the Virology program at the National Environmental Research Center, Cincinnati, Ohio. Comparative data are reported here on the relative resistances of selected RNA and DNA coliphages, animal viruses, and fecal coliforms. The use of coliforms and coliphages as indicators of animal virus pollution of water will also be evaluated.

MATERIALS AND METHODS

The test microbes that were used in these studies were two DNA phages, T2 and T5, two RNA phages, f2 and MS2, *Escherichia coli* ATCC 11229, and two animal viruses, poliovirus 1 and coxsackievirus A9. Chlorine demand-free suspensions of stock viruses were prepared by differential centrifugation and stored in vials at -70° C until used. Bacterial suspensions were prepared in demand-free phosphate buffer after low speed centrifugation on the day of the experiment. The buffer systems consisted of 0.05 *M*, pH 6 phosphate buffer (KH_2PO_4-K_2HPO_4) and 0.05 *M*, pH 10 borate buffer (H_3BO_3-KCl-NaOH). The buffers were made demand-free by the addition of chlorine to provide a residual of 3-5 mg/l. After storage at room temperature for one week, the water was dechlorinated for 24 hrs under ultraviolet light (chlorine is highly photoreactive at ultraviolet wave lengths of about 2600Å and decomposes rapidly) and was thus chlorine demand-free. The amperometric titration method employing the Wallace and Tiernan Amperometric Titrator was used to determine free chlorine during the experiments.

In the test procedure the five 600-ml chlorine demand-free beakers which were prepared are as follows: a titer control beaker containing 400 ml of dechlorinated buffer; a neutralizer control beaker with 400 ml of chlorinated buffer which was neutralized with sodium thiosulfate just before sample inoculation; two test beakers having 400 ml of chlorinated buffer at specified chlorine levels; and, an uninoculated chlorine control beaker containing 400 ml of chlorinated buffer.

All beakers were placed in a 5° C water bath and covered with metal lids through which extended glass stirring rods connected to an overhead variable speed stirring device adjusted to 81 rpm. After temperature equilibration, chlorine levels were determined, and the test microbe was added to the appropriate beaker. Five-ml test samples were removed at frequent intervals and blown rapidly into 5 ml of sodium thiosulfate to yield a final neutralizer concentration of 6 mg/l. The chlorine levels in the test and control beakers were consistent throughout each experiment.

The animal viruses were assayed by the plaque method in Vero (poliovirus 1) and in primary rhesus monkey kidney cells (coxsackievirus A9). The phages were assayed on host bacteria by the plaque method. Bacterial numbers were measured on surface inoculated tergitol-7-agar plates that were supplemented with 40 mg/l of triphenyltetrazolium chloride.

RESULTS AND DISCUSSION

When chlorine is added to water, the following chemical reactions occur:

$$Cl_2 + H_2O \rightleftarrows HOCl + H^+ + Cl^- \qquad (1)$$

and

$$HOCl \rightleftarrows H^+ + OCl^- \qquad (2)$$

Measurable molecular chlorine (Cl_2) does not exist unless the pH is below 3 and free chlorine is over 7.1 mg/l. The hypochlorous acid (HOCl) produced further ionizes to form hypochlorite ion (OCl^-) and hydrogen ion (Equation 2). The dissociation of hypochlorous acid is dependent chiefly upon pH and, to a much lesser extent, temperature, with almost 100 per cent HOCl present at pH 5, and almost 100 per cent OCl^- present at pH 10.

In Figure 15.1, comparisons were made of survival characteristics after chlorination at pH 6 and 5° C of T2 and T5 DNA bacteriophages, f2 and MS2 RNA bacteriophages, *Escherichia coli*, poliovirus 1, and coxsackievirus A-9. Since this study was done at pH 6, greater than 95% of the free chlorine is present as HOCl. The 99% inactivation points used in the construction of the log-log concentration-time plot shown in Figure 15.1 were obtained from separate survival studies in different concentrations of HOCl in the same buffer system (phosphate) at the same pH (6) and temperature (5° C). Four to six different survival studies were made for each microbe at different levels of HOCl. From each survival curve the time required for 99% inactivation of the test organism by each HOCl concentration

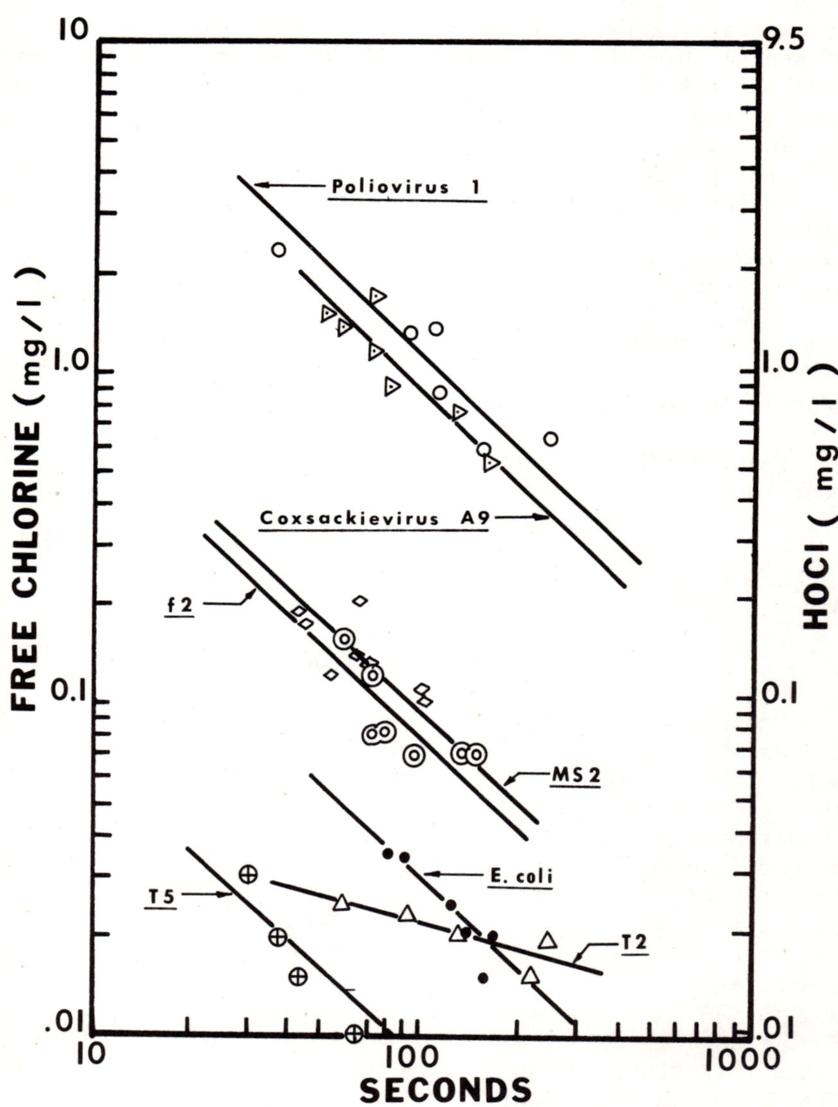

Figure 15.1. Concentration-time relationship for 99% inactivation of animal and bacterial viruses, and the reference bacterium Escherichia coli ATCC 11229, by hypochlorous acid at 5° C and pH 6.

tested was obtained and recorded on the composite figure shown here. Therefore, any given point on this figure represents the time required for 99% inactivation at a given level of HOCl. All slopes, except that for T2, approximate a slope of 1. The animal viruses, poliovirus 1 and coxsackievirus A9, were the most resistant, being 13 times more so than the RNA coliphages f2 and MS2, which were about equally sensitive, about 40 times more resistant than *E. coli*, and about 135 times more resistant than coliphage T5. The small slope of the coliphage T2 curve makes it difficult to compare its inactivation rate with those of the other organisms tested. The slope for coliphage T2 approximated 0.3 and implies a more involved inactivation process. The reaction was similar to that encountered with denaturants, and may have been due to the complex tail structure of coliphage T2.

The greater resistance of the two animal viruses, poliovirus 1 and coxsackievirus A9, compared to the faster kill exhibited by the four bacteriophages and the test bacterium *E. coli* points out that the use of coliphages and even an important member of the total coliform and fecal coliform groups, *i.e.*, *E. coli*, as indicators of enteric virus pollution of disinfected water is of questionable validity. The coliphages and the bacterium *E. coli* were more sensitive than the animal viruses used in these studies to chlorination by the chemical species HOCl. For an organism to be useful as an indicator of sewage contamination of water, and thus indirectly ascertain the presence or absence of enteric pathogens, it should be present whenever enteric pathogens are present, but have a reasonably longer survival time than the hardiest enteric pathogens.[2] The present studies show that these latter criteria, among others, are not met by either the coliphages or the coliform bacterium *E. coli* when they are used to reflect enteric virus presence in water.

The inactivation studies shown in Figure 15.2 were made at pH 10, where 99.7% of the free chlorine was in the form of OCl^-. The 99% inactivation points used to construct the log-log concentration-time plots shown in Figure 15.2 were obtained from individual survival studies at different concentrations of OCl^- in a borate buffer system at pH 10 and a temperature of 5° C. After six to seven survival studies were made for each microbe tested at different levels of OCl^- the time required for 99% inactivation of the test microbe at each OCl^- concentration was determined and plotted on the composite figure shown here. Any given point on this figure represents the time required for 99% inactivation of a test

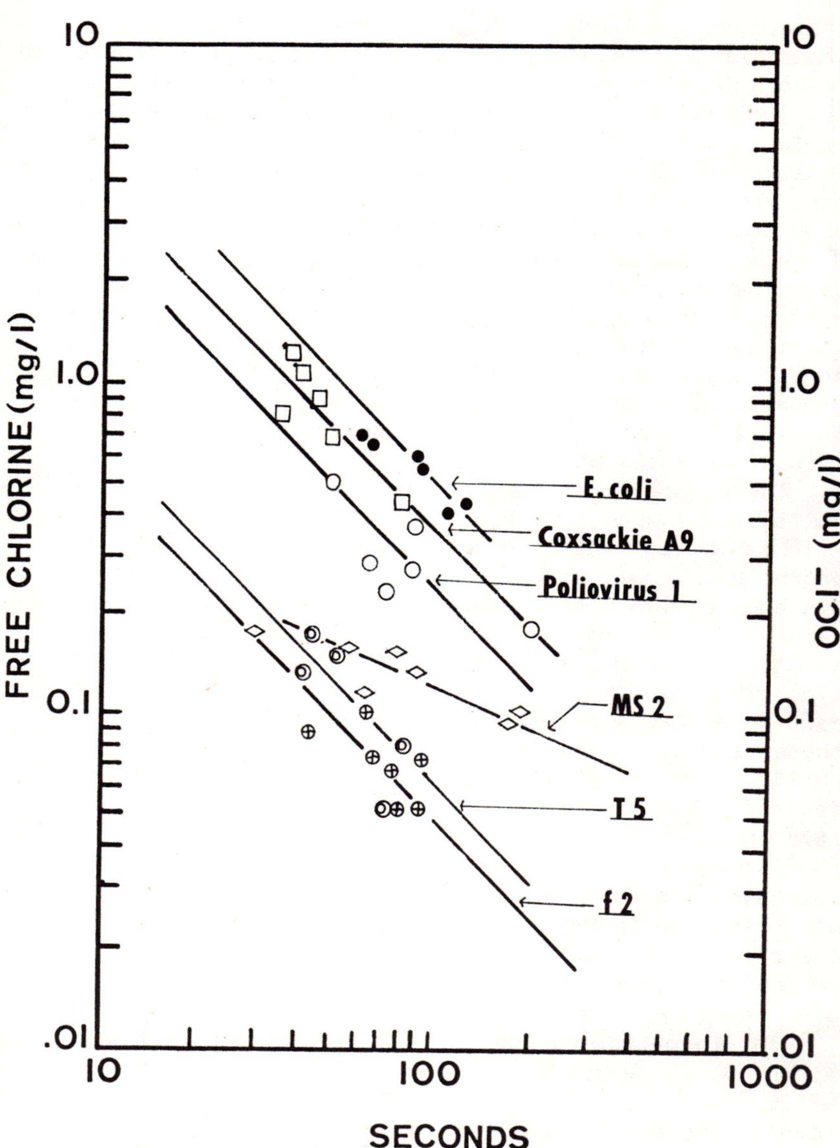

Figure 15.2. *Concentration-time relationship for 99% inactivation of animal and bacterial viruses, and the reference bacterium Escherichia coli ATCC 11229, by hypochlorite ion at 5° C and pH 10.*

microbe at the stated OCl^- level. *E. coli* was the most resistant organism tested, being 1.4 times as resistant as coxsackievirus A9, about twice as resistant as poliovirus 1, about 8 times as resistant as T5, and 10 times as resistant as f2. Since MS2 survival did not exhibit a slope of 1 with OCl^- it was therefore difficult to compare its survival characteristics to the other microbes tested. Although *E. coli* in this experimental system had a slightly longer survival time than the test animal viruses, and could therefore indicate animal viral pollution, the coliphage models would have been of little benefit in the detection of animal virus presence because of their more rapid inactivation by OCl^-.

Hypochlorite ion has been considered to be less microbicidal than HOCl. However, in our studies a more rapid inactivation of both of the test animal viruses occurred at a pH 10 level where the free chlorine was in the form of OCl^- rather than HOCl.[3] This is contrary to the findings of Clarke and Kabler[4] and Weidenkopf.[5] The high pH of the experimental system did not appear to have any effect on the test organisms. The enhanced effect of OCl^- against an animal virus, poliovirus 1, is apparent in Figure 15.3, where the concentration-time relationships for 99% inactivation of poliovirus 1 by HOCl and OCl^- are compared. It was found that OCl^-, based on the times of inactivation, was seven times more effective against the animal virus than HOCl. On the other hand, the predicted microbicidal relationship of the OCl^- to HOCl for *E. coli* was evidenced in the studies shown in Figure 15.4, where the HOCl, based on the times of inactivation, was about 50 times more effective than the OCl^- in destroying the test bacterium. This difference in destruction effectiveness compared favorably to previous literature reports which indicated that HOCl was 70-80 times as bactericidal as OCl^-.[6-8]

Recent on-going research indicates that the borate buffer system used herein has an influence on the OCl^- and HOCl microbicidal relationships. We are presently attempting to reassess the chemistry of HOCl ionization to determine whether the borate buffer system alters the equilibrium of the reaction, suppressing ionization or bringing about the formation of virucidal forms heretofore undescribed. Such research, it is hoped, will clarify the finding that OCl^- in the test system was a more effective virucidal form of free chlorine than HOCl.

In conclusion, there are still many questions in disinfection of water yet unanswered. Good comparative disinfection data are required from which guidelines can

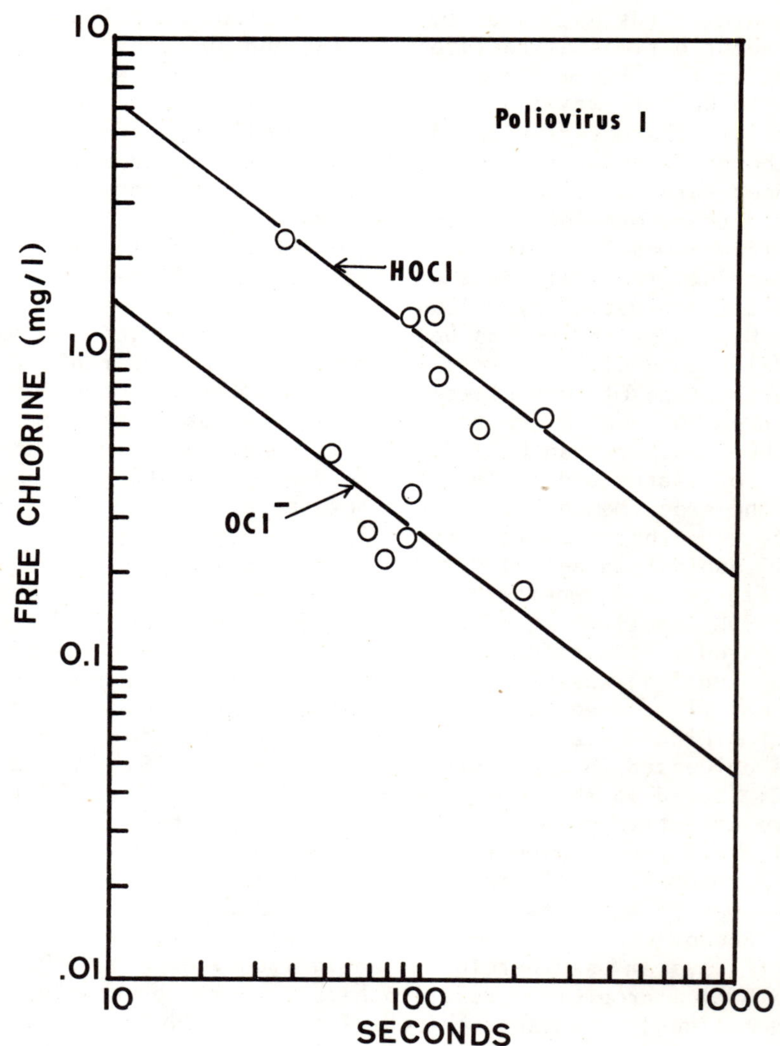

Figure 15.3. Concentration-time relationship for 99% inactivation of poliovirus 1 by hypochlorous acid and hypochlorite ion at 5° C.

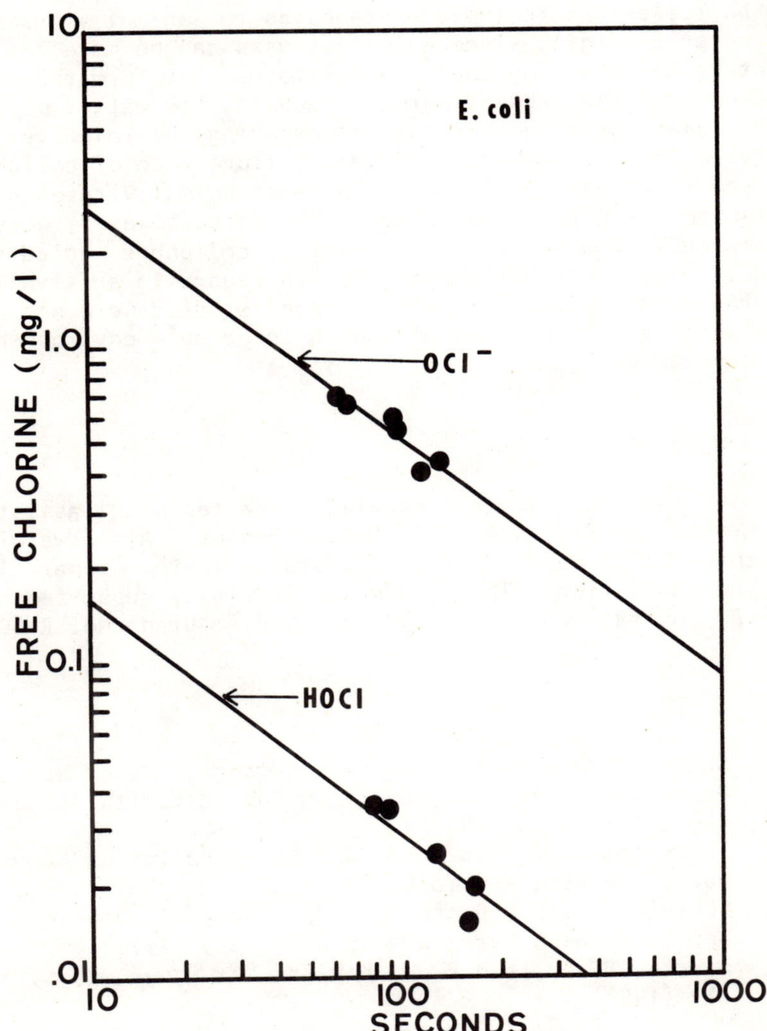

Figure 15.4. *Concentration-time relationship for 99% destruction of Escherichia coli ATCC 11229 by hypochlorous acid and hypochlorite ion at 5° C.*

be drawn for providing safe waters free from pathogens. Present microbiological standards for water quality need to be reassessed to include standards for animal viruses of enteric origin, since animal viruses can be more resistant to chlorination than the coliforms that form the backbone of the present safety standards for water. Additional research is required comparing the relative survival times, numbers, and seasonal presence of coliphages in the water environment with those of animal viruses of pathogenic significance to man. The data, however, does cast doubt upon the ultimate merit of coliphage indicator models of enteric virus presence, in regard to survival characteristics, since the coliphages studied were not as resistant as the viruses of man to comparable environmental circumstances.

ACKNOWLEDGMENTS

The authors wish to acknowledge the technical assistance of Mr. Daniel Siders and Ms. Sandy Cronier Also, we wish to thank Ms. June Schuck for assistance in the preparation of the manuscript. This study was partially supported by E.P.A. Contract No. 14-12-567, and E.P.A. Grant No. R-800370.

REFERENCES

1. *Standard Methods for the Examination of Water and Wastewater*, 13th ed., American Public Health Association, New York, 1971.
2. Scarpino, P. V. In *Water and Water Pollution Handbook.* Marcel Dekker, New York, 1971.
3. Scarpino, P. V., G. Berg, S. L. Chang, D. Dahling, and M. Lucas. *Water Research, 6,* 959 (1972).
4. Clarke, N. A., and P. W. Kabler. *Am. Jour. Hyg., 59,* 119 (1954).
5. Weiderkopf, S. J. *Virology, 5,* 56 (1958).
6. Butterfield, C. T., E. Wattie, S. Megregian, and C. W. Chambers. *Pub. Health Rep., 58,* 1837 (1943).
7. Fair, G. M., J. C. Morris, S. L. Chang, I. Weil, and R. P. Burden. *J. Amer. Water Works Assoc., 40,* 1051 (1948).
8. Berg, G. *Health Lab. Sci., 3,* 170 (1966).

CHAPTER 16

KINETICS OF VIRAL INACTIVATION BY BROMINE

David G. Taylor and J. Donald Johnson*

INTRODUCTION

From the time of the London Broad Street Pump cholera epidemic in 1854, people have been aware of the disease-carrying capability of our potable water supplies. Earlier in this century bacteria were implicated as the primary disease-producing agents in man. Now viruses are the challenge and concern of public health officials. This concern is justifiable for several reasons. First, there is a general ignorance of viruses and the facts that are known are mostly negative in that viruses are only perceived as being harmful. Of course, this is no different from the time when bacteria were discovered as disease organisms. At that time all bacteria were considered harmful. We know today that only a small percentage of the species of bacteria are pathogenic to man. Viruses are probably similar in such respects to bacteria. Secondly, it is much more difficult to isolate and demonstrate viral, as opposed to bacterial, contaminants in our water supplies. The fact that disease agents may be in the drinking water but cannot be adequately measured is of great concern to public health officials and scientists. Clearly, there is some basis for concern since the dose of virus capable of producing infection is extremely low. In fact, the infective dose of Poliovirus I has been shown to be as little

---

*D. G. Taylor, National Institute for Occupational Safety and Health, Cincinnati, Ohio. J. D. Johnson, Department of Environmental Sciences and Engineering, School of Public Health, University of North Carolina, Chapel Hill, North Carolina.

as two plaque-forming viral units.[1] The measurement of this dose even by the largest treatment plant laboratories is not economically feasible.

Viruses are difficult to detect not only analytically but also epidemiologically. Berg[2] has pointed out that epidemiological methods are not sensitive enough to detect the low level spread of viruses. Viral disease may spread from community to community by the water route. Other modes of transport could then spread the disease within the community. Water may be important, therefore, not primarily in the spread of explosive viral epidemics, but in the low level transmission of infectious viral agents. With the continued use and reuse of waters as both potable sources and as wastewater receiving streams, it is important that close attention be paid to the practice of disinfecting water supplies so that the hardy viruses are not recycled along with the water.

Recognition of the need for disinfection of water supplies in this country took place in the 19th century. Continuous chlorination as a means of disinfecting a water supply was begun in the United States in 1908 in Jersey City, New Jersey.[3] Since that time, chlorine has been the chemical agent of choice for the disinfection of water supplies because of its long use, its familiarity, and its low bulk cost. In addition, it is the most investigated chemical disinfectant against waterborne pathogens.[4] Two other halogens, bromine and iodine, have been tested to a lesser extent for their ability to kill pathogens. The utilization of the latter two halogens has been primarily for swimming pool rather than water supply disinfection.[5]

The important human pathogens which may be transmitted by way of the water route include bacteria, amoebic cysts, and viruses. It is this last group of microbes, the viruses, with which this work will be concerned. The virus groups known to be important in the water disinfection field are the polioviruses, the Coxsackie viruses, the Echo viruses, the adenoviruses, and the virus of infectious hepatitis.[6] More than one hundred different viral strains are known to be excreted in large quantities in the feces of infected persons.[3] Two of the very small viruses, polio and infectious hepatitis, have been of such import as to warrant major concern.

Although the processes involved in the chemical inactivation of viruses are not well understood, some progress has been made on halogen devitalization. The disinfection of viruses using chlorine and iodine has been studied by many workers.[7-12] Bromine and its ammonia compounds (monobromamine, dibromamine and nitrogen tribromide) are the least studied of the halogen disinfectants.

Bromine Disinfection

Bromine has been used as a disinfectant for swimming pools and other water systems since the early 1930's.[13] Investigators have found that there are several advantages to using bromine in place of the other halogens. A major disadvantage of the use of chlorine is its tendency to form chloramines. Chloramines have proved to be ineffective as disinfectants for cysts, bacteria and viruses.[10,14,15] Moreover, monochloramine has been reported to be toxic to fish.[16] Brown, *et al.*,[17] reported that the persistence of bromine used in swimming pools is longer than that of chlorine. They also reported that even up to a total concentration of 9 mg/l bromine caused none of the eye irritation problems which have occurred with chlorine. At this concentration there were also no "swimming pool odor" problems.

The halogens have different effects on algae. Kott[18] has found that algae such as *Cosmarium*, which is known to be resistant to chlorine, are readily killed by bromine. Chlorine given in two doses of 0.2 to 1.0 mg/l has an algistatic effect on *Chlorella*, whereas two doses of bromine at the same weight concentration effectively kills *Chlorella*.[19] Black, *et al.*,[20] reported that in the absence of an algicide, iodinated swimming pools must be superchlorinated about once per month to kill the algae. In these studies bromine appears to be the most effective algicide.

Bromine acts well also as a bactericide. Tanner and Pitner[21] found that bromine quickly kills such bacteria as *Staphlococcus*, *E. coli* and *E. typhosa*. They noted that at lower pH (3.5-4.0) bromine is even more effective than at neutrality. Wyss and Stockton[22] reported that the spores of *B. subtilis* and *P. fluorescens* are 500 times more resistant to bromine than vegetative cells. Excess ammonia does not inhibit the killing action of bromine. They also showed that the relative disinfecting efficiency is $Br_2 > HOBr > OBr^-$.

Johannesson,[23] in studying the bromination of swimming pools, found that monobromamine is effective against *E. coli*. Only 0.28 mg/l NH_2Br at pH 8.2 is required to produce greater than 99.9% kill in one minute. Monochloramine, on the other hand, shows negligible kill in 25 minutes at the same concentration.

Kruse[24] has studied the cysticidal properties of bromine and compared them to chlorine and iodine. He found *E. histolytica* cysts to be more resistant to the free halogens than either viruses or bacteria. He concludes that bromine is the most rapid acting halogen cysticide. Amoebic cysts are 99.9% inactivated by 2.1 mg/l bromine in ten minutes at pH 7 and 30°C. Except at high pH, the combined bromine species (mono- and dibromamine) are equally as active as the free species. Whether free or combined bromine was employed, it was shown to be superior to either chlorine or iodine at the same experimental pH which ranged from 4 to 10.

Only two groups have reported definitive studies of viral inactivation utilizing bromine or its compounds. Brown and McLean[5] using bromine, chlorine, and iodine separately, studied viral disinfection in swimming pools. The Poliovirus (Types 1, 2, and 3), Coxsackie B5, Echo 2, Echo 9, and Parainfluenze I viruses were utilized in this study. A concentration of bromine greater than 2 mg/l was found to be effective and the authors concluded that bromination could replace chlorination in treating these waters.

Kruse[25] determined that the virus, RNA f_2 coliphage, was more resistant than *E. coli* cells to bromine at pH 7.5. Five logs of kill were attained in 20 minutes at pH 7.5 and 0°C; added bromide increased the viral inactivation while decreasing the bacterial disinfection; free bromine was a better virucide than the bromamine (mole ratio 200:1, N:Br at pH 7.5); the free nucleic acid (RNA) of the phage was more resistant to either free or combined bromine than the intact virus; and bromamine was a far superior virucide than chloramine.

Chemical Speciation and Disinfection

Whatever disinfectant is used, it is important to realize that individual chemical species kill viruses. For this reason, it is usually not possible to correlate disinfection results with the total analytical concentration of the disinfectant. The concentration of each chemical species must be considered separately since each has its own characteristic ability to inactivate viruses. The rate of inactivation is a function of the ease with which the disinfectant reacts with specific labile molecules of the virus particle. The molecules which are destroyed by the disinfectant must be necessary to maintain the viability and infectivity of the virus. When this destruction occurs, the virus is considered to be killed since it cannot either reproduce, infect cells, or cause overt disease. Disinfection is a complex process by which individual chemical species attack susceptible molecules of the virus. Viral studies, utilizing an individual disinfecting species in order to obtain kinetic data, must be carried out to understand more clearly the mode of action of these chemical disinfectants.

Aqueous Bromine Chemistry

The chemistry of the aqueous bromine system is complex. When introduced into water, bromine dissolves and then hydrolyzes to form hypobromous acid.[26]

$$Br_2(\ell) = Br_2(aq) \qquad (1)$$

$$Br_2 + H_2O = HOBr + H^+ + Br^- \qquad (2)$$

$$K_h = 5.8 \times 10^{-9} \text{ at } 25°C.$$

If the solution is made alkaline (pH > 9), the hypobromous acid dissociates to form the anion as the predominant species.[27]

$$HOBr = OBr^- + H^+ \qquad (3)$$

$$K_d = 2 \times 10^{-9} \text{ at } 25°C.$$

A high concentration of bromide (*ca.* 10^{-1} *M*) in bromine at pH values less than 7 shifts the equilibrium toward the tribromide anion.[28]

$$Br_2 + Br^- = Br_3^- \qquad (4)$$

$$K_{Br} = 15.9 \text{ at } 25°C.$$

From inspection of Equations 1-4, it is readily apparent that the distribution of the oxidizing power among the chemical species Br_2, HOBr, OBr^-, and Br_3^- is greatly influenced by the pH and bromide concentration.

In order to see more readily under what conditions of pH and bromide concentration the individual species exist, these equations are combined in Figure 16.1 into a

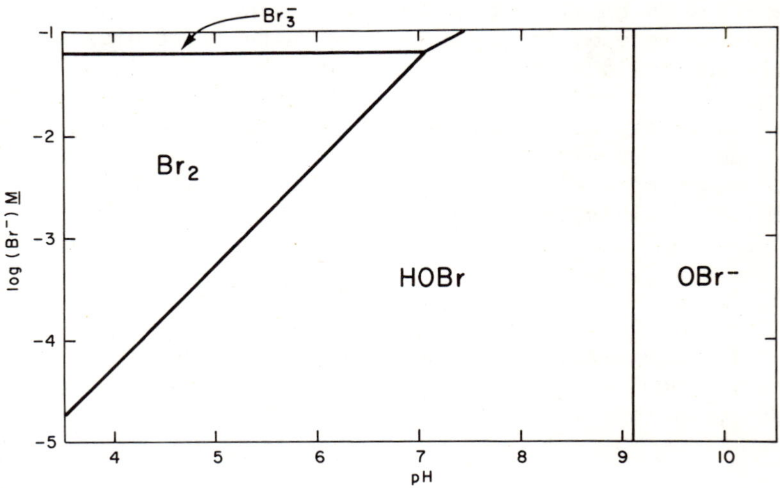

Figure 16.1. Principal species of free bromine in equilibrium with bromide ion and hydrogen ion. Lines are equal mixtures.

$\log[Br^-]$ versus pH diagram in which the lines represent equimolar concentrations of the adjacent species. The dissociation of hypobromous acid is given at 0° C (pK_d = 9.1). The equivalence lines are obtained directly from the equilibria involved. For example, Equation 2 gives the Br_2 - HOBr line

$$K_h = \frac{[HOBr][H^+][Br^-]}{[Br_2]} \qquad (5)$$

Taking logs,

$$\log K_h = \log \frac{[HOBr]}{[Br_2]} + \log [H^+] + \log[Br^-] \qquad (6)$$

and since $pH = -\log[H^+]$ the equation may be rearranged to give

$$\log[Br^-] = pH + \log K_h + \log \frac{[Br_2]}{[HOBr]} \qquad (7)$$

When [HOBr] = [Br_2], $\log \frac{[Br_2]}{[HOBr]} = 0$. Thus, Equation 8 is

$$\log [Br^-] = pH - pK_h \quad (8)$$

Similar manipulations of Equations 2, 3, and 4 give the other lines in Figure 16.1. It is possible, by controlling the pH and bromide concentration, to select the chemical species of interest in large excess over the others. If the bromide concentration is 10^{-7} *M* at pH 4.5 then essentially pure HOBr is formed. Nearly pure Br_2 may be obtained at the same pH when the bromide concentration is 3.0×10^{-3} *M*. Under these conditions about 95% of the bromine oxidizing power exists in the species Br_2 with about 5% as Br_3^- and less than 1% as HOBr. The bromine species distribution is profoundly affected by both the pH of the solution and the bromide concentration.

Figure 16.2 shows the corresponding systems for two other halogens, chlorine and iodine. The lines were calculated from the equilibrium constants just as the bromine system was. In contrast to the bromine system, where only hypobromous acid is present in a significant concentration at neutral pH, chlorine exists as both hypochlorite anion and hypochlorous acid. At neutral pH iodine is distributed between the molecular iodine species and the acid species. Therefore, a direct comparison of the corresponding species in the halogen systems at the same pH is difficult.

Another complicating feature of the halogen systems is the formation of ammonia compounds. For iodine, it is sufficient to say that under the conditions of water disinfection, where low concentrations of iodine are used, there is no significant formation of iodine-nitrogen compounds. This is a result of the low reactivity and oxidation potential of iodine. Bromine does form three bromine-ammonia compounds in water.[29,30]

It has been only recently that data on the distribution of the bromine-ammonia compounds as a function of pH and mixing ratio have been summarized.[31] The data were obtained from UV spectra of mixtures of the N-Br species. The substitution reactions between ammonia and bromine are extremely fast and may be summarized in step-wise fashion.

$$NH_3 + HOBr \rightleftarrows NH_2Br + H_2O \quad (9)$$

$$NH_2Br + HOBr \rightleftarrows NHBr_2 + H_2O \quad (10)$$

$$NHBr_2 + HOBr \rightleftarrows NBr_3 + H_2O \quad (11)$$

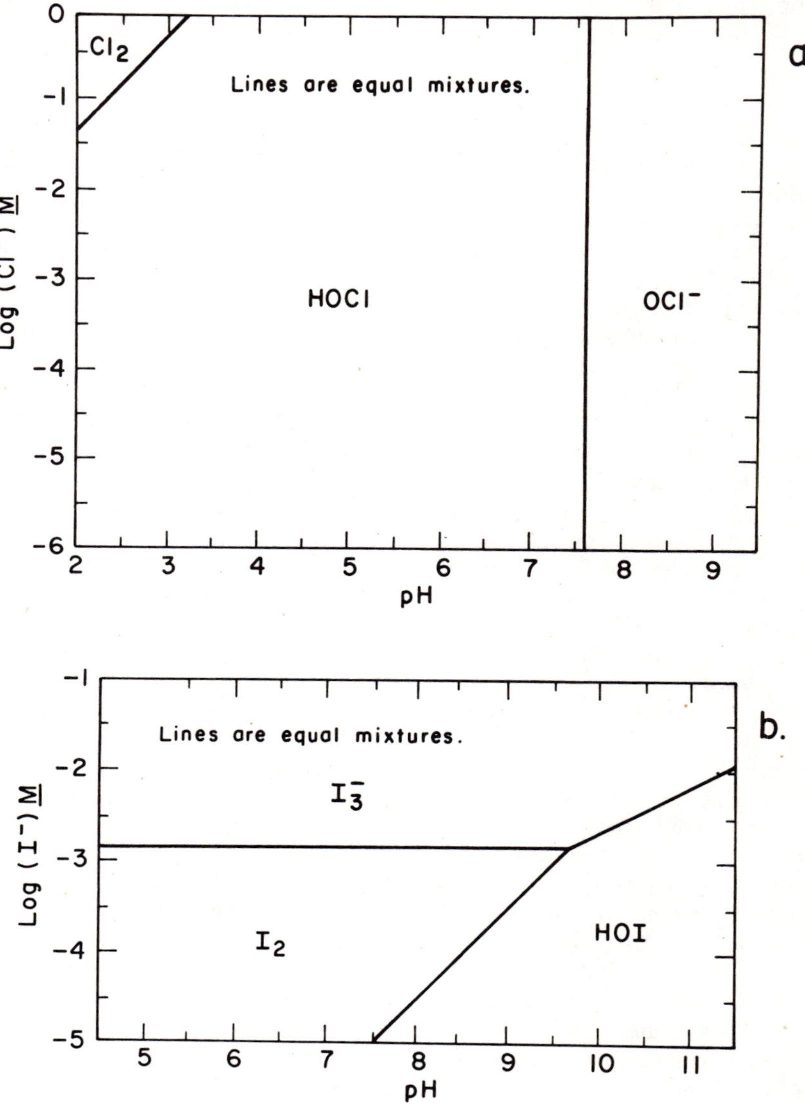

Figure 16.2. a. *Principal species of free chlorine in equilibrium with chloride ion and hydrogen ion.*
b. *Principal species of free iodine in equilibrium with iodide ion and hydrogen ion.*

The initial molar ratio of ammonia to bromine and the solution pH determines which bromamine species is formed. Figure 16.3 presents the bromamine distribution diagram similar to the distribution for the pure halogens in Figures 16.1 and 16.2.[31] The separation between Br_2 and HOBr is shown here for a bromide concentration of 10^{-5} *M*. At pH 4.5 and a mole ratio (N:Br) of 1:3 nitrogen tribromide is formed.

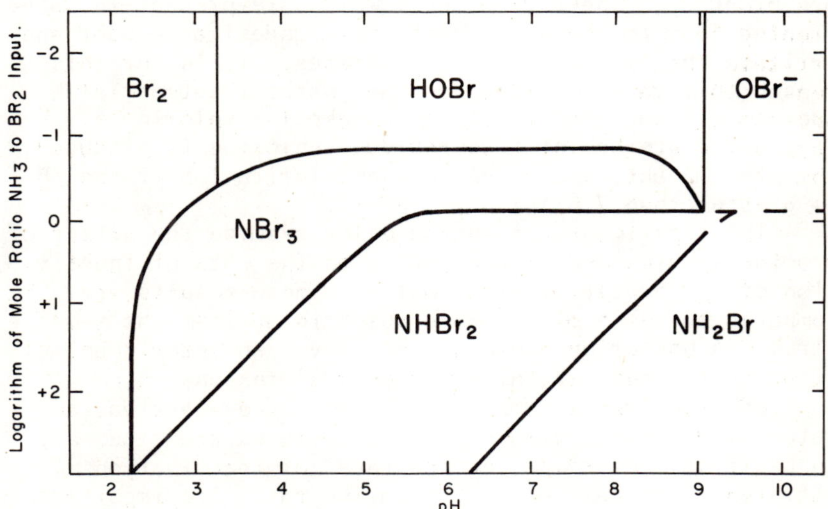

Figure 16.3. Principal species of bromine and bromamine predominating after 1 to 2 minutes at various pH and ammonia to bromine ratios. Lines represent equimolar concentrations. Hypobromous acid separation from bromine given for 10^{-5} M bromide.

The stability of these halamines in water varies, but the Johnson and Overby[31] results show that tribromamine becomes more stable as the pH is decreased from 7.0 to 4.0. Galal-Gorchev and Morris[30] found that NBr_3 decomposes about 5% per hour at pH 4.5

The halamine system of chlorine has quite different characteristics from those of the bromine system discussed above. The ratio of the bromamines produced in water is equilibrium controlled so it is possible to represent the bromamine system in an equilibrium-type diagram (Figure 16.3). The ratios of the chloramines formed after ammonia and chlorine are mixed are not equilibrium controlled. They are rate controlled and do not lend themselves to precise representation in an equilibrium diagram. Above a mole ratio (N:Cl) of 1:1 the major species formed is monochloramine. Ammonia is in excess, and the breakpoint has not been reached. Near the breakpoint, at a mole ratio (N:Cl) of 1.6:1, dichloramine and nitrogen trichloride are produced. These compounds, which are present in water treated just to the breakpoint, have undesirable odor and irritate the eyes and mucous membranes. It is for this reason that many swimming pool and water treatment plant operators find it difficult to breakpoint chlorinate. Beyond the breakpoint free residual chlorine is produced, the predominant species being hypochlorite ion if the pH is greater than 7.6.

The experiments presented below examine the effect of bromine species and concentration on the rate of inactivation of the bacterial virus øX174. These results are compared to those of hypochlorous acid against the same virus. A bacteriophage was chosen over an enteric animal virus as the test organism for several reasons. In applied experiments comparing the absolute inactivation rates of different viruses, animal viruses are necessary. In basic studies comparing the rates of inactivation with different chemical species, relative rates are important, and phage are more appropriate. Phage are easier to work with and require far less analysis time at a fraction of the cost. These bacterial viruses or phage are well-characterized biochemically and may be obtained in high titer. High titers are not attainable with animal viruses and this limits the dynamic range and experimental end-points to a low fractional kill. Under these conditions mechanistic results are more difficult to ascertain. Phage may be prepared free from cellular debris and in a mono-disperse rather than an aggregated state. Debris and viral aggregation can mask the disinfectant kinetics of animal virus inactivation.

The studies reported here were designed to provide information on the importance of bromine chemical speciation in the viral disinfection of water. The relative disinfection rates and the mode of action of the species were sought. This work is particularly important because of renewed interest in the use of bromine for swimming pool disinfection and in any other water system where water is recirculated routinely.

MATERIALS AND METHODS

All water used in this study was filtered through a one micron filter and columns of activated carbon and anion-cation mixed bed ion exchange resins to produce water with a specific resistance of 18 megohms per cubic centimeter. Five-gallon Pyrex carboys, fitted with 50% sulfuric acid traps, were filled and the ion exchanged water chlorinated to 1.5-3.0 mg/l for two to five days to eliminate any final trace of halogen demanding materials. If the chlorine concentration had not decreased more than 0.3 mg/l, the water was dechlorinated with a ten-inch PCQ UV lamp (UltraViolet Products). This water was halogen-demand free and used to prepare chemical reagents and test samples.

All the glassware used with the halogens was stored, soaking in high concentrations of alkaline bromine, and rinsed in halogen-demand free water prior to use.

Bromine stock solutions were prepared at a concentration of about 240 mg/l by adding about 0.15 ml of reagent grade bromine liquid (Allied Chemical, 99.5%) to two liters of deionized water. The pH was adjusted to 10-10.5 with 1 *M* sodium hydroxide while mixing with a Teflon-coated magnetic stirring bar. After solution the pH was checked, the solution placed in a brown two-liter automatic buret (Ace Glass), and left two to three days before standardization. The bromine stock was titrated to the starch endpoint with sodium thiosulfate for total bromine (free plus bromates) at pH 1.3, while free bromine ($HOBr + OBr + Br_2$) was titrated at pH 4.0 by the procedure of Larson and Sollo.[54]

Alkaline chlorine stocks were prepared by bubbling research grade chlorine gas (Mathieson, 99.965% minimum) into two liters of deionized water to obtain a concentration of about 150 mg/l. Enough 1 *M* sodium hydroxide was added to maintain a pH of 10.5. The chlorine stock was then stored just as the bromine stock until standardization. The chlorine stock solutions were standardized by the starch-iodide procedure.[33] Potassium iodate was the primary standard.

The pH buffers used in this study were made up at 0.1 *M* and diluted to 0.05 *M* for each experiment.

Production of Individual Chemical Species

Molecular bromine in the test solution was produced at pH 4.5 by carrying out the experiments at a bromide concentration of 3×10^{-3} *M*. The controls contained the same potassium bromide concentration as the test samples.

The bromine stock (*ca*. 8×10^{-4} *M*) used for the production of pure HOBr was treated with silver nitrate at a concentration of 8.3×10^{-4} *M* at pH 7.0 to remove bromide anion. The stirred solution was allowed to settle in the dark for two days. The supernatant was then drawn off and a portion tested for any further precipitation of AgBr by diluting 1:1 with 0.1 *M* $AgNO_3$. The K_{so} for the precipitation of silver bromide is 7.7×10^{-13}. The final bromide concentration in the test solutions after the dilution of the stock, therefore, was less than 10^{-8} *M* at pH 4.5. Hypobromous acid was at least in a 20,000-fold excess of Br_2. The calculated excess of silver ion was added to the controls for the experiments. No effect on the controls due to silver ion was observed.

Nitrogen tribromide was generated at pH 4.5 by pouring together with stirring, equal volumes of buffered ammonium chloride and standard bromine solution. The final molar ratio (NH_3:Br_2) was 1:3 in the test reactor. Identical ratio solutions were scanned in the UV on a Cary 14 Spectrophotometer against a buffered ammonium chloride reference to check the purity of the NBr_3 formed at 258 and 323 nm.[30]

Since all the free chlorine oxidizing power exists as HOCl at pH 4.5, no special procedure was required. The chlorine was simply diluted to the desired concentration in the buffered reactor solution.

All bromine halogen concentrations in the test system were analyzed by a modification of the SNORT Test for chlorine.[33,34] Fifty-milliliter samples were drawn from the test reactor and mixed with 0.25 ml of 0.4% potassium iodide. Forty seconds after addition of the KI the sample was poured over the magnetically stirred SNORT reagents in a 250-ml beaker. The samples were read two minutes after mixing with the reagents in 10-cm cells in a Cary 14 Spectrophotometer. In some of the early experiments 4-cm cells and an expanded scale slidewire (0 to 0.2 absorbance units) were used. Chlorine was determined by the standard SNORT method.[33]

The Viral System

The virus used in the study was øX174am3, a small (about 22 nm) single-stranded DNA coliphage. This bacterial virus has an icosahedral capsid and no outer membrane. The protein coat has a known amino acid composition.[35-37] Dr.

Clyde L. Hutchison III and Dr. Marshall H. Edgell of the University of North Carolina Department of Bacteriology and Immunology provided the highly purified phage for use in this study. This virus is an amber strain which does not lyse *E. coli* C cells but continues to produce virus in this host. It was possible, therefore, to grow very high concentrations (about 10^{14} PFU/ml final concentration) of this virus on *E. coli* C host cells. The resulting high concentration of virus was purified by a density equilibrium centrifugation and sucrose gradient procedure. For details see Pagano and Hutchison.[38]

The virus is conveniently stored in 0.05 *M* sodium tetraborate in a refrigerator at 5° C. Stock titers can be maintained under these conditions for long periods of time. The virus titers were determined by the plaque assay technique on double-layered tryptone agar plates.[39] Serial dilutions were made in 0.05 *M* borate. The assay cells were *E. coli* X, a genetic recombination of *E. coli* C and *E. coli* K12.

The Experimental System

The experiments were carried out in covered 600-ml beakers. Teflon coated magnetic bars were used for stirring. The chemical system was prereacted before the addition of the virus and was batch operated. The reactors were submerged in a thermostatically-controlled water bath (Lo Temp-Bath, Wilkens-Anderson Co.).

The test procedure was as follows. Buffer and demand-free water were added to the continuously stirred test beaker, and after a twenty minute period for temperature equilibration the halogen was added to the reactor. The halogen concentration was determined after about five minutes. Then 1.0 ml of virus suspension (*ca.* 10^{12} PFU/ml) was added to the remaining 110 ml of buffered halogen, and 5-ml samples were removed. These samples were rapidly blown into neutralizer (2.5 ml 0.05 *M* borate + 2.5 ml 2-4 x 10^{-4} *N* $Na_2S_2O_3$) in sterile screwcap tubes and shaken. After removing the virus samples, a second 50-ml sample was withdrawn for the estimation of the halogen concentration. The pH was checked with an expanded scale pH meter (Leeds and Northrup, Model 7403).

Losses in halogen concentration were experienced from the beginning until the end of the test period, even though precautions were taken to prevent these losses. The spontaneous decomposition of halogens in buffered solutions has been reported by several workers, but the rates reported are extremely slow.[40-42] The halogen concentrations used in the experiments were quite low. For hypobromous acid, the concentrations ran from a low of 8×10^{-7} *M* to a high of 2.6×10^{-6} *M*. On a weight basis this range is 0.13 to 0.42 mg/l. At these concentrations, therefore, a small loss represents a significant proportion of the total halogen present. The most serious losses occurred when the molecular species Br_2 was tested in the stirred reactor.

In order to represent more accurately the mean reactor concentration during the test period, the following procedure was adopted. Preliminary sampling of the reactor indicated that the halogen loss was linear with time. Each part of the experimental procedure, therefore, was timed so that the "time line" between the halogen sampling and the virus sampling was known. Since the losses were linear with time, the concentrations were calculated for the time equal to zero (when the virus was added to the halogen) and the time when the last virus sample was taken from the disinfection reactor. As an example of this procedure, the data from the 0.7 μ*M* nitrogen tribromide experiment is used in the calculations.

Time Line

Interval	Event
	Add bromine and ammonia to reactor
7.0 minutes	
	NBr_3 Sample #1; $[NBr_3]$ = 1.83 μ*M*.
3.5 minutes	
	Add virus; t = 0.
4.0 minutes	
	End virus sampling; t = 4.0 minutes.
2.5 minutes	
	NBr_3 Sample #2; $[NBr_3]$ = 1.53 μ*M*.

The total time between the two nitrogen tribromide concentration samples is ten minutes. The total concentration loss in this period is 0.3 μ*M*. Assuming the linear loss of halogen, the NBr_3 concentration at the time the virus was introduced is 1.83 μ*M* less the loss between Sample #1 and t = 0, *i.e.*,

$$1.83\ \mu M - \frac{3.5\ \text{minutes}}{10.0\ \text{minutes}}\,(0.3\ \mu M) = 1.73\ \mu M.$$

At the end of the virus sampling period (t = 4.0) the loss is

$$\frac{7.5}{10.0}(0.3\ \mu M) = 0.22\ \mu M,$$

and the concentration at t = 4.0 is 1.83 - 0.22 = 1.61 μM. The mean nitrogen tribromide concentration during the experimental sampling time is

$$\frac{1.73 + 1.61}{2} = 1.67\ \mu M.$$

The average halogen loss during the experiment is

$$\frac{(1.73 - 1.61)\ 100}{1.67} = 7\%.$$

RESULTS AND DISCUSSION

The objective of this work was to obtain kinetic data to assist in understanding the mode of interaction between bromine and øX174. The low temperature (0° C) was used in most of the experiments to decrease the fast rates observed in the preliminary experiments. It was assumed that kinetic curves would offer more insight into the disinfection process than would merely the data on the percentage kill for a fixed time interval, as is generally reported in disinfection studies. The time required to inactivate a fixed proportion of microbes is a better relative disinfection indicator than the fixed time-kill.[43] The basic data include the entire disinfection curve shown as the logarithm of the counts of the remaining viable øX174 phage in plaque-forming units per milliliter (PFU/ml) as a function of time. The straight part of each curve is represented by the linear regression of the data.

The experimental results for mean reactor concentration, percentage loss of halogen during the experiment, the time range used in the linear regression, and the regression slope with the 95% confidence interval are presented in Table 16.I.

Figure 16.4 is a composite of the kinetic data for øX174 inactivation at pH 4.5 and 0° C for five hypobromous acid concentrations. Since there are only negligible losses in the control titers in Figure 16.4, no correction has been made in the test data for control loss.

Whatever the shape of the observed kinetic curves as seen in Figure 16.4, it is apparent that even at low temperature small concentrations of the hypobromous acid

Table 16.I

Experimental Data Taken at pH 4.5 and 0° C

Species	*Mean Concentration (μM)*	*% Loss*	*Linear Range (min)*	*Regression Slope ± 95% Confidence Interval (min^{-1})*
	0.8	16	0.5 - 4.5	0.626 ± 0.026
	1.4	18	1.7 - 3.8	1.482 ± 0.059
HOBr	1.5	24	1.5 - 3.5	1.704 ± 0.166
	1.7	18	0.9 - 2.5	2.680 ± 0.129
	2.6	20	0.5 - 1.4	4.772 ± 0.377
	0.6	26	1.0 - 3.0	1.393 ± 0.094
HOCl	1.0	20	1.0 - 2.5	2.742 ± 0.079
	2.0	16	0.5 - 1.5	4.289 ± 0.780
	0.6	50	0.5 - 2.1	2.697 ± 0.113
Br_2	0.8	53	0.5 - 1.4	3.276 ± 0.263
	0.9	40	0.5 - 1.3	3.240 ± 0.793
	0.25	20	0.5 - 3.5	0.201 ± 0.035
NBr_3	0.96	10	0.5 - 3.5	0.527 ± 0.037
	1.67	7	0.5 - 3.5	1.031 ± 0.038

species are extremely effective in destroying øX174. Only 0.4 mg/l (2.6 μM) hypobromous acid is required for six logs of disinfection in 1.5 minutes. This data illustrates that bromine is an extremely efficient virucide at pH 4.5 and 0° C.

Characteristics of Kinetic Curves

There are several features of these kinetic curves common to all test concentrations. Throughout most of the

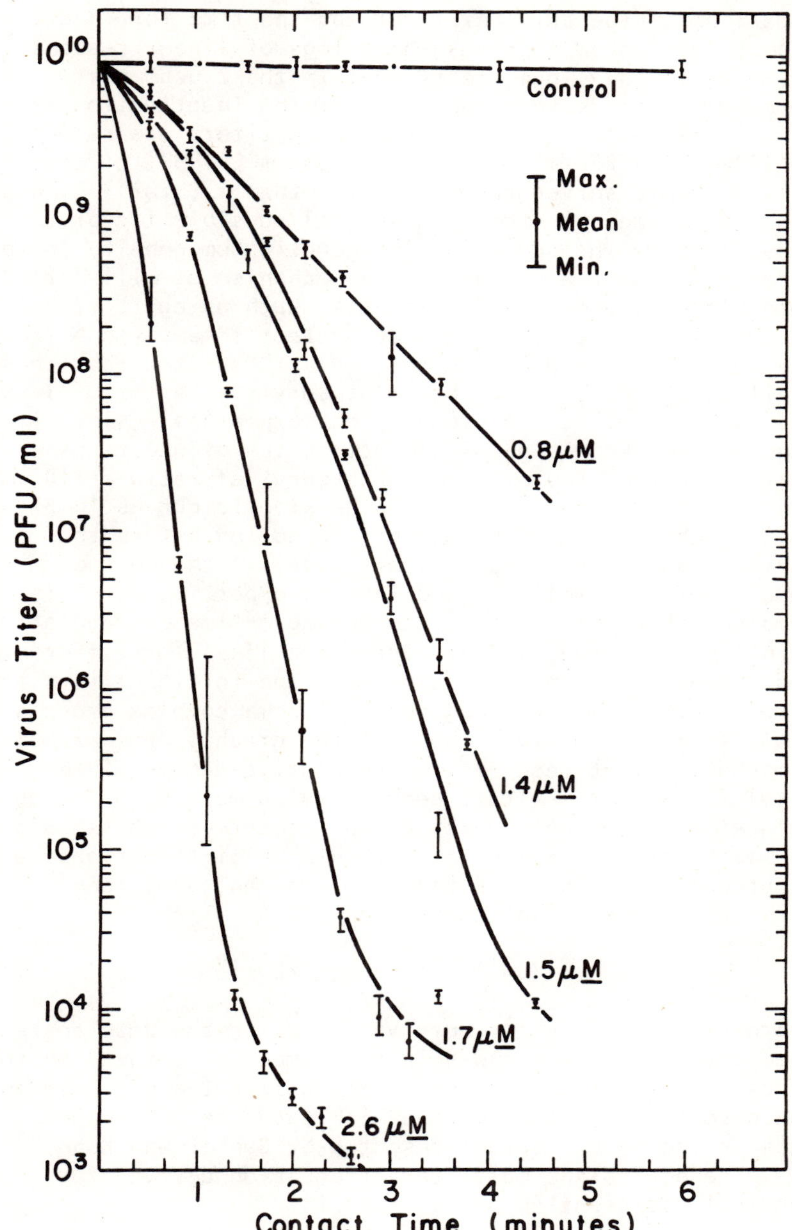

Figure 16.4. Inactivation of øX174 by hypobromous acid at pH 4.5.

time interval there is a linear relationship between the logarithm of the survivor titer and the time for inactivation. The 0.8 μM test has three logs of linear response, while nearly six logs are evident in the 2.6 μM test. However, there is some tendency for the inactivation rate to decrease with time after the phage titers are reduced to 10^4 to 10^5 PFU/ml, *i.e.*, after about 99.999% of the kill. This observed rate decrease, that is, tailing, has been attributed by others to several possible factors. These factors include a lack of genetic homogeneity in the viral particles, a change in the mechanism of kill for some virions, and physical effects, such as aggregation of the virus particles that may protect some virions from the action of the disinfectant. Hiatt[44] stated that these "tailing effects," which occur at survival ratios of less than 10^{-5} to 10^{-6}, are probably not explainable since it requires the knowledge of the attributes of one virus particle out of one million for a survival ratio of 10^{-6}.

Another characteristic of the kinetic curves in Figure 16.4 is the initial shoulder or lag period before the maximum inactivation rate is achieved. Although the shoulder is not well defined in some experiments, it is apparent that one exists since the zero-time extrapolation points of the experimental regression lines for different concentrations do not coincide with the initial titers at zero time. It may be imagined that some complex process results in the observed shape of the graphs. For example, a mechanism that requires the cumulative damage of several vital sites on the virus, each of which must be "hit" once before inactivation occurs, could account for the shoulder. The mathematical expression for this multi-target approach as presented by Fowler[45] and by Oliver and Shepstone[46] for radiation survival curves is

$$S = 1 - [1-\exp(-D/D_o)]^n \tag{13}$$

where D is the dose for survival S, D_o is the dose to reduce the survival by e^{-1}, and n is the number of targets on the virus. Oliver and Shepstone rationalized the e^{-1} dose as the dose required to give one hit in the target volume. This is the Type C curve presented by Gunter and Kohn.[47]

The equivalent multi-target equation derived for chemical disinfection by Morris[43] is

$$N/N_o = 1 - [1-\exp(at/t_r)]^n \tag{14}$$

where t_r is the normalized time for a reduction in the viable organisms by e^{-1}, and "a" is the normalizing constant

$$\ln\,[1 - (1-e^{-1})]^{1/n*}.$$

The dose in the chemical case is the contact time multiplied by the concentration (D = Ct). Dividing this dose by the dose (D_o) to reduce N/N_o by e^{-1} at t_r gives $D/D_o = t/t_r$ because the concentration terms cancel. Figure 16.5 shows the equation plotted on a logarithmic scale for n equal to 1, 2, 3, and 4. The results of the hypobromous acid experiments at pH 4.5 are superimposed on the figure. It

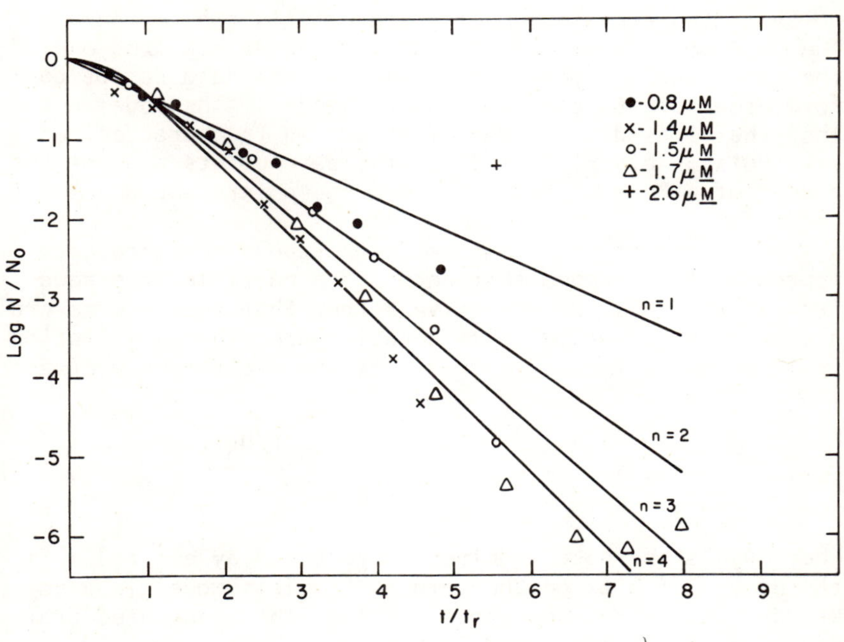

Figure 16.5. Theoretical multi-target survival curves for 1, 2, 3, and 4 targets. Symbols represent the data for five HOBr experiments.

*This is the corrected constant. In Reference 43 "a" appeared as $\ln[1 - (1-e^{-1})]^{-n}$.

is apparent that not only does the shape of the experimental data not closely fit the multi-target equation, but also that n varies between tests. Since n varies for the set of experiments in Figures 16.4 and 16.5 it is unlikely that n has any special significance as to the number of vital sites required for inactivation. Berg, *et al.*[17] also found this number to be variable in their devitalization experiments with elemental iodine. Due to the variance of n in different experiments, Alper, Gillies, and Elkind[48] advocate the use of the term "extrapolation number" for n so that no particular mechanism of inactivation is inferred from the data. The data appears to have a greater concave downward shape than the model equation. The points are scattered around the two- or three-target lines at the beginning and around the four-target line at greater times ($t/t_r > 5$). If one assumes that the model is correct, however, the average number of targets is two to four. The lack of constancy in the fit of the data around one of the multi-target lines indicates that the data do not conform precisely to the multi-target model. The model assumes that the virus-disinfectant interaction is characterized by a Poisson distribution. If the data points are not the result of a Poisson process, then the points would not fit the model.

An alternative mechanism to the above multi-target approach is to assume that each phage particle or plaque-forming unit has one sensitive target that requires several hits for inactivation. The equation used in the radiation field which defines this multi-hit, single target approach is

$$N/N_o = \exp(-D/D_o) \sum_{k=0}^{m-1} \frac{(D/D_o)^k}{k!} \qquad (15)$$

where D_o is the dose required to reduce N by e^{-1} and m is the number of hits on the target.[46] This model is also based upon the Poisson process. The lines generated from Equation 15 are presented in Figure 16.6 for several m-values as adapted from Figure 1 in Oliver and Shepstone.[46]

The HOBr data points are superimposed on this graph. With the exception of the 2.6 μ*M* test data points (+), the experimental results indicate that it takes less than one hit on the single target for inactivation. How can less than one "hit" by the disinfectant kill the virus (*i.e.* prevent plaque formation)? This is an insoluble dilemma. The multi-hit model is based upon one "target" being "hit"

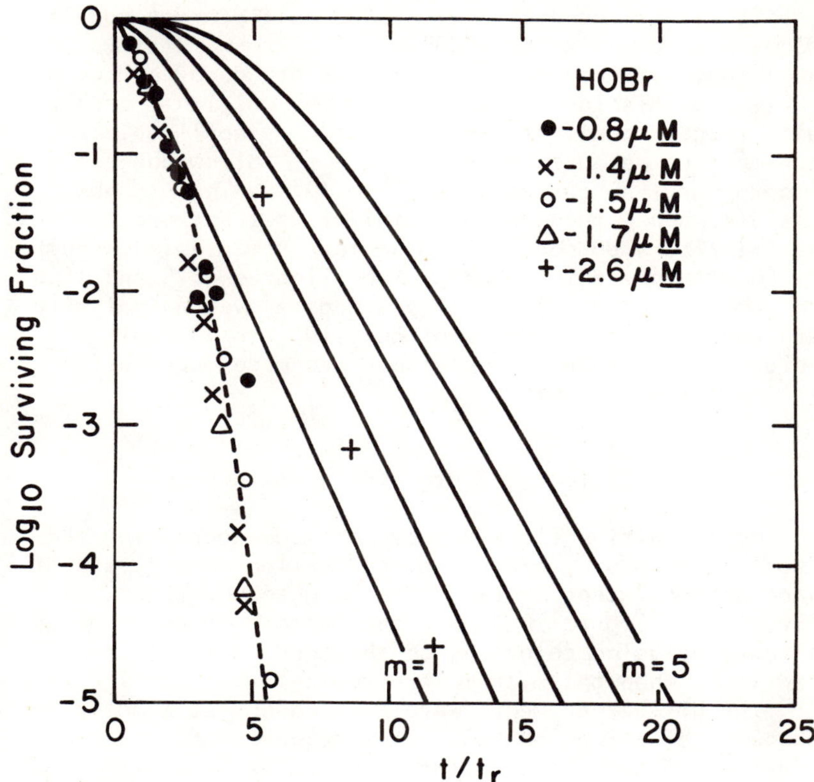

Figure 16.6. Theoretical multi-hit survival curves. ϕX174 data superimposed on graph.

m times, and m clearly is less than one throughout most of the range. On the other hand, the data indicates that for the multi-target-single-hit model (Figure 16.5) the number of targets is two to four. The data fit neither model satisfactorily. Therefore, the data presented here may not be characterized by a given number of targets or hits to explain the initial shoulder on the survival curves.

Other authors have confused the model approach by continuing to use the term multi-hit for a multi-target process (*e.g.*, Reference 4). An attempt has been made here to show the distinction between multi-target and multi-hit models. Fowler and Oliver and Shepstone have made a similar effort in the radiation field.

It is clear that many experimental survival curves do not fit either the multi-hit or the multi-target models proposed to explain the kinetics of microbe inactivation. Phage inactivation experiments, however, do offer the possibility of distinguishing between the multi-hit and multi-target models if one of the two is applicable. This is true because it is possible to follow phage titers down to survival ratios of 10^{-5} to 10^{-8}. This is not possible when working with animal viruses because the low initial titers do not permit the measurement of low numbers of PFU with present methods. From Figures 16.5 and 16.6 one can see the advantage of phage data over animal virus data where survival ratios of only 10^{-1} to 10^{-3} are typically available. Comparisons are made much easier with the greater kills.

Effects of Br_2, NBr_3, and HOCl

No definitive statement can be made concerning the effect of molecular bromine concentration on the rate of inactivation except to say that the observed rates, as indicated in Figure 16.7, are the fastest of any species tested, including chlorine, at the concentration levels used (less than one micromolar bromine).

The NBr_3 experiments were carried out at a molar ratio (NH_3:Br_2) of 1:3 at pH 4.5. It is clear that nitrogen tribromide possesses significant virucidal properties at these low concentrations. The data presented in Figure 16.8 shows that a tribromamine concentration of only 1.7 μM is capable of inactivating 99.99% of the virus present in four minutes at a pH of 4.5.

Unlike HOBr, hypochlorous acid shows little if any lag period at the beginning of the inactivation process (see Figure 16.9). In fact, the shape of the HOCl experimental curve at the 2.0 μM level appears like the molecular bromine graphs shown in Figure 16.7, dropping off dramatically in the first 30 seconds. At this HOCl concentration it takes just under one minute to achieve 99.999% kill.

Comparison of Species Activity

The relative halogen activity comparison is made on a molar basis in order to compare inactivation rates on a particle interaction basis. Figure 16.10 presents the relationship between the observed, first order rate

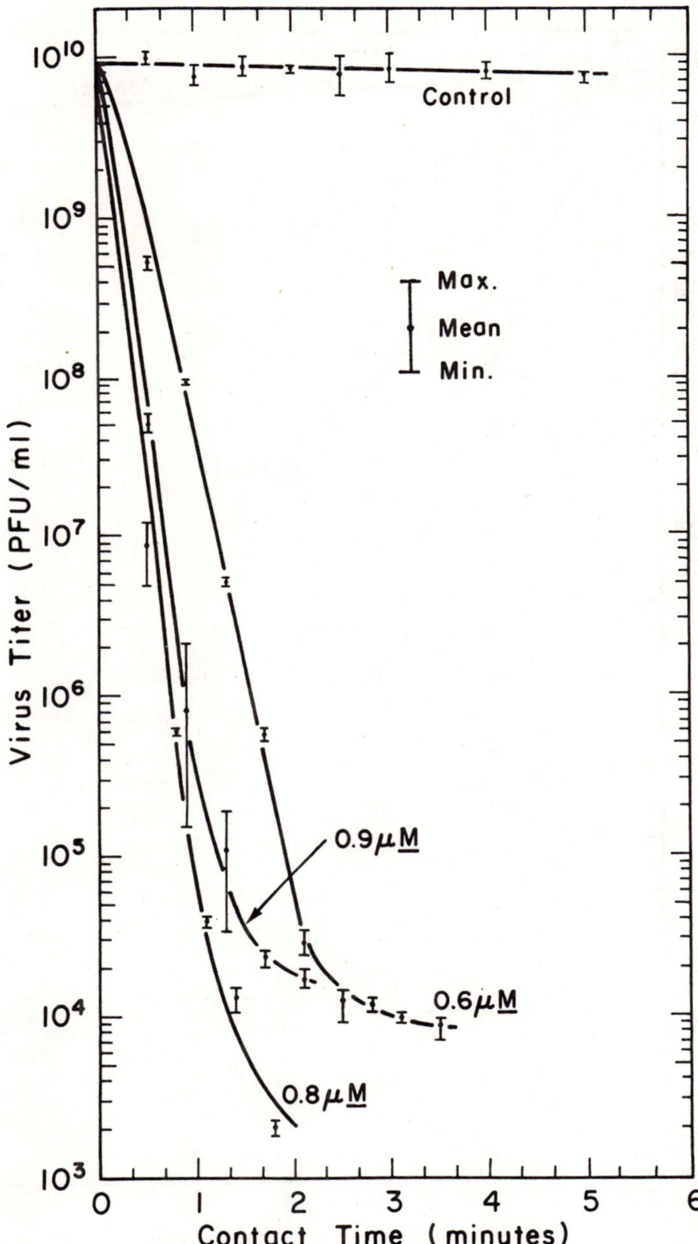

Figure 16.7. Inactivation of ϕX174 by molecular bromine at pH 4.5.

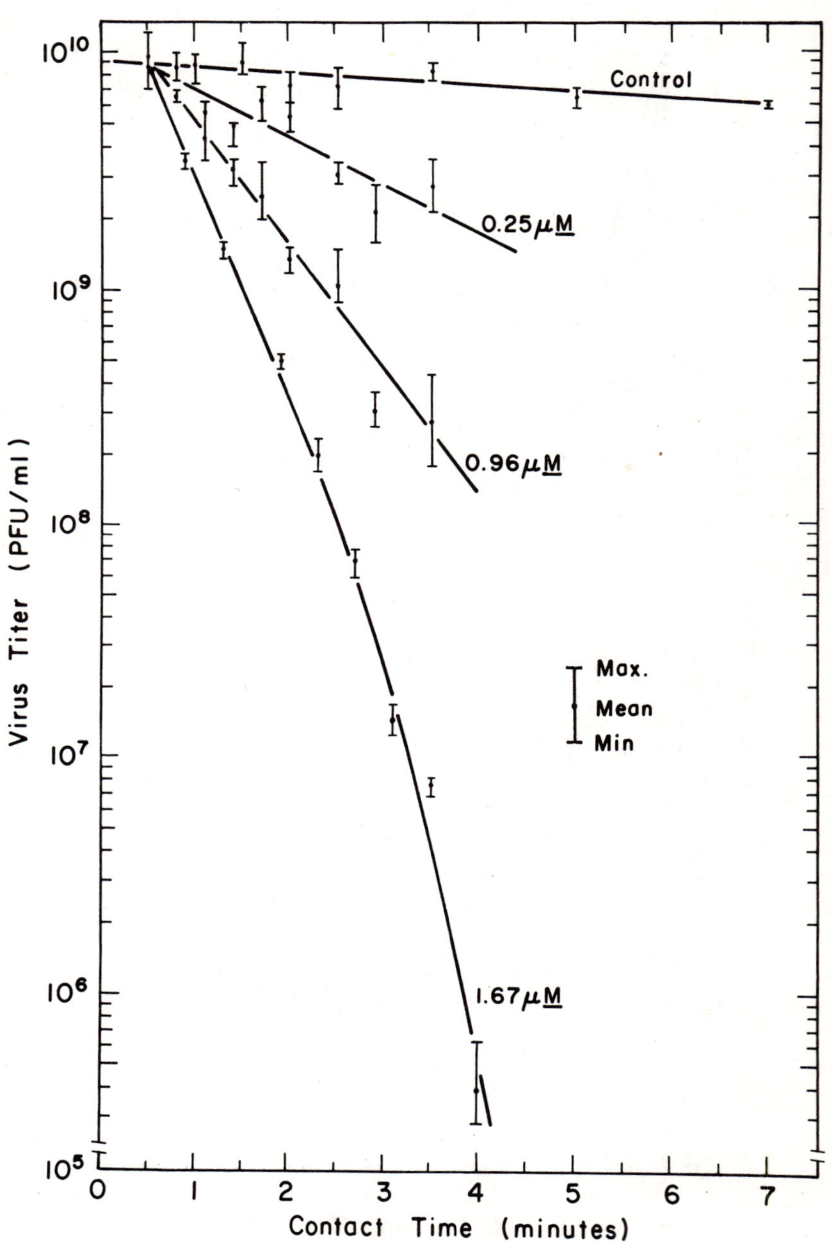

Figure 16.8. Inactivation of ϕX174 by nitrogen tribromide at pH 4.5.

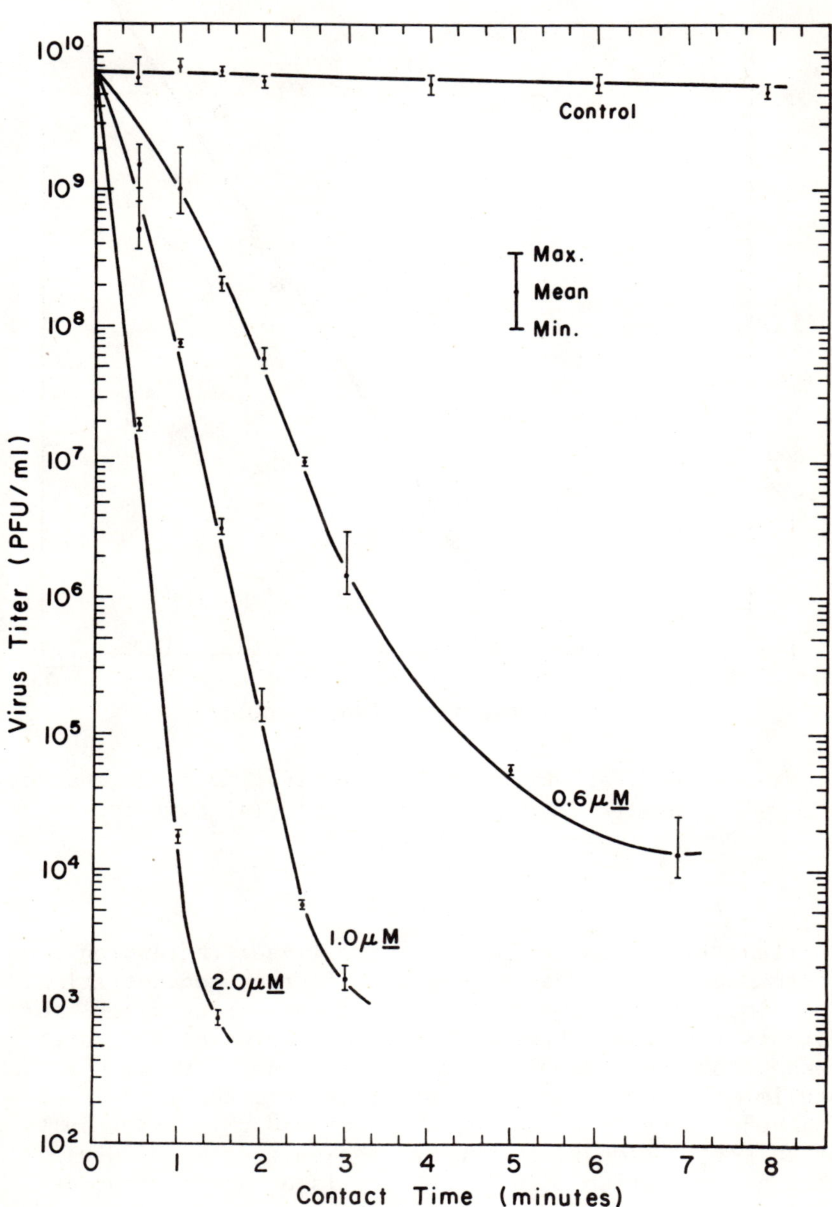

Figure 16.9. Inactivation of ØX174 by hypochlorous acid at pH 4.5.

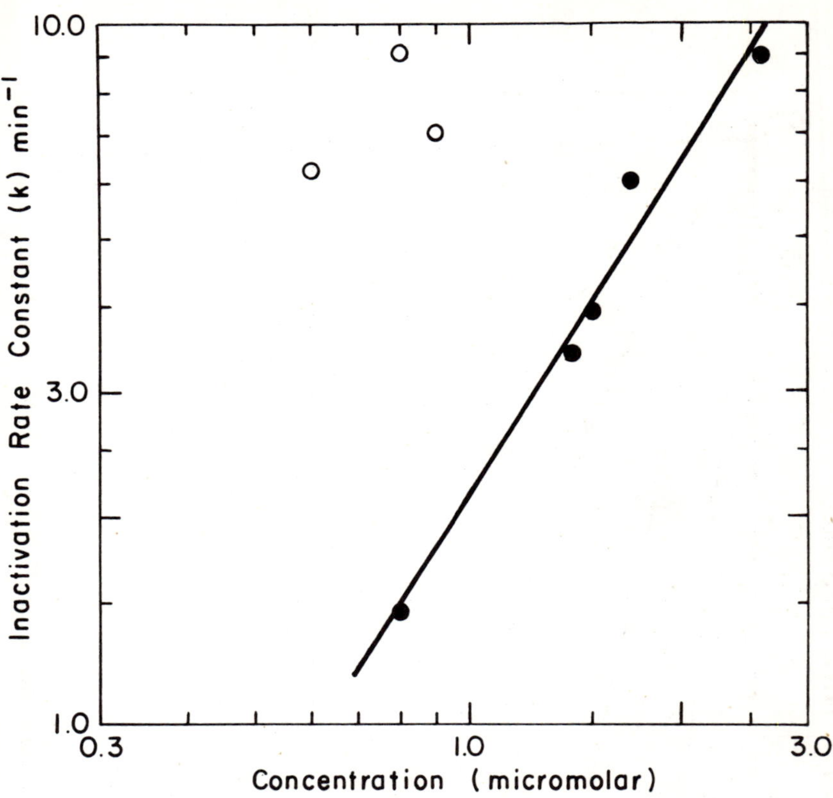

Figure 16.10. *Variation in the inactivation rate constant (k) for hypobromous acid (●) compared to molecular bromine (o).*

constant (k) as determined from the linear regression slope of the survival curve and the disinfectant concentration. The slope of the regression line ($\log k = n \log C + D$) for HOBr is 1.6. The three open circles represent the rates taken from the molecular bromine results. Even with this preliminary comparison, the Br_2 rates are about six times the HOBr rates at a concentration of 0.8 μM. The results of this rate comparison support the suggestion of Kruse[25] that a combination of HOBr and Br_2 is a better virucide than HOBr alone. These experiments show that molecular bromine is clearly superior to HOBr in its ability to kill the virus; and since the two chemical species do not interact to destroy each other's oxidizing ability, it is expected that Br_2 mixed with HOBr would kill the virus more effectively than HOBr alone.

A comparison of the nitrogen tribromide and hypobromous acid results is presented in Figure 16.11. The 0.9 min^{-1} slope of the NBr_3 plot is about one-half that of the free bromine species. This indicates that the inactivation rate constant for nitrogen tribromide is less dependent upon concentration than the inactivation constant for HOBr. The HOBr species is clearly a superior virucide at the higher concentrations, but NBr_3 has significant disinfecting ability. One mg/l of nitrogen tribromide (as bromine) killed 99.999% of the øX174 present in about five minutes.

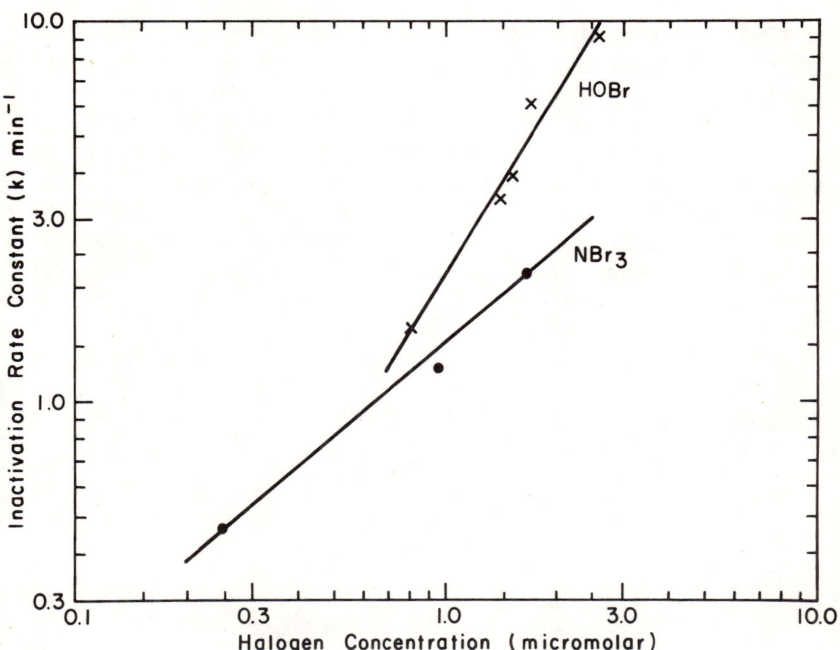

Figure 16.11. Comparison of the inactivation rate constants (k) for hypobromous acid and nitrogen tribromide as a function of concentration.

The above analysis of the concentration dependence of the rates ignores any non-linearity in the survival curves. Morris[43] has stated that even when the slopes of survival curves are time dependent, "comparisons of germicides and evaluations of concentration dependence on the basis of times for fixed fractional kills are still valid as long as the functional relation between log N and t does not change." Making this assumption, it is valid to use times for a given fractional kill to assess concentration dependence. The Watson equation is applicable under these conditions even if there is an initial shoulder on the inactivation curve.

$$C^n t = A \tag{16}$$

or

$$\log t_f = -n \log C + \log A \tag{17}$$

where t_f is the time for a given fractional kill f, C is the disinfectant concentration, n is the so-called coefficient of dilution, and A is a constant. Plots of the logarithmic form of the Watson equation for HOBr and HOCl data at $t_{99.9}$ yielded coefficients of dilution of 1.6 and 1.2, respectively, which compare favorably with the raw data as log k versus log C (1.6 and 1.1). It makes little difference in this case which way the rates are compared, either by rate constant or by t_f. Such data, at $t_{99.0}$, are shown in Figure 16.12.

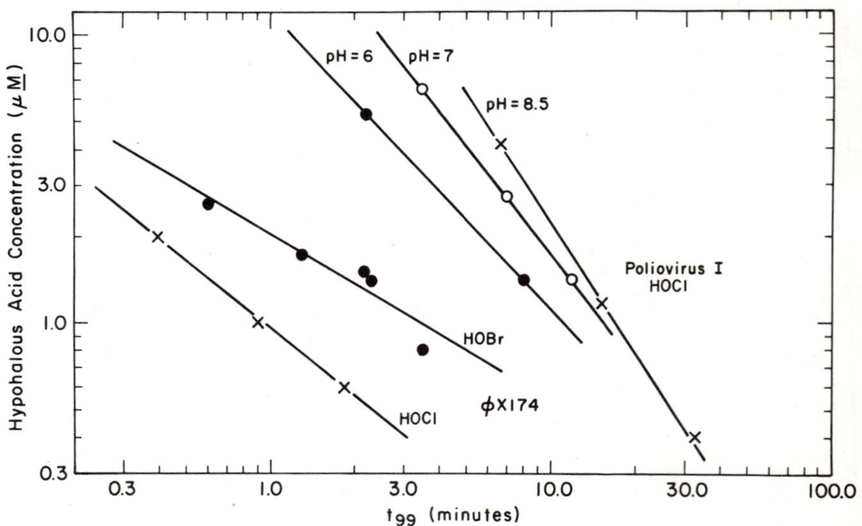

Figure 16.12. The effect of concentration on the time for 99.0% kill of φX174 and Type I Poliovirus.

Because experimental procedure and conditions vary from study to study, it is difficult to compare directly the present results using the bacterial virus øX174 to those of others using a different test organism. An attempt is made to compare the hypohalous acid-øX174 inactivation results here with Weidenkopf's[8] utilizing poliovirus. He studied the devitalization of Type I Poliomyelitis virus by hypochlorous acid at three hydrogen ion concentrations (pH 6, 7 and 8.5). Figure 16.12 shows the effect of concentration on the time required to kill 99% of the poliovirus at the three pH values.[8] The øX174-HOBr and -HOCl results at pH 4.5 are also presented. In order to compare the poliovirus data with the øX174 results, the former must be extrapolated to pH 4.5. It takes about 1.0 to 2.5 minutes to kill 99% of the poliovirus at this pH. The t_{99} for HOCl acting upon øX174 is 0.2 minutes. It follows that an estimate from Weidenkopf's data of the relative effect of HOCl on the enteric animal virus (polio) compared to the effect on the bacterial virus is 5-12 to 1. Poliovirus is about ten times less sensitive than øX174 to HOCl. Scarpino, *et al.*[50] found the Type 1 Poliovirus to be about 130 times more resistant to HOCl than is *E. coli*. The relative resistance of these organisms to HOCl, decreasing about one order of magnitude at each step, is poliovirus > øX174 > *E. coli*.

Certainly, the numbers represent only estimates of the relative halogen effects on these organisms, but they do provide a reference point for comparing this bacterial virus with an enteric virus and an enteric bacterium. It is believed that the time for 99% kill of the two viruses is within about one order of magnitude.

Effect of Temperature

An increase in the temperature affects the rate of inactivation of øX174 by hypobromous acid at pH 7.0. The bromine results shown above in Table 16.I were carried out at 0° C and pH 4.5. A summary of the temperature results at pH 7.0 are given in Table 16.II. As an example of the kinetic curves at pH 7, only 1.2 μM HOBr is required to obtain seven logs of disinfection in less than 3.5 minutes at 30° C (Figure 16.13).

The temperature results are summarized in Figure 16.14. Although the 15° and 30° C lines are nearly parallel, the 0° C data deviate markedly from this pattern. A concentration threshold effect at low temperature would account for this slope change. A possible explanation for

Table 16.II

Temperature Data Taken at pH 7.0 Using Hypobromous Acid

Temperature	*Concentration* (μ*M*)	*% Loss*	*Range (min)*	*Regression Slope ± 95% Confidence Interval* (min^{-1})
0° C	1.0	13	0.5 - 8.1	0.002 ± 0.006
	1.3	15	0.5 - 6.0	0.002 ± 0.008
	2.1	11[a]	0.5 - 5.0	0.638 ± 0.025
	3.6	29[a]	0.5 - 2.0	2.278 ± 0.163
	3.8	26[a]	0.5 - 2.5	2.153 ± 0.078
15° C	0.6	30	0.5 - 3.5	0.045 ± 0.024
	1.0	24	0.5 - 5.0	0.091 ± 0.011
	1.1	21	0.5 - 4.0	0.038 ± 0.013
	2.0	13	0.5 - 4.0	0.659 ± 0.036
30° C	0.7	33	0.8 - 3.5	0.848 ± 0.021
	1.2	38	0.5 - 3.5	2.253 ± 0.059

[a]Loss during entire time period from Sample No. 1 to Sample No. 2.

the deviation is that there is a change in the overall reaction mechanism between 0° and 15° C. A change in the slope of the log C - log k plots has been observed in the devitalization of other viruses. For example, Bachrach[44] has observed an apparent two-mechanism process in the thermal inactivation of the virus of foot-and-mouth disease. Certainly, more temperature data is necessary to characterize fully the possible change in the overall inactivation mechanism here.

The Arrhenius activation energy is calculated at 1.0 μM HOBr, around which most of the data lies. Figure 16.15 is a plot of the logarithm of the rate constant as

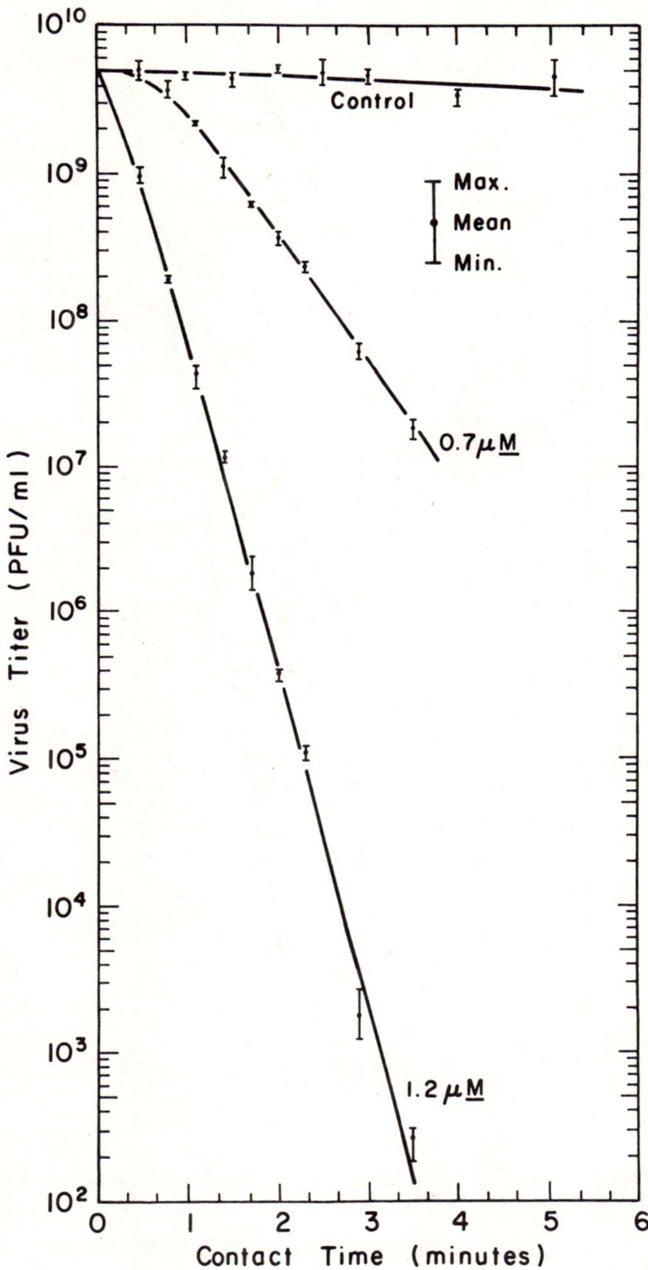

Figure 16.13. Inactivation of ØX174 by hypobromous acid at pH 7.0 and 30° C.

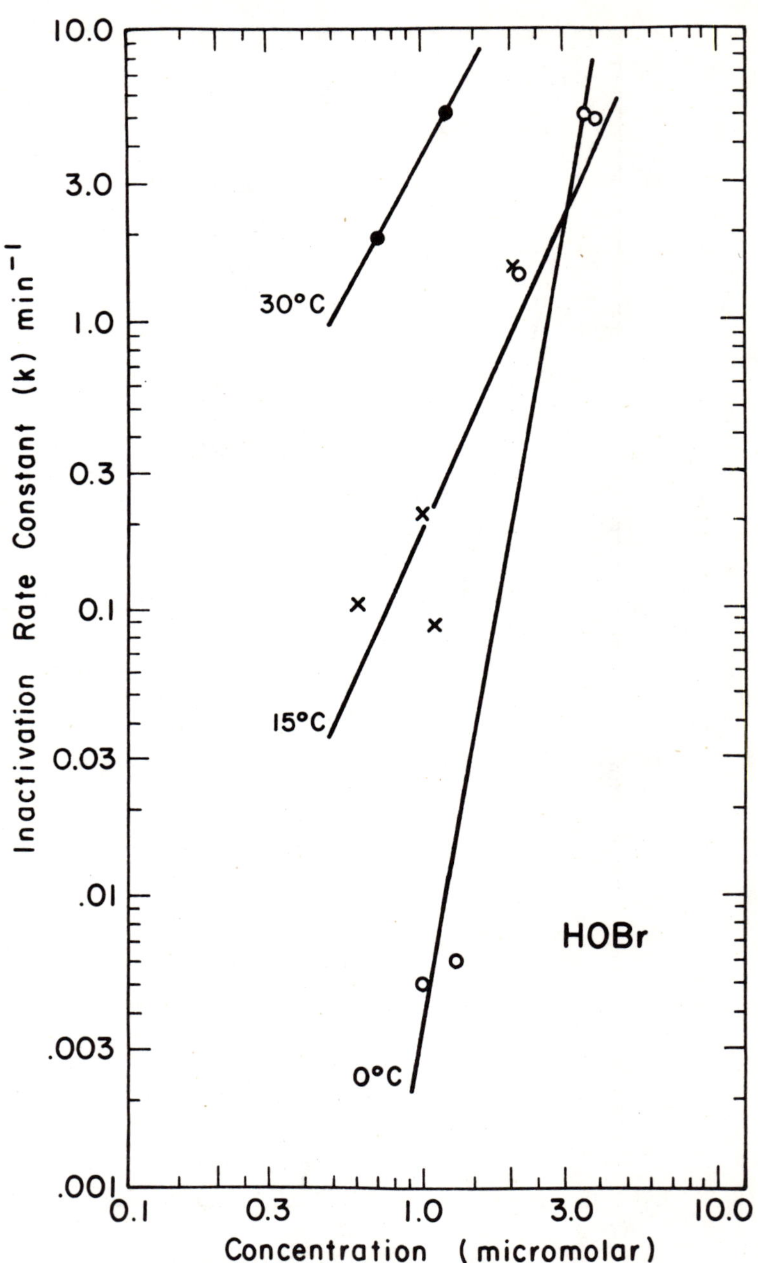

Figure 16.14. Variation in the inactivation rate constant (k) at three temperatures.

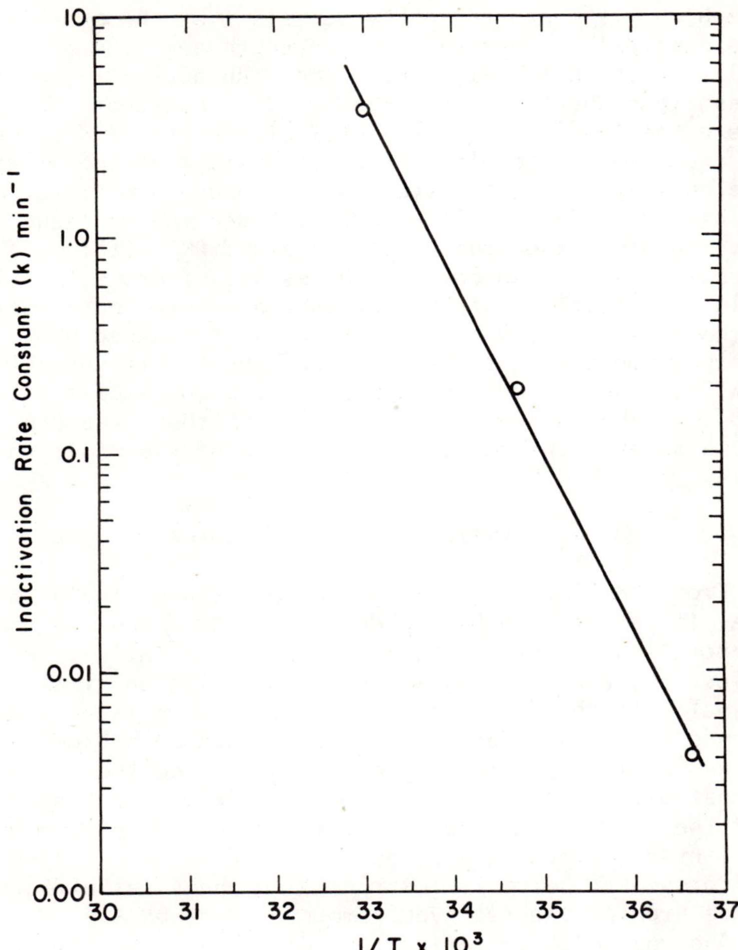

Figure 16.15. Arrhenius plot of the inactivation of ØX174 by 1.0 μM hypobromous acid at pH 7.0.

a function of the reciprocal of temperature as required by the Arrhenius equation

$$\log k = \frac{-Ea}{2.303}\left[\frac{1}{T}\right] + \frac{\log A}{2.303} \,. \tag{18}$$

From the slope of the Arrhenius plot, the activation energy is calculated to be 37 Kcal/mole. Since this value is quite high, it supports the hypothesis that the overall process of ØX174 inactivation by hypobromous acid at pH 7 and 1.0 μM is chemically controlled. An activation energy of less than about 10 Kcal/mole would be necessary to explain the inactivation as a diffusion controlled process. This experimental result further suggests that the virus is being inactivated by attack on the protein coat rather than by the diffusion of the HOBr through the coat and subsequent attack on the single-stranded DNA within. This suggestion is in accordance with results of Kruse,[25] using bromine to inactivate f_2 bacteriophage, where inactivation for the whole virus was faster than for the naked DNA. The experiments of Lytle and Ginoza[51] using nitrous acid on ØX174, however, do not support this proposal since they found that the rate of devitalization of the naked DNA was the same as the devitalization rate of the whole phage.

Inactivation Model for Bromine

From the limited data on bromine disinfection given above, it is not possible to develop a comprehensive model incorporating the effects of bromine concentration, bromine species, pH, temperature, and viral clumping on viral survival. If the shape of the survival curve is an indication of the inactivation mechanism, then each bromine species is different in its mode of attack on the virus. The most pragmatic and useful model would be one that, under the conditions used, predicts the effect of the most prevalent disinfecting species on viruses. The fitting model presented below is restricted to the viral effects of free bromine species hypobromous acid at pH 4.5.

One way to characterize the HOBr data is to fit it to the logarithmic form of the Watson equation (Equation 17). Table 16.III shows the values to be used for A and n at a given percentage of kill (f). The time required to inactivate a chosen fraction (t_f) may be obtained when the HOBr concentration is known. For example, the time required to kill 99% of the virus using 2.0 μM HOBr at pH 4.5 is calculated to be 1.07 minutes. Although the procedure of fitting the data to a mathematical expression is simplistic, this equation may be used to predict kill times under the given conditions. It should be emphasized that the concentration refers to the measured residual of hypobromous acid in the aqueous system.

Table 16.III

Constants for Use in Equation 17 to Determine Times for a Given Percentage (f) of ØX174 Kill by Hypobromous Acid

f	*n*	*log A*
90	1.27	0.242
99	1.51	0.486
99.9	1.57	0.621
99.99	2.09	0.849
99.999	2.01	0.897

Although a more complete model is not possible at this time, the characteristics of such a model can be outlined. The logarithmic nature of the survival curves is constant and should be incorporated into the model. Since the inactivation rate constant for ØX174 is greatly affected by the pH and temperature, they should be included. Therefore, the equation

$$\log N/N_o = -kC^nt \tag{19}$$

where C^nt equals a constant (A) for a given percentage of kill and can be determined from HOBr concentration studies at several pH values. The rate constant would be chosen at a given pH value on the basis of temperature, *i.e.*, using the Arrhenius Equation (Equation 18). The causes of the observed shoulders and tailing effects of survival curves cannot be resolved here. No consistent pattern in the shape of these curves has been found regardless of whether species, concentration, temperature or pH are considered.

SUMMARY AND CONCLUSIONS

The aqueous bromine system offers a unique opportunity to study five chemical species as biocides at one pH. The study of individual species is not possible with the chlorine or iodine systems at a constant pH. The chemical species hypobromous acid, molecular bromine and nitrogen tribromide were tested at pH 4.5 for their virucidal activity and compared to the activity of hypochlorous acid against the bacterial virus ØX174.

Bromine was studied as an alternative to chlorine because bromine does not ionize at neutral pH, has more desirable characteristics as a swimming pool disinfectant, and is active even in the form of bromamine. Chloramines, on the other hand, possess little virucidal activity.

Bromine, in contrast to chlorine, is found to be an effective virucide at low concentration in all the chemical forms tested. Five logs of viral inactivation (99.999%) are achieved in 1.2 minutes at pH 4.5 and 0° C using 0.4 mg/l (2.6 μM) HOBr. At 0.8 μM, molecular bromine is about six times and hypochlorous acid three times more effective than hypobromous acid as a virucide under the test conditions. The relative rates of øX174 inactivation in descending order are Br_2 > HOCl > HOBr > NBr_3 at a concentration of 1.0 μM.

Although hypochlorous acid is more effective than HOBr and NBr_3, it forms chloramines in waters containing ammonia. According to the literature, these chloramine species are ineffective as virucides. Therefore, in waters where ammonia is present, chlorine should be inferior to bromine as a disinfectant. Nitrogen tribromide, which is formed under breakpoint bromination conditions, has significant virucidal activity. Though it was the least active species tested, NBr_3 devitalized 99.99% of the virus in four minutes at 0.8 mg/l (as bromine). This fact is significant because this is the first time that NBr_3 has been tested for its virucidal effectiveness.

Comparison of the results using øX174 with those of Weidenkopf[8] extrapolated back to pH 4.5 using Poliovirus I indicates that Poliovirus is five to twelve times less sensitive than øX174 to 0.21 mg/l HOCl.

Bromine inactivation of øX174 is chemically controlled, not mass transport controlled. From temperature studies on the variation in the inactivation rate constant, the Arrhenius activation energy calculated at 1.0 μM HOBr is found to be 37 Kcal/mole. This high activation energy rules out the possibility that the overall disinfection kinetics are mass transport controlled. The overall disinfection rate processes must be limited by a chemical reaction. A concentration threshold effect noted at low temperature may account for the slope change of the 0° C log-log plot. More data is needed for the development of a bromine model for viral inactivation.

The survival curves of HOBr at pH 4.5 show an initial lag period before the maximum first order rate is attained. Although good fits of the data are not obtained when the points are superimposed on graphs of either the multi-target or the multi-hit models of disinfection, the data most closely approximates the three-target, single-hit model. Combination multi-target, multi-hit models have been presented elsewhere.[45,52] These models may be a better approximation of the process. More data, however, are required to verify the multiple target and hit hypothesis. It is possible that the assumption of a Poisson distribution may be in error, or the inactivation process may be too complex to be described by the simple, multi-target or the multi-hit models. A further consideration that has been neglected is the size distribution of the viral particles themselves in determining the shape of the survival curves. Moreover, if there is a distribution in the number of viral particles per plaque-forming unit, then this is an additional variable in the shape of the survival curve.

Although the economics of bromine was not a part of this study, bromine cost limitations on its use cannot be ignored. Bromine is presently twice the cost of chlorine. The use of bromine chloride to generate free bromine in water is feasible at a lower cost.[53] Bromine chloride may be fed and metered as a gas at low pressure rather than as a liquid, and therefore, promises to be easier to dose than liquid bromine. Moreover, since hypobromous acid is only ionized 6% at pH 7.6, as compared to the 50% ionization of hypochlorous acid, only about one-half of the amount of bromine is required to achieve the same effective molar concentration of active halogen at that pH.

REFERENCES

1. Polotkin, S. A. and M. Katz. In *Transmission of Viruses by the Water Route*, G. Berg, Ed., Interscience, New York, 1967, pp. 151-166.
2. Grabow, W. O. K. *Water Research, 2,* 675 (1968).
3. Fair, G. M., J. C. Geyer, and D. A. Okun. *Water and Wastewater Engineering*, Vol. 2, John Wiley, New York, 1968, Chapters 1, 19 and 31.
4. Chang, S. L. *J. San. Eng. Div. ASCE, 97*, No. SA5, 689 (1971).
5. Brown, J. R. and D. M. McLean. *J. Appl. Bacteriol. 29,* 559 (1966).
6. Berg, G. *J. New England Water Works Assoc., 78,* 79 (1964).
7. Berg, G., S. L. Chang, and E. K. Harris. *Virology, 22,* 469 (1964).

8. Weidenkopf, S. J. *Virology, 5,* 56 (1958).
9. Kelly, S. M. and W. W. Sanderson. *J. Public Health, 50,* 14 (1960).
10. Kruse, C. W., *et al.* In *National Specialty Conference on Disinfection,* American Society of Civil Engineers, New York, 1970, pp. 113-136.
11. Olivieri, V. P., T. K. Donovan, and K. Kawata. In *National Specialty Conference on Disinfection,* American Society of Civil Engineers, New York, 1970, pp. 365-384.
12. Sproul, O. J., *et al.* In *National Specialty Conference on Disinfection,* American Society of Civil Engineers, New York, 1970, pp. 385-396.
13. Farkas-Himsley, H. In *Bromine and Its Compounds,* Z. E. Jolles, Ed., Academic Press, New York, 1966, Part 6, Chapter 2, pp. 554-562.
14. Stringer, R. and C. W. Kruse. *J. San. Eng. Div., ASCE, 97,* No. SA6, 801 (1971).
15. Butterfield, C. T. *Public Health Reports, 63,* 934 (1948).
16. Spelman, P. T. paper presented at the Water Pollution Control Federation Meeting, Atlanta, Georgia, 1972.
17. Brown, J. R., D. M. McLean, and M. C. Nixon. *Canadian J. Public Health, 54,* 267 (1963).
18. Kott, Y. *Water Research, 3,* 265 (1969).
19. Kott, Y. and J. Edlir. *Water Research, 3,* 251 (1968).
20. Black, A. P., *et al. J. Amer. Water Works Assoc., 60,* 69 (1968).
21. Tanner, F. W. and G. Pitner. *Proc. Soc. Exp. Biol. Med., 40,* 143 (1939).
22. Wyss, O. and J. R. Stockton. *Biochem., 12,* 267 (1947).
23. Johannesson, J. K. *Nature, 181,* 1799 (1958).
24. Kruse, C. W. "Etiology of *E. Histolytica* Transmission," Final Technical Report, Contract No. DADA 17070-C-0025, U. S. Army Medical Research and Development Command, Washington, D.C., 1971.
25. Kruse, C. W. "Mode of Action of Halogens on Bacteria and Viruses and Protozoa in Water Systems," Final Technical Report Contract No. DA-49-193-MD-2314, U.S. Army Medical Research and Development Command, Washington, D.C., 1969.
26. Liebhafsky, H. A. *J. Am. Chem. Soc., 56,* 1500 (1934).
27. Farkas, L. and M. Lewin. *J. Am. Chem. Soc., 72,* 5766 (1950).
28. Griffith, R. O., A. McKeown, and A. G. Winn. *Trans. Faraday Soc., 28,* 101 (1932).

29. Johannesson, J. K. *J. Chem. Soc.*, 2998 (1959).
30. Galal-Gorchev, H. and J. C. Morris. *Inorg. Chem.*, *4*, 899 (1965).
31. Johnson, J. D. and R. Overby. *J. San. Eng. Div., ASCE*, *97*, No. SA5, 617 (1971).
32. Johnson, J. D. "Bromine Residual Chemistry," Annual Progress Report, Contract No. DADA-17-72-C-2053, U.S. Army Medical Research and Development Command, Washington, D.C., 1972.
33. *Standard Methods for the Examination of Water and Wastewater.* 13th ed. American Public Health Association, New York, 1971.
34. Johnson, J. D. and R. Overby. *Anal. Chem.*, *41*, 1744 (1969).
35. Edgell, M. H., C. A. Hutchison, III, and R. L. Sinsheimer. *J. Molecular Biol.*, *43*, 547 (1969).
36. Carusi, E. A. and R. L. Sinsheimer. *J. Molecular Biol.*, *7*, 388 (1963).
37. Davis, B. D., *et al. Microbiology.* Harper and Row, New York, 1967, Chapter 42, pp. 1078-1082.
38. Pagano, J. S. and C. A. Hutchison, III. *Methods in Virology*, Vol. V, K. Maramorosch and H. Koprowski, Eds., Academic Press, New York, 1972, pp. 79-123.
39. Adams, M. H. *Bacteriophages*, Interscience, New York, 1959.
40. Chapin, R. M. *J. Am. Chem. Soc.*, *56*, 2211 (1934).
41. Prutton, C. F. and S. H. Maron. *J. Am. Chem. Soc.*, *57*, 1652 (1935).
42. Engle, P., A. Oplatka, and B. Perlmutter-Hayman. *J. Am. Chem. Soc.*, *76*, 2010 (1954).
43. Morris, J. C. In *National Specialty Conference on Disinfection*, American Society of Civil Engineers, New York, 1970, pp. 609-633.
44. Hiatt, C. W. *Bacteriol. Rev.*, *28*, 150 (1964).
45. Fowler, J. F. *Phys. Med. Biol.*, *9*, 177 (1964).
46. Oliver, M. A. and B. J. Shepstone. *Phys. Med. Biol.*, *9*, 167 (1964).
47. Gunter, S. F. and H. I. Kohn. *J. Bacteriol.*, *71*, 571 (1956).
48. Alper, T., N. E. Gillies, and M. M. Elkind. *Nature*, *183*, 1062 (1960).
49. Watson, H. E. *J. Hygiene*, *8*, 536 (1908).
50. Scarpino, P. V., *et al.* See Chapter 15, *Chemistry of Water Supply, Treatment and Distribution.*
51. Lytle, D. C. and W. Ginoza. *Virology*, *38*, 152 (1969).
52. Atwater, K. C. and A. Norman. *Proc. Nat. Acad. Sci.*, *35*, 696 (1949).

53. Schmeisser, M. and E. Schurter. In *Bromine and Its Compounds*, Z. E. Jolles, Ed., Academic Press, New York, 1966, Part 2, Chapter 4, pp. 179-252.
54. Larsen, T. E. and F. Sollo. "Determination of Free Bromine in Water," Annual Progress Report, Contract No. DA-49-193-MD-2909, U.S. Army Medical Research and Development Command, Washington, D.C., 1967.

CHAPTER 17

INACTIVATION OF VIRUSES AND BACTERIA BY OZONE

E. Katzenelson, B. Kletter,
Hanna Schechter and H. I. Shuval*

INTRODUCTION

Ozone, O_3, which was first suggested as a disinfectant for drinking water in the 19th century, is presently used for this purpose in many European countries. A review of the literature[1] shows that it is superior to other chemical disinfectants. It acts quicker and in lower concentrations, and has no side effects such as taste and odor which are characteristic of other chemicals, *e.g.*, chlorine.[2] These positive qualities have led to a renewed interest in ozone as a disinfectant of water, especially in those countries where it is not yet utilized.

It is striking that no basic research has been done on the inactivation kinetics of microorganisms by O_3 concentrations while at constant levels. In particular, research with viruses has been neglected. Such research is important in order to understand the correlation between concentration and killing ability of the disinfectant, and would make possible the calculation of the quantities of ozone needed for different kinds of water. In this study efforts were made to develop methods by which the disinfection abilities of ozone at constant levels of concentration can be probed. Various microorganisms, found in polluted waters were chosen. These included *Escherichia coli*, a well-known indicator of fecal contamination of water, coliphage T_2, and poliovirus type I. The main weight of the research was placed on viruses since these are known to be more resistant to disinfectants than bacteria.

---

*E. Katzenelson, B. Kletter, H. Schechter, and H. I. Shuval, Environmental Health Laboratory, Hebrew University Hadassah Medical School, Jerusalem, Israel.

MATERIALS AND METHODS

Microorganisms

Poliovirus type I (Brunhilda) was used throughout this study. The virus was grown in Vero cells (Flow Laboratories), accumulated, and concentrated by the phase separation method[3] and by two subsequent purification cycles of the concentrated virus. A cycle consisted of centrifugation of the suspension for 30 min at 12,000 x g. The supernatant was then centrifuged for one hour at 100,000 x g. The pellet was resuspended in 27 ml of 0.005 *M* phosphate buffer, pH 7.0, and ultrasonicated at 20 Mc/sec for 2 minutes. Finally, the purified virus suspension was centrifuged at 4,000 x g for 30 minutes and stored in glass vials, 0.25 ml in each, at -70° C. Titration of the virus was done on Vero cells, which resulted in 5.0×10^8 pfu/ml.

Coliphage T_2 was grown on *E. coli*. The bacteria were inoculated in an Erlenmeyer flask containing nutrient broth and incubated in a shaker bath at 37° C for 5 hr, at the end of which period 10^9 pfu bacteriophage were added. The suspension was incubated again for 3 hours. At the end of the entire 8-hr incubation period, the suspension became almost clear. This solution was centrifuged for 45 min at 12,000 x g; the supernatant was centrifuged for an additional 3-4 hr at 40,000 x g. The sediment was resuspended in phosphate buffer, pH 7.5, and again centrifuged. The cycle was repeated four times. After the last washing the sediment was resuspended in phosphate buffer, diluted to 5×10^8 pfu/ml and kept under refrigeration.

Before each experiment *E. coli* were inoculated into 30 ml of nutrient broth and incubated overnight at room temperature, followed by 4-5 hr in a shaker bath at 37° C. The bacteria were harvested and washed four times by centrifugation in 50 ml of 0.5 *M* phosphate buffer, pH 7.0. After the fourth washing the sediment was resuspended in 50 ml of the same buffer solution. The suspension was then filtered through a 0.3μ Millipore filter to remove possible clumps. The final solution usually contained 2×10^9 bacteria/ml.

Ozone

Ozone was generated from the oxygen in air in a Fisher-Lab Ozonizer (OZ III). A suction pump (Charles Austen Pumps, Ltd.) was used to pass the air through the Ozonizer. The airstream was dried by means of a Koty "Senior" air purifier and flow equalizer. The ozone-rich air was bubbled through a 0.05 *M* phosphate buffer, pH 7.0, until a high concentration of O_3 was attained. The O_3 excess remaining in the air was neutralized by passing the airstream through two traps, the first of which contained a concentrated solution of KI, and the second, thiosulfate. Before each experiment, the concentrated solution was diluted until it reached the appropriate concentration.

Ozone concentration was determined by utilizing a spectrophotometric method developed in the authors' laboratory.[4] Water for the solutions used in the ozone experiments was made ozone-demand-free by distillation in an all-glass still in the presence of alkaline potassium permanganate. This was followed by ozonating the bidistilled water and by dissipation of the ozone by boiling.

In the course of this study it was found that ozone-demand-free glassware was a prerequisite for maintaining a constant level of ozone concentrations throughout the experiments. Therefore, all glassware was washed with chromic acid cleaning solution, after which it was rinsed in tap water and distilled water, soaked in a strong ozone solution, and dried at 180° C.

PROCEDURE

One ml of poliovirus stock suspension diluted 1:10 in ozone-demand-free water was put into a beaker containing 400 ml of 0.05 *M* phosphate buffer, pH 7.2, and dissolved ozone in the desired concentration. Before and during the experiment the beakers were kept in a thermostatically controlled water bath. The contents of the beakers were constantly stirred at 81 rpm with glass paddles connected to a Phipps and Bird mechanical stirrer. The beakers were covered to prevent the escape of ozone. Samples for virus determination were taken through holes in the covers. The first sample of 5 ml was taken 8 seconds after the introduction of the virus. Further samples (5 ml each) were taken at 8-second intervals for the first two minutes, and at 3, 5, 10, and 15 minutes. The samples were introduced into a 5-ml neutralizer solution of 0.002 *M* tetrasodium pyrophosphate and 0.0004 *M* sodium sulphite immediately upon withdrawal. "Sampling time" was the very moment the sample was put into the neutralizer. Pipetting never required more than 2 seconds. A beaker containing 200 ml of phosphate buffer and 0.5 ml of the diluted virus stock was stirred simultaneously in the same water bath to act as a control. Samples from the latter were withdrawn immediately upon introduction of the virus, and 5 and 15 min afterwards. Samples for ozone determination were drawn immediately before and 1, 2, 5, 10, and 15 min after virus addition. Neutralized samples were assayed for virus in tissue cultures on the day of the experiment. The preparation of dilution and the inoculation into tissue cultures have been described elsewhere.[5]

The procedures followed for the studies with the coliphage were the same as those for poliovirus, with the exception of the sampling frequency which was done every 10 seconds. For the assays of bacteriophages, samples were diluted in nutrient broth. One ml of each dilution was mixed with 1 ml of 1.5% liquified agar at 52° C containing *E. coli* bacteria. After stirring, the mixture was poured onto petri dishes containing agar. Plaques were counted after an 18-hr incubation period at 37° C.

Inactivation experiments with *E. coli* were essentially the same as with coliphage T_2. Bacterial assays were done by the pour plate method in nutrient agar.

Redox potentials were measured with a potentiometer (Radiometer Type PHM 26C) with calomel and platinum electrodes. The potentials are given as the uncorrected values in millivolts, using the calomel electrode as the reference cell.

RESULTS

Inactivation Kinetics of Poliovirus I

Concentrations of 0.07, 0.1, 0.15, 0.2, 0.3, 0.5, 0.8, 1.5 and 2.5 mg/l ozone were used. Figures 17.1, 17.2 and 17.3 depict the inactivation kinetics of the poliovirus with ozone concentrations ranging from 0.3 to 1.5 mg/l at 5° C. Each graph sums up at least three experiments. A significant feature is the 2-stage action of ozone. The first stage is short, less than 8 seconds, with a virus kill

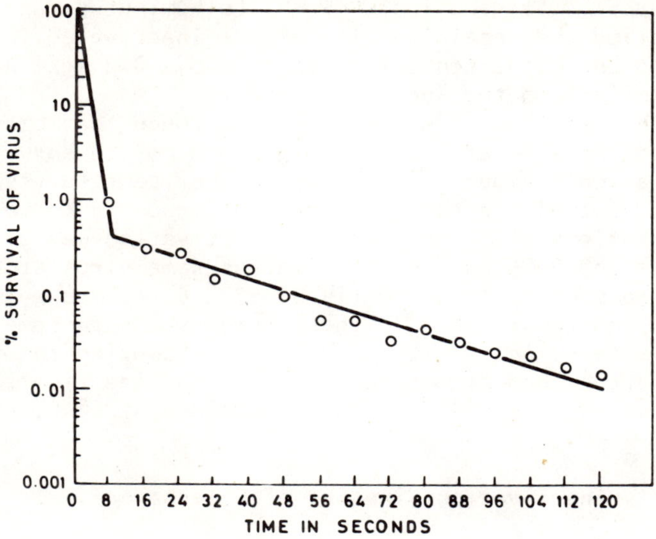

Figure 17.1. Inactivation kinetics of poliovirus I by 0.3 mg/l ozone at 5° C.

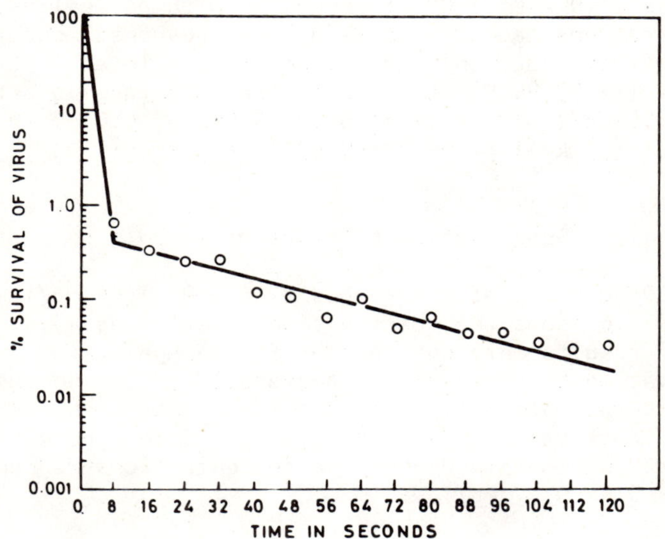

Figure 17.2. Inactivation kinetics of poliovirus I by 0.8 mg/l ozone at 5° C.

of 99-99.5%. Stage 2 lasts from 1 to 5 minutes, and in this period the remaining viruses are inactivated. Increasing the ozone concentration to above 0.2 mg/l had very little effect on the inactivation rate.

At the start of this study, it was found that ozone concentrations lower than 0.15 mg/l did not cause virus inactivation. Figure 17.4 shows the per cent survival of poliovirus type I after 40 seconds of exposure to various concentrations of ozone. Later, inactivation was obtained with 0.1 and 0.05 mg/l ozone with the same virus stock, which had been kept constantly at -70° C. There was, however, no consistency in these results and in some of the experiments no inactivation was seen. Lowering the temperature further had almost no effect on the inactivation rate; the curves were similar to those obtained at 5° C.

Inactivation Kinetics of Coliphage T_2

O_3 concentrations ranging from 0.01-1.3 mg/l were used. The inactivation kinetics of coliphage T_2 with 0.01, 0.09, and 0.26 mg/l O_3 at 1° C are depicted in Figure 17.5. Similar results were obtained at 5° C. Both the 2-stage inactivation curve and the no-dose-response above concentrations of 0.2 mg/l O_3 are seen, as with poliovirus. Lower concentrations caused a certain "dose response" occurring in the first stage only, and is expressed in an initial kill of (1) 99-99.9% at O_3 concentrations ranging between 0.01-0.06 mg/l; (2) kill of 99.9-99.99% at 0.07-0.19 mg/l O_3; and, (3) kill of 99.99-99.999% at 0.2 mg/l O_3.

Inactivation Kinetics of *E. coli*

The O_3 concentrations ranged from 0.02-1.55 mg/l. Figure 17.6 shows the inactivation kinetics at 1° C. Similar results were obtained at 5° C. Again, the 2-stage inactivation curve is seen. Above 0.1 mg/l O_3 no dose response was obtained. The lowest concentration that caused inactivation was 0.04 mg/l O_3. Some type of "dose response" was obtained when the concentrations ranged between 0.04-0.07 mg/l O_3.

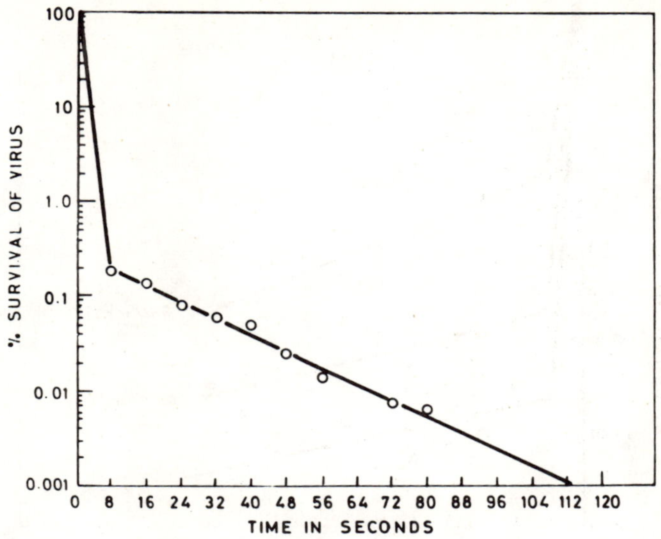

Figure 17.3. Inactivation kinetics of poliovirus I by 1.5 mg/l ozone at 5° C.

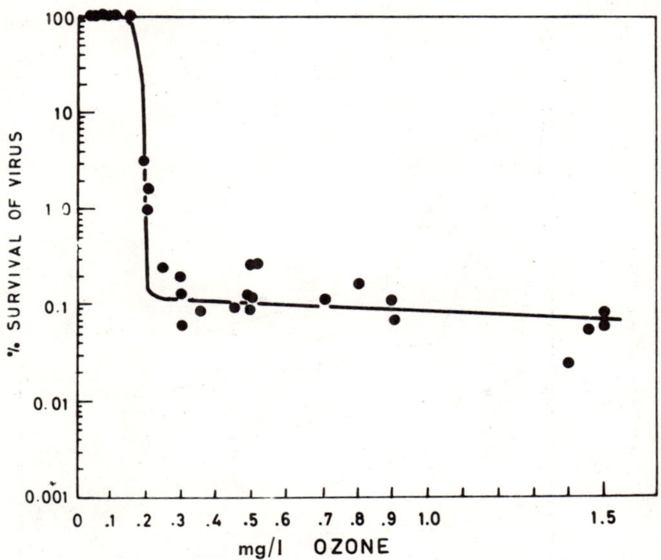

Figure 17.4. Poliovirus I survival after 40" of exposure to various concentrations of ozone.

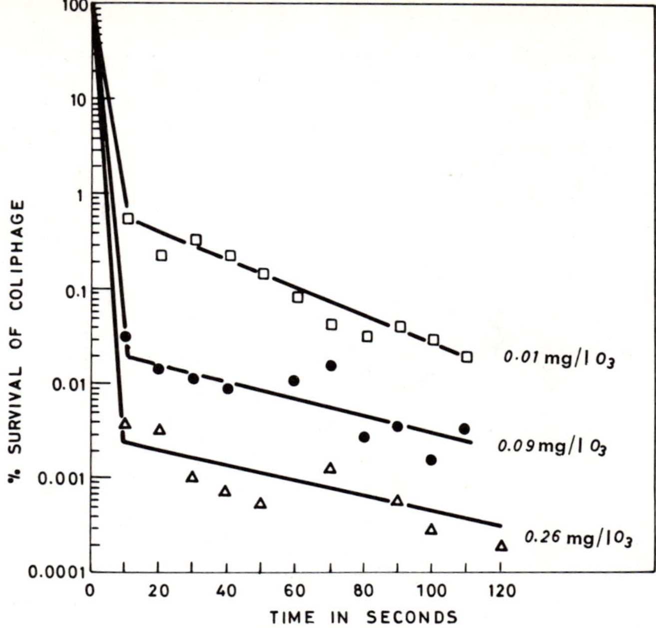

Figure 17.5. Inactivation kinetics of coliphage T_2 by various concentrations of ozone at 1° C.

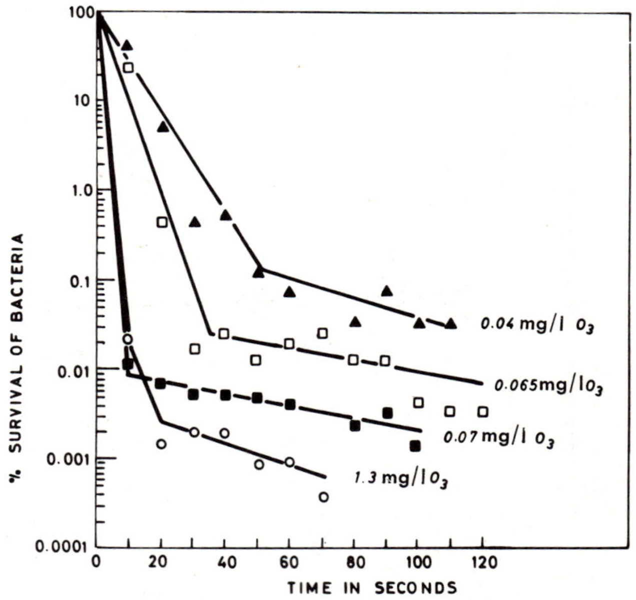

Figure 17.6. Inactivation kinetics of *E. coli* by various concentrations of ozone at 1° C.

Introduction of Virus at Two Different Times

A possible explanation for the 2-stage inactivation curve could be a chemical change of ozone, *e.g.*, the appearance of some intermediate decomposition species triggered by the microorganism. Therefore, a second portion of poliovirus was added, introduced 56 seconds after the first administration. As shown in Figure 17.7, the inactivation rate of this additional virus inoculum was the same as that of the first portion, which implies that the inactivating power of the ozone was not affected by the experimental conditions.

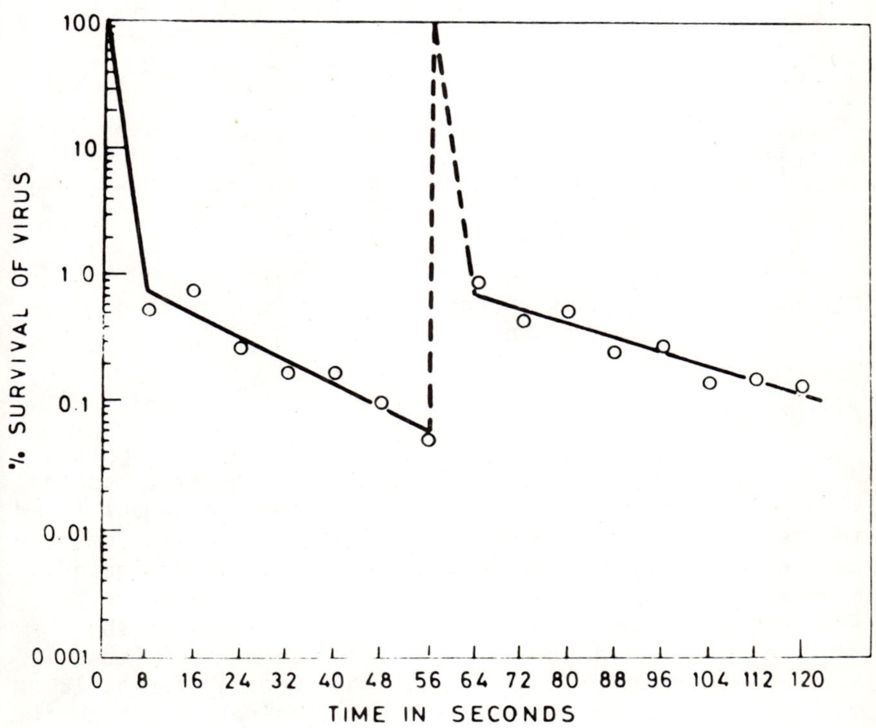

Figure 17.7. Inactivation kinetics of poliovirus I, added in two portions, at 0 time and after 56 seconds, by 1.5 mg/l ozone.

Redox Potentials of Ozone Solutions

To verify a possible connection between the redox potentials and the disinfection ability of ozone, the redox potentials of different O_3 concentrations were measured. Figure 17.8 shows the results of many such measurements. Redox potentials increase with increasing O_3 concentrations, the latter ranging from 0.03-0.2 mg/l, after which there is a leveling off.

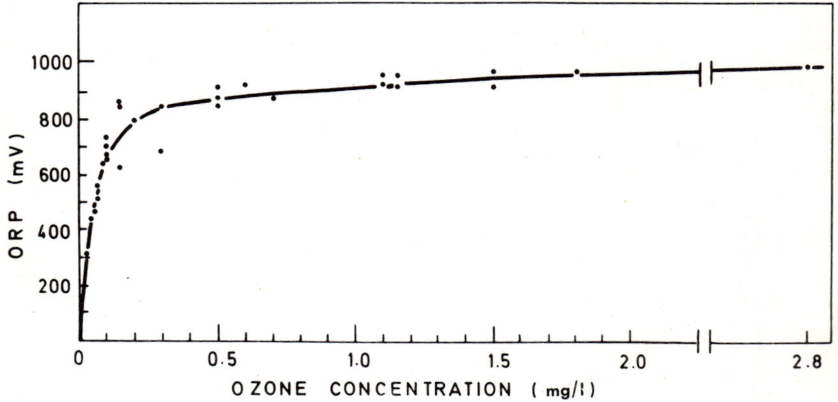

Figure 17.8. Correlation between ozone concentration and redox potential values.

DISCUSSION

During the search for disinfectants of bacterial contamination in water, ozone was the first chemical agent found to possess germicidal properties.[1] Its mode of action, particularly against viruses, has been less investigated, however, than other disinfection agents. Since there is a dearth of data on its mode of action, a study of the behavior of ozone in clean water was undertaken. Ozone is extremely labile and even a very small amount of organic matter in the water causes its rapid dissipation. Much time and ingenuity were required to build an experimental system in which ozone loss could be kept at a minimum. Toward this purpose, ozone-demand-free glassware, microorganism stock, and water had to be prepared. Further measures taken were covering the beakers, keeping the temperature at 5° C or lower during the experiments, and stirring at a rate not exceeding 81 rpm. In spite of these meticulous preparations, it was not always possible to keep the ozone at a constant level. In this study results are given for those experiments in which ozone disappearance did not exceed 20%.

The main object of this work was to elucidate the inactivation kinetics of microorganisms in water. Since poliovirus type I is ubiquitous in sewage it was chosen as a prototype for viral contamination in water. Another contaminator used in this work was the bacterial virus coliphage T_2. Finally, *E. coli*, an indicator of fecal contamination of water, was also included.

Outstanding characteristics of ozone action are its rapid inactivation of the microorganisms tested and the low concentrations of ozone necessary. Comparing the results obtained in this study with those of Scarpino *et al.*,[6] who used chlorine, and with those of Berg *et al.*,[7] who used iodine, the superiority of ozone is striking. With the same concentrations, a 99% kill was reached in less than 10 seconds when using ozone as compared to several minutes needed with the other two agents. In fact, ozone acts so quickly that it became impossible to measure the time required for 99% kill, nor was it possible to obtain curves showing the inactivation rate as a first order reaction according to Chick's Laws.[8]

A noteworthy phenomenon was the 2-stage shape of the inactivation curve. Stage 1 lasted less than 10 seconds with an inactivation of more than 99%, followed by stage 2 of several minutes' duration, during which period final inactivation took place. This phenomenon cannot be ascribed to changes in the composition of ozone, since poliovirus, added 56 seconds after the first virus inoculum, was inactivated at the same rate. A possible explanation for the 2-stage inactivation kinetics would be that 0.5-1.0% of the poliovirus consists of clumps. Galasso and Sharp[9] were the first to mention such clumps, and Berg *et al.*[7] described virus clumps as impeding the inactivation rate. It is interesting to note that this 2-stage action also appeared in experiments with coliphage T_2 and *E. coli*. It may be assumed that here, too, clumps were the cause of this phenomenon. An attempt was made to obliterate the clumps and an accelerated rate of kill with these two microorganisms, as compared to poliovirus, was seen during the first stage. A further explanation could be that these microorganisms are more sensitive to ozone than is poliovirus. O_3 concentrations as low as 0.01 mg/l (the lowest concentration that could be measured) resulted in a rapid inactivation of the coliphage, while *E. coli* was inactivated by 0.04 mg/l O_3.

Increasing the O_3 concentration to above 0.2 mg/l had very little effect on the inactivation rate as far as could be discerned. Since, for technical reasons, the earliest sample could be taken only 8 to 10 seconds after the introduction of the organisms, it was impossible to determine the processes that took place during 0-10 seconds. Dose response during this period may have taken place.

With concentrations lower than 0.2 mg/l O_3, the inactivation curves of poliovirus were not sufficiently reproducible to allow an accurate interpretation of the results. In some experiments these ozone concentrations had no effect, while in others concentrations even lower than 0.05 mg/l caused rapid inactivation. Experiments now in progress in the authors' laboratory point to the fact that storage of poliovirus at -15° C causes a change in its sensitivity to ozone, which expresses itself in a lower rate of inactivation with low O_3 concentrations. Storage may, therefore, be the reason of this lack of consistency in results.

A connection may exist between the mode of action of O_3 and its redox potentials.[10] Redox potentials increased with increasing concentrations of ozone from 0.03 through 0.2 mg/l; above 0.2 mg/l O_3 the potentials leveled off. The fact that increase in redox potentials was not seen above this dose correlates well with the lack of dose response in the inactivation curves. However, redox potential values do not explain the results of ozone action on the microorganisms with concentrations lower than 0.2 mg/l O_3.

Contrary to results with chlorine, lowering the temperature from 5° C to 1° C had almost no effect on the inactivation kinetics of the microorganisms. It should be noted that O_3 is chemically very active, even at temperatures as low as -70° C.

The work described here should be regarded as preliminary. Special techniques have yet to be developed which will enable the measurement of ozone inactivation rates during the first 10 seconds (first stage). Further research is needed on the clump phenomenon and the effect of storage on the disinfection ability of ozone. Whatever remains to be done, the rapidity with which ozone acts at low concentrations, as was demonstrated here, promises positive utilization of ozone in the disinfection of water and wastewater.

SUMMARY

The inactivation kinetics of various microorganisms, which serve as indicators for water pollution, against ozone were studied. The inactivation reactor system and experimental protocol that enable ozone concentrations to remain stable in water, and reproducibility of the results, are described. Inactivation kinetics curves of poliovirus type I, coliphage T_2, and *E. coli*, by various concentrations of ozone and at two different temperatures, are presented. The results clearly show the effectiveness of ozone in inactivating microorganisms. Lower concentrations of ozone and shorter contact times are required than are needed when using chlorine or any other agent presently utilized for the disinfection of water.

ACKNOWLEDGMENTS

Thanks are expressed to Mrs. Nehama Biederman, Yael Rotem, Rivka Kalbo, and Miss Jacqueline Sabbagh for their excellent technical assistance.

REFERENCES

1. Venosa, A. D. In *Ozone in Water and Wastewater Treatment*, F. L. Evans, Ed. Ann Arbor Science Publishers, Ann Arbor, Michigan, 1972, pp. 83-100.
2. Berg, G. In *Proc. National Specialty Conference on Disinfection*. American Society of Civil Engineers Publication, 1970, pp. 339-364.
3. Shuval, H. I., B. Fattal, S. Cymbalista, and N. Goldblum. *Water Res., 3,* 225 (1969).
4. Schechter, H. *Water Res., 7* (1973).
5. Shuval, H. I., A. Thompson, B. Fattal, S. Cymbalista, and Y. Winter. *J. Sanitary Eng.,* October, 587 (1971).
6. Scarpino, P. V., G. Berg, S. L. Chang, D. Dahling, and M. Lucas. *Water Res., 6,* 959 (1972).
7. Berg, G., S. L. Chang, and E. K. Harris. *Virology, 22,* 469 (1964).
8. Chick, H. *J. Hyg., 8,* 92 (1908).
9. Galasso, G. L., and D. G. Sharp. *J. Immunol., 88,* 339 (1962).
10. Chang, S. L. In *Proc. National Specialty Conference on Disinfection*. American Society of Civil Engineers Publication, 1970, pp. 635-680.

CHAPTER 18

THE EFFECT OF WATER TREATMENT AND DISTRIBUTION ON TRACE ELEMENT CONCENTRATIONS

Julian B. Andelman*

INTRODUCTION

There is a growing body of epidemiological and other evidence indicating an association between water quality and mortality from cardiovascular and other chronic diseases. One of the earliest studies, done in Japan by Kobayashi, showed a high correlation between death rates from cerebral hemorrhage and the acidity of river water, particularly those waters with high sulfate-bicarbonate ratios.[1] This was followed by other studies by Schroeder in the United States,[2-4] and by Crawford, Gardner, Morris, and Heady in Britain,[5-8] that developed statistical correlations of cardiovascular and other chronic disease mortality with various water quality parameters. Similar analyses were made by Biorck and co-workers in Sweden,[9] Biersteker in the Netherlands,[10] Anderson and co-workers in Canada,[11] and Dudley, Beldin, and Johnson in the United States.[12] These and other similar studies have been reviewed by Winton and McCabe[13] and Masironi *et al.*[14] who have considered the possible role of trace elements in municipal waters in relation to cardiovascular diseases. Although the most common correlation has been an inverse one between cardiovascular mortality and water hardness, some correlations have also been found with certain trace element concentrations.

---

*J. B. Andelman, Professor of Water Chemistry, Graduate School of Public Health, University of Pittsburgh, Pittsburgh, Pennsylvania.

There are speculations that the concentration levels of trace elements are influenced by or related to the water hardness.[14] Evidence for such a relationship is shown in Table 18.I, which is taken from Masironi,[15] and indicates that there is a high degree of positive correlation between water hardness and the concentration of many trace elements in river waters of the United States. It has also been

Table 18.I

Correlations of Hardness with Various Trace Elements in River Waters of the U.S. 1961-1962 (116 samples)[15]

Element	*r*	*Element*	*r*
Bi	0.95	Zn	0.90
Sb	0.95	Pb	0.87
Sn	0.94	B	0.85
Co	0.94	Mn	0.65
Cd	0.93	F	0.57
Cr	0.93	Be	0.56
V	0.92	Ag	0.45
Ni	0.92	Ba	0.44
Mo	0.91	Fe	0.30
		Cu	0.21

Levels of Significance

r	*Prob. %*
0.24	1
0.3	0.1

suggested that hardness and/or calcium might secondarily act to control the exchange of trace elements between soils or distribution system pipes and their contacting waters.[14]

The mechanisms by which such trace elements could influence cardiovascular diseases are for the most part unclear. Masironi[16] has reviewed the possible mechanisms for several trace metals, and has noted that cadmium has been the one element most extensively studied, particularly in relation to arterial hypertension. It has been suggested that approximately 30 trace elements might be usefully studied in terms of their uptake in man in relation to cardiovascular diseases.[14]

It should be noted that diseases other than cardiovascular have been statistically associated with trace elements in water. For example, Berg and Burbank[17] have shown statistically significant positive correlations between mortality from various types of cancer and concentrations of several trace elements in water supplies, including arsenic, beryllium, cadmium, lead, and nickel. Thus, aside from chemical poisoning associated with unusually high concentrations of trace metals in a water supply, it is apparent that the more usual levels encountered are also of interest in that they may have some influence on chronic disease in the populations exposed. Nevertheless, to put into perspective the concentrations that are of some concern from the point of view of possible acute effects. Table 18.II lists the concentrations of those trace elements indicated in the U.S. Public Health Service Drinking Water Standards of 1962 which, if exceeded, constitute grounds for rejection of the supply.[18] In the same table are the results from a survey of community water supplies of the U.S. indicating the maximum concentrations found for such trace elements in samples obtained from the distribution systems.[19] In

Table 18.II

U.S. Public Health Service Mandatory Standards for Trace Elements in Drinking Water[18] *Compared with Concentrations Found in Distribution System Samples in a U.S. Community Water Supply Survey*[19]

Element	*Concentration (mg/l)*		*% Samples Exceeding Limit*
	Standard	*Max. Found*	
As	0.05	0.1	0.2
Ba	1.0	1.55	<0.1
Cd	0.01	3.94	0.2
Cr(VI)	0.05	0.08	0.2
Pb	0.05	0.64	1.4
Se	0.01	0.07	0.4
Ag	0.05	0.03	0

general, the frequency with which the standard was exceeded was relatively small, but the maximum concentrations found were quite high for some elements. Another point of perspective in this regard is the extremely low incidence of reported outbreaks of waterborne disease from chemical poisoning in the U.S.[20] It should be noted, however, that outbreaks from such chemical poisonings were those most frequently found not to be reported to the National Center for Disease Control, since such reports are not required.

The focus of this paper shall be the concentrations of trace elements to which humans are exposed in their community water supplies, principally and ultimately to provide the kind of data base for assessing possible relationships with chronic disease. Thus, the concern is not whether the U.S. Public Health Service Standards shown in Table 18.II are met. Rather, what kinds of variability in time are encountered and what effects do treatment and distribution have on the trace element concentrations as they reach the consumer? Do raw and finished water supplies vary significantly on a macro- and microgeographic scale? Is it necessary to measure trace element concentrations in community water supplies frequently to assess their average concentrations? In order to aid the epidemiologist in assessing man's exposure to trace elements from his water supplies and other sources, all of these important questions need to be addressed.

RAW WATER INFLUENCE ON FINISHED WATER QUALITY

The concentrations of trace elements and their variations in raw water supplies are of prime importance in relation to the ultimate quality of the finished water reaching the consumer. One should then expect that raw water variabilities, both geographically and with time, would be reflected in the finished waters. One indication of such variability on a macrogeographic basis is shown in Table 18.III, which shows the mean positive concentrations for several trace elements in three U.S. river basins and compares them with U.S. mean values. The underlined values are above the U S mean concentrations, and the data are taken from the comprehensive survey by Kopp and Kroner.[21] It is apparent that, except for copper, concentrations in the Lake Erie basin are significantly higher than the U.S. averages, in Alaska they are significantly lower, and in the Northeast there is a mixed behavior.

Table 18.III

Mean Positive Trace Element Concentrations in U.S. River Basins (μg/l), 1962-1967[21]

Element	*All U.S.*	*Northeast*	*Lake Erie*	*Alaska*
Zn	64	<u>96</u>	<u>205</u>	28
As	64	34	<u>308</u>	34
B	101	32	<u>210</u>	28
Cu	15	<u>15</u>	11	9
Ni	19	8	<u>56</u>	5
Pb	23	17	<u>39</u>	12
V	40	9	<u>54</u>	32

One can also encounter such geographic variability on a microgeographic scale. For example, in Lake Michigan in one relatively short period zinc concentrations varied from 3 to 80 μg/l, depending on location within the lake. The comparable concentrations for chromium ranged from 0.55 to 1.1 μg/l.[22] Similarly, Table 18.IV shows the mean concentrations for several trace elements in the two major source tributaries of the Ohio River, the Allegheny and the Monongahela, just before they join at Pittsburgh to form the Ohio.[23] These data show that in almost every instance the mean monthly concentrations varied significantly between the two rivers at sampling points less than a mile apart.

Another important feature of the data in Table 18.IV is the changing concentrations with time for the trace elements tabulated there. Such changes are striking and are an indication of the need for frequent sampling in assessing trace element concentrations.

Such time variability can also be seen, as shown in Table 18.V, in the concentrations of boron and nickel in finished waters at treatment plants using these two rivers as their raw water supplies. These concentrations also show marked variation with time for each of these elements at each of the three treatment plants, and incidentally indicate significant differences in the finished waters of the two plants taking their water from nearby points on the Allegheny River. Also shown in that table are the

Table 18.IV

Variation in Time of Concentration (μg/l) of Several Trace Elements in the Allegheny and Monongahela Rivers at Pittsburgh[23]

Element	Nov. 1964		Jan. 1965		March 1965		May 1965		July 1965		Sept. 1965	
	A	M	A	M	A	M	A	M	A	M	A	M
Barium	27	34	59	15	40	19	30	25	22	12	30	24
Boron	100	129	118	74	111	130	100	105	155	205	69	85
Copper	21	29	97	6	28	27	27	50	26	25	24	32
Nickel	20	<8	<8	<8	<13	<10	25	15	40	20	44	20
Strontium	118	143	45	53	28	47	50	65	100	90	120	130
Zinc	74	118	50	26	81	44	140	120	180	160	120	150

Table 18.V

Variation of Concentration for Boron and Nickel in Finished Water of Three Municipal Treatment Plants at Pittsburgh

		Concentration - μg/l		
		Allegheny		*Monong.*
Element	*Period*[a]	*Plant A*	*Plant B*	*Plant C*
Boron	1	20-200	50-300	90-500
	2	50-100	50-200	30-300
	3	20- 60	7- 70	8- 60
	a	24	-	21
	River - '64-'65[b]	70-160		70-210
Nickel	1	4- 40	0.2-5	0.3-0.5
	2	50-130	0.5- 2	3-10
	3	200-500	0.7- 5	0.7- 2
	a	31	-	2
	River - '64-'65	<8-44		<8-20

[a]Periods -
1 - 6/14-7/13/60 }
2 - 11/22-12/20/60 } Data of Shapiro, *et al.*, 1961[24]
3 - 2/15-3/14/61 }
a - 3/26/62 Data of Durfor and Becker, 1962[25]

[b]River Data - from Shapiro, Andelman and Morgan, 1966.[23]

concentrations for these two elements in single samples taken within the following year in an independent study reported by Durfor and Becker.[25] It is apparent from a comparison of these two sets of data that if one attempts to use single samples as a measure of trace element concentrations at the treatment plant the results can be greatly misleading. Additional evidence to this effect may be seen in Table 18.VI, which, for several trace elements in the same three plants, compares the

Table 18.VI

Comparison of Concentration (μg/l) of Several Trace Elements from Single Samples of Finished Municipal Waters from Allegheny and Monongahela Rivers at Pittsburgh

		Allegheny R.		*Mon. R.*	*U.S.*
Element	*Study*	*Plant A*	*Plant B*	*Plant C*	*Median*
Ag	a	0.4	Trace	<0.3	0.23
	b	<0.12	-	<0.19	
B	a	10	10	8	31
	b	24	-	21	
Ba	a	100	100	30	43
	b	63	-	48	
Cu	a	10	5	8	8.3
	b	18	-	12	
Fe	a	10	10-40	8	43
	b	150	-	52	
Ni	a	400	5	0.8	<2.7
	b	31	-	2.1	
Sr	a	1000	100-500	8-300	110
	b	57	-	150	

[a]Study a - Composited sample, week of 2/15/61.[24]

[b]Study b - Sample of 3/26/62, and U.S. Median.[25]

concentrations composited during one week with the single sample values obtained one year later in the Durfor and Becker study. The concentrations of nickel and strontium varied significantly among the three plants, and for all elements the composited samples varied markedly from those reported in the Durfor and Becker study.

It can then be concluded that both macro- and micro-geographic as well as time variabilities can be encountered for trace elements in both raw and finished water supplies. Without even considering the detailed effects of treatment or distribution this should be recognized, especially in designing sampling schemes to assess their concentrations.

EFFECTS OF TREATMENT ON TRACE ELEMENT CONTENT

It is important to recognize that conventional municipal water treatment can have an effect on trace element content. This can be seen from the data of Kopp (Table 18.VII) comparing the mean concentrations for several trace elements in surface waters of the U.S. with the comparable values for finished municipal waters at treatment plants. The

Table 18.VII

Comparison of Mean Observed Concentrations of Trace Elements in Surface and Finished Waters of the United States[26]

Element[a]	*Concentration (μg/l)*	
	Surface	*Finished*
Zinc	64	79
Boron	101	92
Iron	52	69
Manganese	58	26
Copper	74	43
Barium	43	29
Strontium	217	105

[a]All elements detected in at least 50% of samples.[26]

zinc and iron concentrations for the latter were higher, while all the other elements listed were on the average lower in concentration in the finished waters. This rather broad comparison is only an indication of possible changes, and more detailed studies of specific plants or processes are required.

One such study (Table 18.VIII) reported by Andelman and Shapiro,[27] and based on a student research project directed by this author in England, considered the changes in concentrations for five trace elements during various stages of treatment at a modern municipal plant taking its water from the Thames River and utilizing pre-chlorination, flocculation with alum, rapid sand filtration, and post-chlorination. Analyses were performed on the unfiltered

Table 18.VIII

Variation in Concentration (μg/l) of Several Trace Elements at Various Stages of Treatment of Water from Thames River, England[27]
(Sample A - Midwinter 1971; Sample B - Late Winter 1971)

	Symbol[a]	Cobalt		Copper		Manganese		Nickel		Cadmium
		A	B	A	B	A	B	A	B	B
Raw	T	6.1	5.1	9.6	12.0	39.2	49.5	28.5	23.3	3.5
	F	5.2	4.6	28.4	6.2	32.8	8.3	15.3	9.8	3.1
Post Sed.	T	4.4	5.2	7.1	12.7	13.4	26.8	2.1	21.6	3.5
	F	-	5.1	-	5.0	-	11.5	-	10.0	2.8
Post Filt.	T	5.2	4.6	6.2	15.5	12.1	21.2	10.4	13.9	3.5
	F	4.4	4.4	5.1	10.9	7.9	24.9	10.4	12.8	4.1
Post Chl.	T	-	5.2	-	21.7	-	33.3	-	17.0	5.8
	F(1)	4.8	4.4	14.7	10.1	12.9	22.3	27.8	10.0	2.6
	F(2)	-	4.4	-	10.1	-	22.3	-	10.0	2.4
	F(3)	-	4.4	-	11.7	-	21.6	-	10.0	2.8
Ratio	Chlor. (T)/Raw (T)		1.0		2.0		0.7		0.7	1.7

[a]Symbols: T - Total concentration; F - Filtrate; F(1), F(2), F(3) - Replicates

samples and on the filtrates, reported as "T" and "F" respectively in Table 18.VIII, using a 0.45 micron membrane filter. Samples were taken on two days in winter (samples A and B). Comparing the raw water and final effluent, there was no significant effect on cobalt, and the total concentration of manganese and nickel decreased while that for copper and cadmium increased. In three samples there was an unexplained discrepancy in that the filtered sample had a higher concentration than the unfiltered one. This was probably not due to analytical imprecision, since the reproducibility shown for the post-chlorinated filtrates was quite good. The discrepancies do emphasize the need for replicate samples in performing these analyses to assess trace element concentrations in water supplies. In addition, if one is interested in the actual water to which the consumer is exposed, membrane filtration prior to analyses will in many cases lead to a low and incorrect result.

Softening by ion exchange would also be expected to reduce the concentrations of many trace elements. This was shown to occur in a study by Kopp of the effect of a cation exchange home water softener (Table 18.IX). For those elements whose analyses were sufficiently sensitive, such reductions were found, except for boron, which was in the form of an anion and would not be expected to be removed.

Table 18.IX

Trace Metals (μg/l) in Finished Water Before and After Passage Through a Home Water Softener[26]

Element	*Unsoftened*	*Softened*	*After 24 hour Contact With Copper Tubing*	
			Unsoftened	*Softened*
Zinc	<15	<15	290	100
Boron	140	140	77	137
Iron	79	< 8	77	< 8
Copper	37	< 8	245	126
Lead	<30	<30	3780	2250
Barium	29	< 2	2	< 2
Strontium	93	< 2	129	< 2

Incidentally, the effect of exposure of both the softened and unsoftened waters to contact with soldered copper tubing was to increase the concentrations of zinc, copper and lead. Interestingly, the concentrations of these elements in the unsoftened samples increased more than the softened ones. Boron and barium in the unsoftened samples decreased, while strontium increased. Iron in either water showed no changes on contact with the pipe. Such changes are indicative of what may be encountered in a distribution system, as will be discussed later.

Both the chemical form and physical state of a trace element are likely to have an effect on its removal in treatment. Certainly the fact that it is in a particulate or colloidal form is important in that regard. The data in Table 18.VIII indicate, for example, that sedimentation reduced a large portion of the total manganese, and filtration did the same for nickel. By contrast, in both cases the filterable concentrations increased slightly.

Longsdon and Symons[28] studied the removal of both inorganic and methyl mercury by a variety of treatment processes. They found that typically 40 to 60 per cent of the inorganic mercury was removed by iron coagulation. Such removal was relatively independent of the turbidity of the untreated water. In contrast, the methyl mercury removal was considerably less and increased with increases in the turbidity of the untreated water. They also observed that coagulation with alum reduced the inorganic mercury concentration, and that when the coagulation was followed by increasing doses of powdered activated carbon the removal also increased. In their measurements on mercury removal by coagulation, the dose of iron, the pH of the treated water, and the hardness all affected the fraction removal.

It is thus apparent that various water treatment processes can have an important effect on the concentration of trace elements in finished waters. However, considerably more research, similar to that of Logsdon and Symons, is necessary in order to predict or optimize such removal.

COMBINED EFFECTS OF TREATMENT AND DISTRIBUTION

The combined effects of treatment and distribution on trace element concentrations can be illustrated in three studies of municipal treatment systems. The first, the system represented by plant A in Tables 18.V and 18.VI, is that of the City of Pittsburgh, which at the time of the study was utilizing sedimentation, slow sand filtration, and post-chlorination, with soda ash added for pH control. The plant treated about 90 mgd and obtained its raw water from the Allegheny River. Samples were collected at four-hour intervals and composited to form one sample each week. The results of the analyses on five successive weekly composite samples for barium, copper and nickel are reported in Table 18.X. In each case samples were obtained at the plant prior to and after chlorination, as well as at a tap at one remote point in the distribution system.[24] The results for these three elements are presented here because

Table 18.X

Effect of Chlorination and Distribution of Barium, Copper, and Nickel in a Municipal Water Supply System at Pittsburgh[24]

		Concentration - μg/l		
Element	*Week*[a]	*Before Chlor.*	*After Chlor.*	*Tap*
Barium	1	300	90	40
	2	90	80	40
	3	90	100	40
	4	100	90	30
	5	100	90	30
Copper	1	30	90	4000
	2	30	< 90	4000
	3	30	<100	4000-10,000
	4	40	90	9000
	5	30	< 90	>3000
Nickel	1	7-30	30	100
	2	30	8-30	100
	3	9-30	10-40	> 100
	4	10	9-30	90
	5	30	9-30	30

[a]Composite weekly samples obtained in April and May, 1961.

they show relatively large changes. It is apparent that the nickel and copper occurred in significantly higher concentrations in the tap water as compared with that at the treatment plant following chlorination; the barium concentrations were lower. Of the three, copper uniformly showed increases as a result of chlorination. It should also be noted that a relatively large increase in copper content following chlorination was also found in the Thames River study reported in Table 18.VIII. The copper concentrations at the tap reported in Table 18.X are unusually high and are probably due to the fact that the building in which the tap was located was newly constructed.

The combined effects of treatment and distribution were also investigated by Boström and Wester[29] in their study of trace elements in the municipal treatment systems of three cities in Sweden--Malmo, Stockholm, and Göteberg. A comparison of raw and tap water concentrations for seven of the elements from that study is shown in Table 18.XI. These data indicate that for copper there was a consistent decrease in concentration in the tap as compared with the raw water, an increase for zinc, and no changes for mercury. In contrast, barium, selenium, cobalt and cadmium were variable in their reductions or increases, depending on the city.

Table 18.XI

Comparison of Several Trace Elements in Raw and Tap Water of Three Cities in Sweden[29]

Element	*Raw Water Conc. Range μg/l*	*Concentration Ratio, Tap/Raw*		
		Malmo	*Stockholm*	*Göteberg*
Cu	2-13	0.5	0.2	0.4
Zn	8-28	4.5	2.8	1.4
Hg	0.09-0.4	1.1	1.0	1.0
Ba	1-3	6.7	4.0	0.5
Se	0.05-0.5	1.8	0.5	1.6
Co	0.1	0.7	1.0	3.0
Cd	0.02-0.3	2.5	0.5	1.0

Finally, a rather large study was done by Barnett, Skougstad and Miller[30] of macro and trace elements in raw, finished, and tap water of the Denver municipal system. The results for the trace elements are reported in Table 18.XII in terms of ratios of concentrations as an indication of either variability in the raw or tap water, or changes in the tap water compared with that of the finished water at the treatment plants. In the raw water the maximum/minimum ratio for most trace elements studied varied from

Table 18.XII

Variations in Concentration of Several Trace Elements in Raw and Tap Water, and Between Finished and Tap Water in the Denver Municipal System[30]

	Ratio of Concentrations			
	Raw Water	*Tap/Finished*		*Tap Water*
Element	*Greatest Max/Min*	*Max. Reduction*	*Max. Increase*	*Greatest Max/Min*
Al	6.5	0.53	57	4.3
Ba	1.5	0.61	1.5	1.5
B	2.0	0.53	1.5	2.0
Cu	4.0	0.45	5.5	5.3
Fe	30	0.037	25	20
Li	2.4	0.57	3.1	4.0
Mn	6.2	0.5	4.0	2.5
Mo	130	0.41	4.0	32
Rb	4.4	0.5	1.7	2.0
Sr	1.6	0.59	2.1	2.1
Zn	10.5	0.56	5.0	2.7

1.5 to 6.5, but higher values were obtained for iron, zinc and aluminum. Although it was noted that, except perhaps for iron and copper, the water at domestic taps was quite similar to that at the treatment plants, there were at times significant differences, as indicated by the maximum reductions (as high as 30, but typically 2) or increases (as high as 57, but typically 3 to 5) in the ratio of the concentrations at the taps compared with those at the treatment plants.

CONCLUSIONS

There is some epidemiological evidence relating the trace element content of municipal water supplies to chronic disease, especially cardiovascular diseases. At this point the evidence is quite inconclusive and the possible mechanisms are generally uncertain. Acute illness from exposure to high concentrations of these trace elements in the United States is negligible. However, a survey of community water supplies has shown that for all those trace elements for which there are mandatory U.S. Public Health Service standards, only silver was not found in distribution system samples above the maximum permissable concentrations.

In order to properly assess the concentration levels of trace elements in municipal waters at the point of human consumption, it is necessary to consider geographic and time variabilities, as well as changes with treatment and distribution. Large variations in the concentration of many trace elements have been observed in different regions for surface waters of the United States, as well as on a microgeographic scale, even for different water sources serving the same municipality. In addition, the time variabilities at a given water intake have been found to be considerable.

Water treatment can have a significant effect on trace element content. Although for many elements concentrations are reduced in treatment, in some cases increases occur. It has also been shown that chemical state and physical form will influence the extent of removal, as will adjusting variables in the treatment operation. Considerably more investigation is required in the latter area.

Finally, the distribution process itself can affect the concentration of trace elements, with both increases and decreases having been observed. All these variabilities and changes emphasize the need for a carefully designed sampling scheme in attempting to assess the levels of concentrations of trace elements to which humans are exposed through their drinking water. In particular, the need to sample at the terminals of the distribution system, rather than at the treatment plant, is especially important.

REFERENCES

1. Kobayashi, J. *Ber. Ohara Inst. Landw. Fosch., 11,* 12 (1957).
2. Schroeder, H. A. *J. Am. Med. Assoc., 172,* 1902 (1960).
3. Schroeder, H. A. *J. Chron. Dis., 12,* 586 (1961).
4. Schroeder, H. A. *J. Am. Med. Assoc., 195,* 125 (1966).
5. Morris, J. N., M. D. Crawford, and J. A. Heady. *Lancet, 1,* 860 (1961).
6. Morris, J. N., M. D. Crawford, and J. A. Heady. *Lancet, 2,* 506 (1962).
7. Crawford, M. D., M. J. Gardner, and J. N. Morris. *Lancet, 1,* 827 (1968).
8. Crawford, M. D., M. J. Gardner, and P. A. Sedgwick. *Lancet, 1,* 988 (1972).
9. Biorck, G., H. Bostrom, and A. Widstrom. *Acta Med. Scand., 178,* 239 (1965).
10. Biersteker, K. *T. soc. Geneesk., 45,* 658 (1967).
11. Anderson, T. W., W. H. le Riche, and J. S. Mackay. *New Eng. J. Med., 280,* 805 (1969).
12. Dudley, E. F., R. A. Beldin, and B. C. Johnson. *J. Chron. Dis., 22,* 25 (1969).
13. Winton, E. F. and L. J. McCabe. *J. Amer. Water Works Assoc., 62,* 26 (1970).
14. Masironi, R., A. T. Miesch, M. D. Crawford, and E. I. Hamilton. *Bull. World Health Org., 47,* 139 (1972).
15. Masironi, R. *Bull. World Health Org., 43,* 687 (1970).
16. Masironi, R. *Bull. World Health Org., 40,* 305 (1969).
17. Berg, J. W. and F. Burbank. In *Geochemical Environment in Relation to Health and Disease,* H. C. Hopps and H. L. Cannon, eds., *Ann. N.Y. Acad. Sci., 199,* 249 (1972).
18. "Drinking Water Standards, 1962," U.S. Public Health Service Publication No. 956, Washington, D.C., 1962.
19. McCabe, L. J., J. M. Symons, R. D. Lee, and G. G. Robeck. *J. Amer. Water Works Assoc., 62,* 670 (1970).
20. Craun, G. F. and L. J. McCabe. *J. Amer. Water Works Assoc., 65,* 74 (1973).
21. Kopp, J. F. and R. C. Kroner. "Trace Metals in Waters of the United States," Federal Water Pollution Control Administration, Cincinnati, 1970.
22. Copeland, R. A. and J. C. Ayers. "Trace Element Distributions in Water, Sediment, Phytoplankton, Zooplankton and Benthos of Lake Michigan," Environmental Research Group, Inc., Ann Arbor, Michigan, 1972.
23. Shapiro, M. A., J. B. Andelman, and P. V. Morgan. "Intensive Study of the Water at Critical Points on

the Monongahela, Allegheny, and Ohio Rivers in the Pittsburgh, Pennsylvania Area," University of Pittsburgh, 1967.

24. Shapiro, M. A., W. H. Hill, P. K. Chin, and Y. Kobayashi. "Physiological Aspects of Water Quality," Progress Report, University of Pittsburgh, October, 1961.

25. Durfor, C. N. and E. Becker. "Public Water Supplies of the 100 Largest Cities in the United States, 1962," U.S. Geol. Survey Water-Supply Paper 1812, 1964.

26. Kopp, J. F. In *Trace Substances in Environmental Health,* D. D. Hemphill, ed., Vol. III, Univ. of Missouri, Columbia, Missouri, 1970.

27. Andelman, J. B. and M. A. Shapiro. In *Trace Substances in Environmental Health,* D. D. Hemphill, ed., Vol. VI, Univ. of Missouri, Columbia, Missouri, 1973.

28. Logsdon, G. S. and J. M. Symons. "Mercury Removal by Conventional Water Treatment Methods," presented at the American Water Works Association, Chicago, June 1972.

29. Bostrom, H. and P. O. Wester. *Acta Med. Scand., 181,* 465 (1967).

30. Barnett, P. R., M. W. Skougstad, and K. J. Miller. *J. Amer. Water Works Assoc., 61,* 61 (1969).

INDEX

INDEX